Natural Language Processing

Raymond Lee

Natural Language Processing

A Textbook with Python Implementation

Second Edition

 Springer

Raymond Lee
Faculty of Science and Technology
Beijing Normal-Hong Kong Baptist University
Zhuhai, China

ISBN 978-981-96-3207-7   ISBN 978-981-96-3208-4   (eBook)
https://doi.org/10.1007/978-981-96-3208-4

None

This Springer imprint is published by the registered company Springer Nature Singapore Pte Ltd.
The registered company address is: 152 Beach Road, #21-01/04 Gateway East, Singapore 189721,
Singapore

If disposing of this product, please recycle the paper.

This book is dedicated to all readers and students taking my undergraduate and postgraduate courses in natural language processing; your enthusiasm in seeking knowledge motivated me to write this book.

Preface

Motivation of This Book

Natural language processing (NLP) and its associated applications have flourished due to advancements in artificial intelligence (AI) over the past few decades. NLP applications include information retrieval (IR) systems, text summarization (TS) systems, question-and-answering (chatbot) systems, as well as recent developments in large language models (LLMs) and generative AI (GenAI). These topics are prevalent in both industry and academia, offering varied routines that significantly enhance a wide range of everyday services.

The objective of this book is to provide readers with foundational NLP concepts and knowledge through 14 h of step-by-step workshops. These workshops will guide participants in practicing various core Python-based NLP tools, including NLTK, spaCy, TensorFlow, Keras, transformer, and BERT technology, to build their own Python-based NLP applications.

Organization and Structure of This Book

This book is structured into two main parts: Part I—Concepts and Technology, and Part II—Natural Language Processing Workshops with Python Implementation in 14 Hours. In Part I, the first ten chapters lay a solid foundation in natural language processing (NLP) concepts, exploring key topics such as N-gram language models, part-of-speech tagging, syntax analysis, semantic representations, and latest advancements in transfer learning and transformer technology. Each chapter builds on the preceding one, leading to an understanding of major NLP applications and the evolution of large language models and generative AI, while also addressing ethical considerations in AI.

Part II consists of seven practical workshops that provide hands-on experience with Python implementations relevant to the concepts discussed in the first part.

Starting with an introduction to the Natural Language Toolkit (NLTK), the workshops guide readers through various NLP tasks such as N-gram modeling, sentiment analysis, and use of transformers like BERT. Each workshop offers step-by-step instructions, empowering readers to apply their knowledge in real-world scenarios, including the creation and deployment of a chatbot system. This structured approach balances foundational theory with practical application, making it suitable for learners eager to advance their skills in NLP.

Major Enhancement in 2nd Edition

The second edition includes the following major updates:

1. An overview of the development of BERT, transformer models, ChatGPT, and large language models (LLMs) from the 2000s to the present, covered in Sect. 1.5, "A Brief History of NLP," in Chap. 1
2. A new chapter focusing on the latest advancements in NLP, specifically in LLMs and generative AI (GenAI), included in Chap. 10
3. Revised and updated NLP workshops (originally Chaps. 10, 11, 12, 13, 14, 15, and 16, now Chaps. 11, 12, 13, 14, 15, 16, and 17) to align with the latest versions of Python-based NLP packages and tools

Readers of This Book

This book serves as both an NLP textbook and a practical guide for NLP Python implementation, tailored for:

- Undergraduates and postgraduates across various disciplines, including AI, computer science, IT, data science, and related fields
- Lecturers and tutors teaching NLP or AI-related courses
- NLP and AI scientists, as well as developers, who wish to learn the fundamental concepts of NLP and apply them through Python-based workshops
- Readers interested in learning NLP concepts and practicing Python-based NLP techniques using tools such as NLTK, spaCy, TensorFlow, Keras, BERT, transformer technology, and latest developments in LLMs and GenAI

How to Use This Book?

This book can be used as a textbook for undergraduate and postgraduate courses on natural language processing (NLP) and as a reference for general readers who want to learn key NLP technologies and implement NLP applications using contemporary tools such as NLTK, spaCy, TensorFlow, BERT, and transformer technology.

Part I (Chaps. 1, 2, 3, 4, 5, 6, 7, 8, 9, and 10) covers the foundational concepts and key technologies in NLP, including the N-gram language model, part-of-speech tagging, syntax and parsing, meaning representation, semantic analysis, pragmatic analysis, transfer learning, and transformer technology. It also discusses major NLP applications and the latest developments in BERT, ChatGPT, large language models (LLMs), and generative AI (GenAI).

Part II (Chaps. 11, 12, 13, 14, 15, 16, and 17) consists of materials for a 14-h, step-by-step Python-based NLP implementation spread across seven workshops.

For readers and AI scientists, this book serves as a reference for learning NLP and applying Python-based NLP tools and libraries, using the latest development tools and platforms.

For the seven NLP workshops in Part II (Chaps. 11, 12, 13, 14, 15, 16, and 17), readers can download all Jupyter Notebook files and data from my NLP GitHub directory: https://github.com/raymondshtlee/nlp/. For any queries, please feel free to contact me via email at raymondshtlee@uic.edu.cn.

Zhuhai, China   Raymond Lee

Acknowledgments

April 2025

I would like to express my gratitude:

To my wife, Iris, for her patience, encouragement, and understanding at times spent on research and writing.

To Ms. Celine Cheng, executive editor of Springer Nature, and her professional editorial and book production team for their support, advice, and valuable comments.

To Prof. Zhi Chen, President of Beijing Normal-Hong Kong Baptist University (BNBU), for the provision of excellent environment for research, teaching, and writing this book.

To Prof. Jianxin Pan, Vice President (Research and Development) of BNBU, for his support for R&D of NLP and related AI projects.

To Prof. Terry Huajun Ye, Acting Dean of Faculty of Science and Technology of BNBU, and Prof. Weifeng Su, Head of Department of Computer Science of BNBU, for their continuous support for AI and NLP courses.

To research assistant Mr. Zihao Huang for the help of NLP workshop preparation. To research student Ms. Clarissa Shi and student helpers Ms. Siqi Liu, Mr. Yingjie Wang, and Ms. Jie Lie for their help with literature review on major NLP applications and transformer technology and Mr. Zhuohui Chen for help in bug fixing and version update for the workshop programs.

To Beijing Normal-Hong Kong Baptist University for the prominent support by the Guangdong Provincial Key Laboratory IRADS (2022B1212010006).

Faculty of Science and Technology   Raymond Lee
Beijing Normal-Hong Kong Baptist University
Zhuhai, China

Competing Interests The author has no competing interests to declare that are relevant to the content of this manuscript.

Contents

Abbreviations

AI	Artificial intelligence
ASR	Automatic speech recognition
BERT	Bidirectional encoder representations from transformers
CDD	Conceptual dependency diagram
CFG	Context-free grammar
CFL	Context-free language
CNN	Convolutional neural networks
CR	Coreference resolution
DT	Determiner
FOPC	First-order predicate calculus
GenAI	Generative artificial intelligence
GPT	Generative pre-trained transformer
GRU	Gate recurrent unit
HMM	Hidden Markov model
IE	Information extraction
IR	Information retrieval
LLM	Large language model
LSTM	Long short-term memory
MEMM	Maximum entropy Markov model
MeSH	Medical Subject Headings
ML	Machine learning
NER	Named entity recognition
NLP	Natural language processing
NLTK	Natural Language Toolkit
NLU	Natural language understanding
NN	Noun
NNP	Proper noun
Nom	Nominal
NP	Noun phrase
PCFG	Probabilistic context-free grammar
PMI	Point-wise mutual information

POS	Part of speech
POST	Part-of-speech tagging
PPMI	Positive point-wise mutual information
Q&A	Question and answering
RNN	Recurrent neural network
TBL	Transformation-based learning
VB	Verb
VP	Verb phrase
WSD	Word sense disambiguation

Part I
Concepts and Technology

Chapter 1
Natural Language Processing

Consider this scenario: Late in the evening, Jack starts a mobile app and talks with AI Tutor Max.

1.1 Introduction

There are many chatbots available today that enable humans to communicate with devices using natural language. Figure 1.1 illustrates a dialogue between a student, who has returned to the dormitory after a full day of classes, and a mobile application called *AI Tutor 2.0* (Cui et al. 2020), developed as part of our research on AI Tutor chatbots. The goal is to allow the user (Jack) not only to learn from reading books but also to engage in candid conversations with AI Tutor 2.0 (Max), which provides knowledge-based responses in natural language.

This differs from traditional chatbots that respond only to basic commands. Instead, it represents *human-computer interaction*, demonstrating how a student might converse with a tutor about subject knowledge in the physical world. The process is dynamic and involves two main components: (1) *World Knowledge*: This includes simple handshaking dialogues such as greetings and general discussions. Developing this is a complex task, as it requires knowledge and common sense to construct a functional chatbot capable of engaging in everyday dialogues; (2) *Domain Knowledge*: This refers to the technical expertise within a specific subject area. To achieve this, the chatbot must first learn from the author's book, *AI in Daily Life* (Lee 2020), which provides foundational knowledge on the subject. This knowledge is then organized into a *knowledge tree* or *ontology graph*. The goal is to create a new type of publication and interactive device, facilitating learning between humans and computers.

Natural Language Processing (NLP) intersects with various disciplines, including *linguistics, computational linguistics, statistical engineering, AI and machine*

R. Lee, *Natural Language Processing*,
https://doi.org/10.1007/978-981-96-3208-4_1

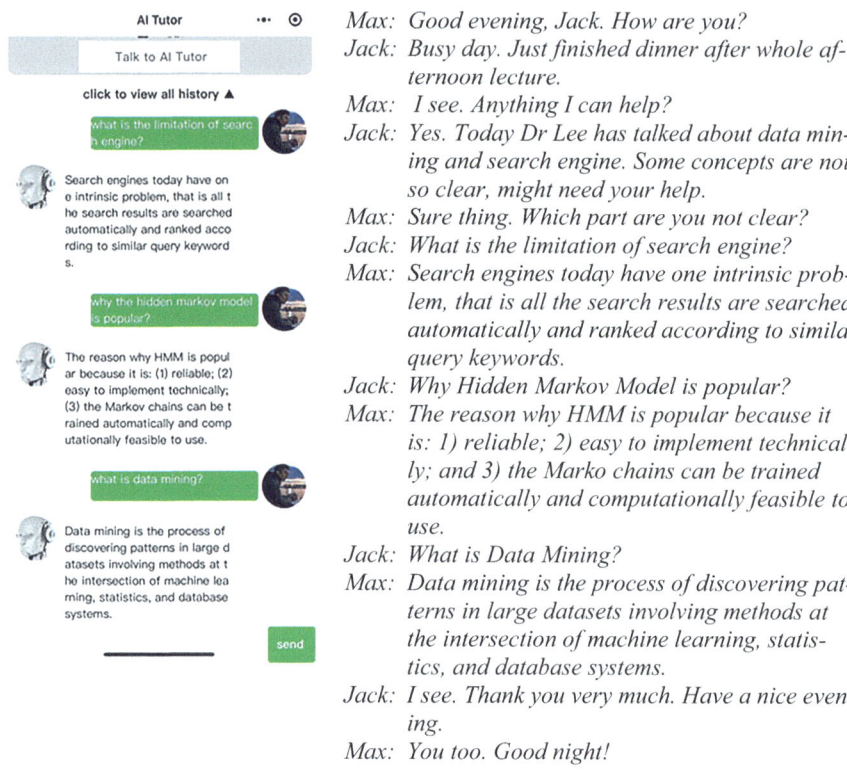

Fig. 1.1 A snapshot of the AI Tutor chatbot

learning, *data mining*, and *voice processing and synthesis*. Over the past few decades, many innovative chatbots developed by NLP and AI scientists have become commercial products. This chapter will introduce the core technology and its components, with more detailed discussions of related technologies in the following chapters.

1.2   Human Language and Intelligence

There is an old saying: *Your actions define who you are*. It is because we never know what people think, the only method is to evaluate and judge their actions and behaviors.

NLP core technologies and methodologies arose from the famous *Turing Test* (Eisenstein 2019; Bender 2013; Turing 1936, 1950) proposed by Sir Alan Turing

Fig. 1.2   Turing test (Tuchong 2020a)

(1912–1954) the in 1950s, the father of AI. Figure 1.2 shows a human judge conversing with two individuals in two rooms. One is a human, the other is either a robot, a chatbot, or an NLP application. During a 20-min conversation, the judge can ask human/machine technical/non-technical questions and require a response to every question so that the judge can decide whether the respondent is a human or a machine. NLP in the *Turing Test* is to recognize, understand questions, and respond in human language. It remains a popular topic in AI today because we can't see and judge people's thinking to define intelligence. It is the ultimate challenge in AI.

Human language is a significant component of human behavior and civilization. Generally, it can be categorized into (1) written and (2) oral aspects. Written language undertakes to process, store, and pass human/natural language knowledge to the next generations. Oral or spoken language acts as a communication medium among individuals.

NLP has examined the basic effects on philosophy such as meaning and knowledge, psychology in word meanings, linguistics in phrase and sentence formation, and computational linguistics in language models. Hence, NLP is cross-disciplinary integration of disciplines such as philosophy in human language ontology models, psychological behavior between natural and human language, linguistics in mathematical and language models, computational linguistics in agents, and ontology trees technology as shown in Table 1.1.

Table 1.1 Various disciplines related to NLP

Discipline	Problems to tackle with	Solutions and tools
Philosophy	What is meaning and knowledge?	Ontology and epistemology
	How do words and sentences acquire meaning?	Natural language argumentation using intuition
	How can we relate ideas and concept into words and meanings	Mathematical models such as logic theory and model theory
Psychology	How can we identify the structure of sentences?	Psychological experiments to measure the performance
	How the meaning of words can be identified?	Statistical analysis of observations
	When does understanding take place?	
Linguistics	How to form phrases and sentences with words?	Mathematical model of language structure
	How can we represent the meaning of a sentence?	Logical model for the representation of language structure and patterns
Computational linguists and NLP	How to model different types of human languages?	Agent ontology and ontological tree modeling
	How to model knowledge and meaning?	NLP techniques discussed in this chapter
	How to use human language for human–machine direct communication?	

1.3   Linguistic Levels of Human Language

Linguistic levels (Hausser 2014) are regarded as functional analyses of human written and spoken languages. There are six levels of linguistics analysis: (1) *phonetics*, (2) *phonology*, (3) *morphology*, (4) *syntax*, (5) *semantics*, and (6) *pragmatics (discourse)* classified in basic sound linguistics. These six levels of linguistics are shown in Fig. 1.3.

The basic linguistic structure of spoken language includes *phonetics* and *phonology*. *Phonetics* refers to the physical aspects of sound, the study of the production and perception of sounds called *phones*. *Phonetics* governs the production of human speech often without preceding knowledge of spoken language, organizes sounds, and studies the *phonemes* of languages that can provide various meanings between words and phrases.

Direct language structure is related to morphological and *syntactic levels*. *Morphology* is the *form* and *word level* determined by grammar and syntax generally. It refers to the smallest form in linguistic analysis, consisting of sounds, to combine words with grammatical or lexical functions.

Lexicology is the study of vocabulary from a word form to a derived form. *Syntax* represents the primary level of clauses and sentences to organize the meaning of different words order, that is, addition and subtraction of spoken language, and deals with related sentence patterns and ambiguous analysis.

The advanced structure deals with actual language meaning at *semantic* and *pragmatic levels*. *Semantic level* is the domain of meaning that consists of

Fig. 1.3 Linguistic levels
of human languages

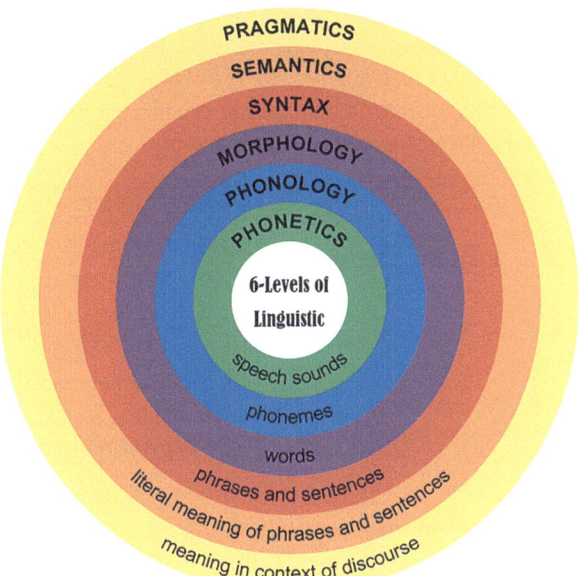

morphology and *syntax* but is regarded as a level that requires one's own learning to assign correct meaning promptly with vocabulary, terminology form, grammar, sentence, and discourse perspective. *Pragmatics* is the use of language in definitive settings. The meaning of *discourse* does not have to be the same as abstract form in actual use. It is largely based on the concept of speech acts and the contents of statements with intent and effect analysis of language performance.

1.4 Human Language Ambiguity

In many language models, cultural differences often produce identical utterances with more than a single meaning in conversation. *Ambiguity* is the capability to understand sentence structures in many ways. There are (1) *lexical*, (2) *syntactic*, (3) *semantic*, and (4) *pragmatics ambiguities* in NLP.

Lexical ambiguity arises from words where a word's meaning depends on contextual utterance. For instance, the word *green* is normally a noun for color. But it can be an adjective or even a verb in different situations.

Syntactic ambiguity arises from sentences and is parsed differently, for example, *Jack watched Helen with a telescope*. It can be described as either Jack watched Helen <u>by using</u> a telescope or Jack watched Helen <u>holding</u> a telescope.

Semantic ambiguity arises from word meaning and can be misinterpreted, or a sentence has ambiguous words or phrases, for example, *The van hits the boar while it is moving*. It can be described as either *the van hits the boar while the van is moving*, or *the van hits the boar while the boar is moving*. It has more than a simple syntactic meaning and requires to work out the correct interpretation.

Pragmatic ambiguity arises from a statement and is not clearly defined when the context of a sentence provides multiple interpretations such as *I like that too*. It can describe what *I like that too*, *others like that too* but the description of *that* is uncertain.

NLP analyzes sentences ambiguity incessantly. If they can be identified earlier, it will be easier to define proper meanings.

1.5   A Brief History of NLP

There are several major NLP transformation stages in NLP history (Santilal 2020).

1.5.1   First Stage: Machine Translation (Before the 1960s)

The concept of NLP was introduced in seventeenth century by philosopher and mathematician Gottfried Wilhelm Leibniz (1646–1716) and polymath René Descartes (1596–1650). Their studies of the relationships between *words* and *languages* formed the basis for language translation engine development (Santilal 2020).

The first patent for an invention related to machine translation was filed by inventor and engineer Georges Artsrouni in 1933, but formal study and research was rendered by Sir Alan Turing from his remarkable article *Computing Machinery and Intelligence* published in 1950 (Turing 1936, 1950) and his famous *Turing Test* officially used as an evaluation criterion for machine intelligence since NLP research and development were mainly focused on language translation at that time.

The first and second International Conference on Machine Translation held in 1952 and 1956 used basic *rule-based* and *stochastic* techniques. The 1954 Georgetown-IBM experiment engaged in wholly automatic machine translation of more than 60 Russian sentences into English and was over-optimistic that the whole machine translation problem could be solved within a few years. However, a breakthrough in NLP was achieved by Emeritus Professor Noam Chomsky on universal grammar for linguistics in 1957, but the ALPAC report published in 1966 revealed deficient progress for AI and machine translation in the past 10 years signifying the first winter of AI.

1.5.2   Second Stage: Early AI on NLP (1960s–1970s)

NLP's major development was focused on how it can be used in different areas such as knowledge engineering called *agent ontology* (Climiano et al. 2014) to shape meaning representations following AI grew popular over time. The BASEBALL system (Green et al. 1961) was a typical example of a Q&A-based domain expert system of human and computer interaction developed in the 1960s, but inputs were restrictive and language processing techniques remained in basic language processing.

In 1968, Professor Marvin Minsky (1927–2016) developed a more powerful NLP system. This advanced system used an AI-based question-answering inference engine between humans and computers to provide knowledge-based interpretations of questions and answers. Further, Professor William A. Woods proposed an *augmented translation network (ATN)* to represent natural language input in 1970. During this period, many programmers started to transcribe codes in different AI languages to conceptualize natural language ontology knowledge of real-world structural information into human understanding mode status. Yet these expert systems were unable to meet expectations signified the second winter of AI.

1.5.3   Third Stage: Grammatical Logic on NLP (1970s–1980s)

NLP research turned to knowledge representation, programming logic, and reasoning in AI. This period was regarded as the grammatical logic phase of NLP in which powerful sentence processing techniques such as SRI's core language engine and discourse representation theory emerged. These innovations introduced new pragmatic representations and discourse interpretation with practical resources and tools such as parsers and Q&A chatbots. Although R&D was hampered by computational power, the lexicon in 1980s aimed to expand NLP.

1.5.4   Fourth Stage: AI and Machine Learning (1980s–2000s)

The revolutionary success of the Hopfield Network in the field of machine learning proposed by Professor Emeritus John Hopfield activated a new era of NLP research using machine learning techniques as an alternative to complex rule-based and stochastic methods in the past decades.

Computational technology upgrades in computational power and memory complemented Chomsky's theory of linguistics had augmented language processing from machine learning methods of corpus linguistics. This development stage was also known as NLP lexical, and corpus referred to grammar emergence in lexicalization method in the late 1980s, which signified the IBM DeepQA project led by Dr. David Ferrucci for their remarkable question-answering system developed in 2006.

1.5.5   Fifth Stage: Rise of BERT, Transformer, ChatGPT, and LLMs (2000s–Present)

NLP has significant advancements over the past two decades fueled by innovations in neural networks, especially deep learning architectures. The timeline of this evolution began in the early 2000s with the rise of *Recurrent Neural Networks (RNNs)*.

RNNs were designed to handle sequential data by maintaining a hidden state that could "*remember*" past information. This allowed them to model temporal dependencies in tasks like speech recognition, translation, and text generation (Hochreiter and Schmidhuber 1997). However, RNNs had limitations in retaining information across long sequences leading to performance degradation (Mikolov et al. 2010).

To address this, *Long Short-Term Memory (LSTM)* networks were introduced as a specialized type of RNN. LSTMs incorporated a memory mechanism to retain information over longer periods, which significantly improved performance in tasks requiring long-range context such as machine translation and time-series prediction (Hochreiter and Schmidhuber 1997). They were more effective than basic RNNs regardless of scalability and computational efficiency (LeCun et al. 2015).

The real breakthrough in NLP came with the introduction of the *Transformer* architecture in 2017 outlined in the landmark paper "*Attention is All You Need*" (Vaswani et al. 2017). Transformers replaced the sequential processing of RNNs and LSTMs with a *self-attention mechanism* to weigh the importance of different words in a sentence relative to each other in a parallel fashion. This design not only improved the speed and accuracy but also paved the way for *Large Language Models (LLMs)* by enabling scalable architectures to handle vast amounts of text (Devlin et al. 2018).

BERT (Bidirectional Encoder Representations from Transformers) developed by Google in 2018 took it further by introducing a *bidirectional* approach to understanding context. BERT read text in both directions simultaneously unlike previous models that processed text in one direction. This innovation allowed the model to achieve state-of-the-art results in tasks like question-answering, sentiment analysis, and natural language understanding (Devlin et al. 2018).

Generative Pre-trained Transformers (GPT) by OpenAI was a major leap in NLP. This technological evolution has demonstrated significant model size scaling development for other robust LLMs. *GPT-1* released in 2018 had 117 million parameters focused on understanding unsupervised learning. *GPT-2* released in 2019 had 1.5 billion parameters that validated the large-scale unsupervised learning capability to generate coherent and contextually relevant text across a wide range of tasks (Radford et al. 2019). However, it raised concerns about the potential for AI misuse to generate misinformation and deepfake text (Solaiman et al. 2019).

GPT-3 released in 2020 had 175 billion parameters and was 100 times more vigorous than *GPT-2* significantly increasing the size of the language model. This extension excelled in few-shot and zero-shot learning to perform tasks with minimal examples that have not been explicitly trained (Brown et al. 2020). Its sheer size could mimic human-like reasoning and perform complex tasks such as programming, writing essays, and engaging in sophisticated dialogue (Floridi and Chiriatti 2020).

OpenAI released *ChatGPT* based on *GPT-4* in 2023 was the next *LLM* evolution. While OpenAI did not disclose the precise parameters number, *GPT-4* advanced in coherence, reasoning, and factual accuracy realizing it a versatile tool for conversational AI, content generation, and real-time problem-solving (OpenAI 2023).

The progression from *GPT-1* to *GPT-4* has demonstrated exponential growth in *parameters and computational power*, bringing AI closer to human-like language understanding and generation. Each *GPT* generation has expanded the boundaries of what AI models could achieve to challenge traditional notions of human intelligence (Marcus and Davis 2020). LLMs continued to exert AI's capability limits to perform tasks that were once considered uniquely human but raised important ethical and societal questions about usage and impact on the future (Bender et al. 2021).

1.6   NLP and AI

NLP can be regarded as automatic or semi-automatic processing of human language (Eisenstein 2019). It requires extensive knowledge of linguistics and logical theory in theoretical mathematics, also known as *computational linguistics*. It is a multidisciplinary study of epistemology, philosophy, psychology, cognitive science, and agent ontology.

NLP is an AI area in which computer machines can analyze and interpret human speech for human-computer interaction (HCI) to generate structural knowledge for information retrieval operations, text and automatic text summarization, sentiment and speech recognition analysis, data mining, deep learning, and machine translation agent ontologies at different levels of Q&A chatbots, as shown in Fig. 1.4.

Fig. 1.4 NLP and AI (Tuchong 2020b)

Fig. 1.5 NLP main
components

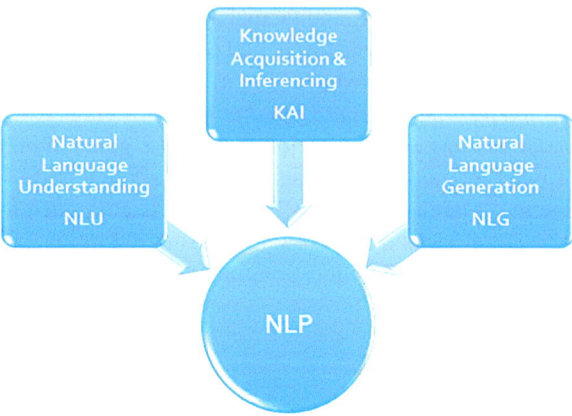

1.7 Main Components of NLP

NLP consists of (1) *Natural Language Understanding (NLU)*, (2) *Knowledge Acquisition and Inferencing (KAI)*, (3) *Natural Language Generation (NLG)* components as shown in Fig. 1.5.

NLU is a technique and method devised to understand the meanings of human-spoken languages by syntax, semantic, and pragmatic analyses.

KAI is a system to generate proper responses after spoken languages are fully recognized by NLU. It is an unresolved knowledge acquisition and inferencing problem in machine learning and AI by the conventional rule-based system due to the intricacies of natural language and conversation, that is, an *if-then-else* type of query-response used in expert systems. Most KAI systems strive to regulate knowledge domain at a specific industry for resolution, that is, customer service knowledge for insurance, medical, etc. Further, agent ontology has achieved a favorable outcome.

NLG includes *answer*, *response*, and *feedback* generation in human-machine dialogue. It is a multi-facet machine translation process that converts responses into text and sentences to perform text-to-speech synthesis from the target language and produce near-human speech responses.

1.8 Natural Language Understanding (NLU)

Natural Language Understanding (NLU) is a process of recognizing and understanding spoken language in four stages (Allen 1994): (1) *speech recognition*, (2) *syntactic (syntax) analysis*, (3) *semantic analysis*, and (4) *pragmatic analysis* as shown in Fig. 1.6.

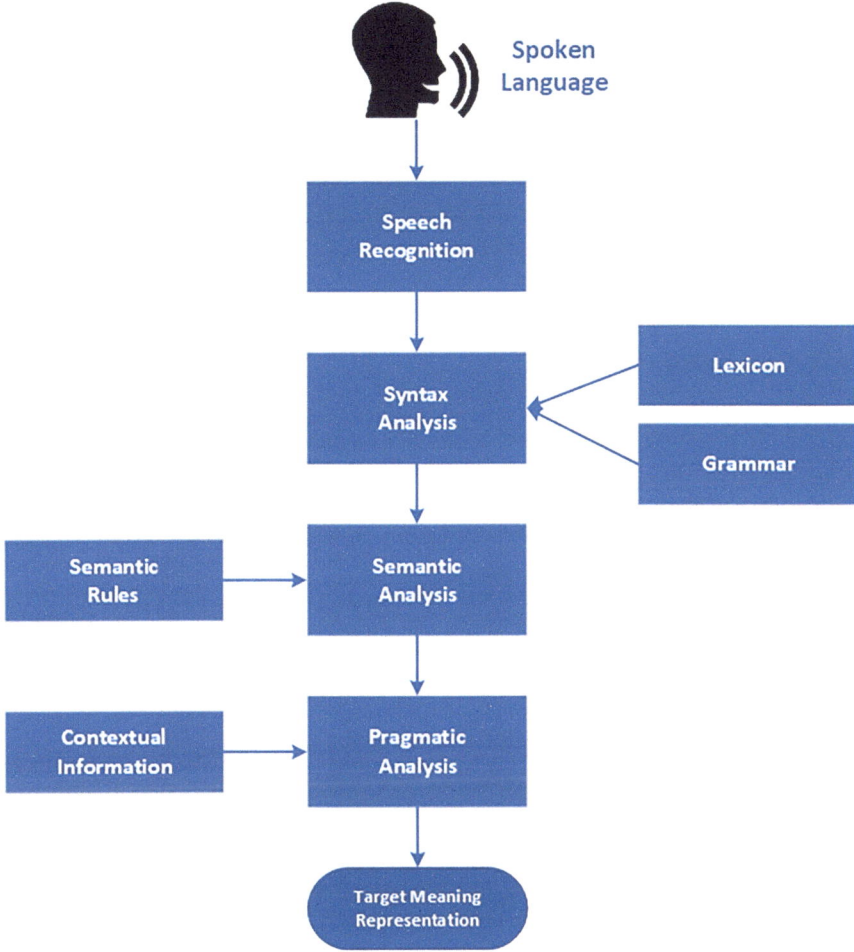

Fig. 1.6 NLU systematic diagram

1.8.1   Speech Recognition

Speech recognition (Li et al. 2015) is the first stage in NLU that performs p*honetic*, *phonological*, and *morphological* processing to analyze spoken language. The task involves breaking down the stems of spoken words called *utterances*, into distinct *tokens* representing paragraphs, sentences, and words in different parts. Current speech recognition models apply *spectrogram analysis* to extract distinct frequencies, for example, the word *uncanny* can be split into two-word tokens *un* and *canny*. Different languages have different spectrogram analyses.

1.8.2   Syntax Analysis

Syntax analysis (Sportier et al. 2013) is the second stage of NLU direct response *speech recognition* to analyze the structural meaning of spoken sentences. This task has two purposes: (1) check syntax correctness of the sentence/utterance and (2) break down spoken sentences into syntactic structures to reflect syntactic relationship between words. For instance, the utterance *oranges to the boys* will be rejected by syntax parser because of syntactic errors.

1.8.3   Semantic Analysis

Semantic analysis (Goddard 1998) is the third stage in NLU and corresponds to *syntax analysis*. This task is to extract the precise *meaning* of a sentence/utterance, or dictionary meanings defined by the text and reject meaningless, for example, semantic analyzer rejects word phrase like *hot snowflakes* despite correct syntactic words meaning but incorrect semantic meaning.

1.8.4   Pragmatic Analysis

Pragmatic analysis (Ibileye 2018) is the fourth stage in NLU and stringent spoken language analysis involving high level or expert knowledge with common sense, for example, *will you crack open* the *door? I'm getting hot.* This sentence/utterance requires extra knowledge in the second clause to understand *crack* is to break in semantic meaning, but it should be interpreted as to open in pragmatic meaning.

1.9   Potential Applications of NLP

After years of R&D from machine translation and rule-based systems to data mining and deep networks, NLP technology has a wide range of applications in everyday activities such as *machine translation, information retrieval, sentiment analysis, information extraction*, and *question-answering chatbots* as in Fig. 1.7.

1.9.1   Machine Translation (MT)

Machine translation (Koehn, 2009; Scott 2018) is the earliest application in NLP since 1950s. Although it is not difficult to translate one language to another yet there are dilemmas for (1) *naturalness* (or fluency) means different languages

Fig. 1.7 Potential NLP
applications

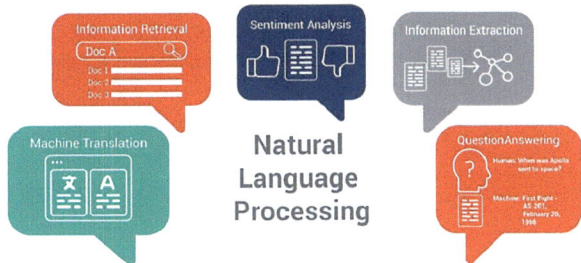

have different styles and usages and (2) *adequacy* (or accuracy) means different languages may present independent ideas in different languages. Experienced human translators address this trade-off in creative ways such as statistical methods, or *case-by-case rule-based systems* in the past but since there have been many ambiguous scenarios in language translation, the goal of machine translation R&D nowadays strive for several AI techniques applications for recurrent networks, or deep networks backbox systems to enhance machine learning capabilities.

1.9.2   Information Extraction (IE)

Information extraction (Hemdev 2011) is an application task to extract key language information from texts or utterances automatically. It can be structural, semi-structural machine-readable documents or from users' languages of NLP in most cases. The recent activities in complex formats such as audio, video, and even interactive dialogue can be extracted from multiple media. Hence, many commercial IE programs become domain-specific such as medical science, law or AI Tutor-specified AI knowledge in our case. By doing so, it is easier to set up an ontology graph and ontology knowledge base to contain all the retrieved information that can be referenced to these domain knowledge graphs to extract useful knowledge.

1.9.3   Information Retrieval (IR)

Information retrieval (Peters et al. 2012) is an application for organizing, retrieving, storing, and evaluating information from documents, source repositories, especially textual information, and multimedia such as video and audio knowledge bases. It helps users to locate relevant documents without answering any questions explicitly. The user must request the IR system to retrieve the relevant output and respond in document form.

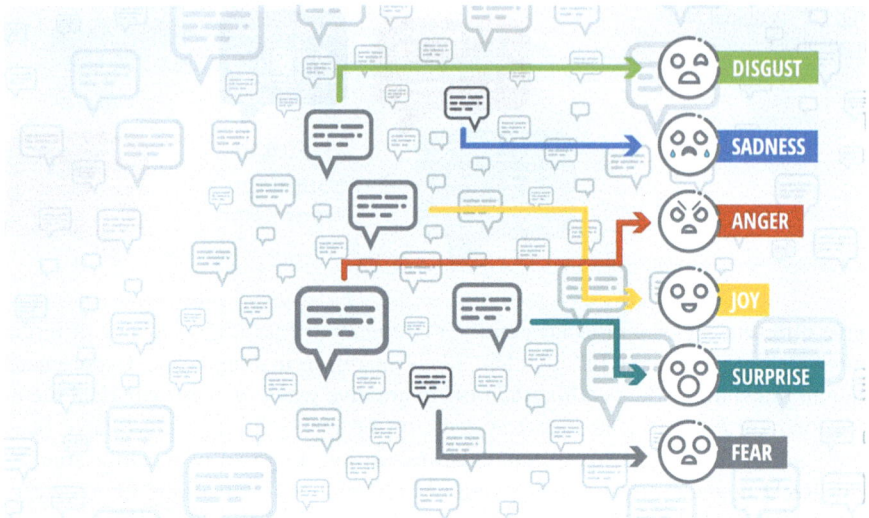

Fig. 1.8 NLP on *sentiment analysis*

1.9.4   Sentiment Analysis

Sentiment analysis (Liu 2012) is a kind of data mining system in NLP to analyze user sentiment toward products, people, and ideas from social media, forums, and online platforms. It is an important application for data extraction from messages, comments, and conversations published on these platforms and assigns a labeled sentiment classification as in Fig. 1.8 to interpret natural language and utterances. *Deep networks* are ways to analyze large amounts of data. In Part 2, the NLP Implementation Workshop will explore how to implement sentiment analysis in detail using *Python spaCy* and *Transformer Technology*.

1.9.5   Question-Answering (Q&A) Chatbots

Q&A systems is the objective in NLP (Raj 2018). A process flow is necessary to implement a Q&A chatbot. It includes voice recognition to convert into a list of tokens in sentences/utterances, syntactic grammatical analysis, semantic meaning analysis of whole sentences, and pragmatic analysis for embedded or complex meanings. When an enquirer's utterance meaning is generated, it is necessary to search from knowledgebase for the most appropriate answer or response through inferencing either by a *rule-based system*, *statistical system*, or *deep network*, for example, *Google BERT system*. Once a response is available, reverse engineering is required to generate a natural voice from a verbal language called voice synthesis. Hence, the Q&A system in NLP is an important technology that can be applied to daily activities such as human-computer interaction in auto-driving, customer services support, and language skills improvement.

The final workshop will discuss how to integrate various Python NLP implementation tools including *NLTK, spaCy, TensorFlow Keras,* and *Transformer Technology* to implement a Q&A movies chatbot system.

References

Allen, J. (1994) Natural Language Understanding (2nd edition). Pearson

Bender, E. M. (2013) Linguistic Fundamentals for Natural Language Processing: 100 Essentials from Morphology and Syntax (Synthesis Lectures on Human Language Technologies). Morgan & Claypool Publishers

Bender, E. M., Gebru, T., McMillan-Major, A., & Shmitchell, S. (2021). On the Dangers of Stochastic Parrots: Can Language Models Be Too Big? Proceedings of the 2021 ACM Conference on Fairness, Accountability, and Transparency, 610-623. https://doi.org/10.1145/3442188.3445922

Brown, T. B., Mann, B., Ryder, N., Subbiah, M., Kaplan, J., Dhariwal, P., Neelakantan, A., Shyam, P., Sastry, G., Askell, A., Agarwal, S., Herbert-Voss, A., Krueger, G., Henighan, T., Child, R., Ramesh, A., Ziegler, D. M., Wu, J., ... & Amodei, D. (2020). Language Models are Few-Shot Learners. arXiv preprint arXiv:2005.14165.

Climiano et al. (2014). Advancements in agent ontology for knowledge engineering. Journal of Natural Language Processing, 10(2), 123–145.

Cui, Y., Huang, C., Lee, Raymond (2020). AI Tutor: A Computer Science Domain Knowledge Graph-Based QA System on JADE platform. World Academy of Science, Engineering and Technology, Open Science Index 168, International Journal of Industrial and Manufacturing Engineering, 14(12), 543 - 553.

Devlin, J., Chang, M. W., Lee, K., & Toutanova, K. (2018). BERT: Pre-training of Deep Bidirectional Transformers for Language Understanding. arXiv preprint arXiv:1810.04805.

Eisenstein, J. (2019) Introduction to Natural Language Processing (Adaptive Computation and Machine Learning series). The MIT Press.

Floridi, L., & Chiriatti, M. (2020). GPT-3: Its Nature, Scope, Limits, and Consequences. Minds and Machines, 30(4), 681-694. https://doi.org/10.1007/s11023-020-09548-1

Goddard, C. (1998) Semantic Analysis: A Practical Introduction (Oxford Textbooks in Linguistics). Oxford University Press.

Green, B., Wolf, A., Chomsky, C. and Laughery, K. (1961). BASEBALL: an automatic question-answerer. In Papers presented at the May 9-11, 1961, western joint IRE-AIEE-ACM computer conference (IRE-AIEE-ACM'61 (Western)). Association for Computing Machinery, New York, NY, USA, 219–224.

Hausser, R. (2014) Foundations of Computational Linguistics: Human-Computer Communication in Natural Language (3rd edition). Springer.

Hemdev, P. (2011) Information Extraction: A Smart Calendar Application: Using NLP, Computational Linguistics, Machine Learning and Information Retrieval Techniques. VDM Verlag Dr. Müller.

Hochreiter, S., & Schmidhuber, J. (1997). Long short-term memory. Neural Computation, 9(8), 1735-1780.

Ibileye, G. (2018) Discourse Analysis and Pragmatics: Issues in Theory and Practice. Malthouse Press.

LeCun, Y., Bengio, Y., & Hinton, G. (2015). Deep learning. Nature, 521(7553), 436-444. https://doi.org/10.1038/nature14539

Lee, R. S. T. (2020). AI in Daily Life. Springer.

Li, J. et al. (2015) Robust Automatic Speech Recognition: A Bridge to Practical Applications. Academic Press.

Liu, B. (2012) Sentiment Analysis and Opinion Mining. Morgan & Claypool Publishers.

Marcus, G., & Davis, E. (2020). GPT-3, Bloviator: OpenAI's Language Generator Has No Idea What It's Talking About. MIT Technology Review. https://www.technologyreview. com/2020/08/22/1007539/gpt3-openai-language-generator-artificial-intelligence-ai-opinion/

Mikolov, T., Karafiát, M., Burget, L., Cernocký, J., & Khudanpur, S. (2010). Recurrent neural network-based language model. In Eleventh Annual Conference of the International Speech Communication Association.

OpenAI. (2023). GPT-4 Technical Report. https://openai.com/research/gpt-4

Peters, C. et al. (2012) Multilingual Information Retrieval: From Research To Practice. Springer.

Radford, A., Wu, J., Child, R., Luan, D., Amodei, D., Sutskever, I., & OpenAI. (2019). Language Models are Unsupervised Multitask Learners. OpenAI Blog, 1(8), 9.

Raj, S. (2018) Building Chatbots with Python: Using Natural Language Processing and Machine Learning. Apress.

Santilal, U. (2020) Natural Language Processing: NLP & its History (Kindle edition). Amazon.com.

Scott, B. (2018) Translation, Brains and the Computer: A Neurolinguistic Solution to Ambiguity and Complexity in Machine Translation (Machine Translation: Technologies and Applications Book 2). Springer.

Solaiman, I., Brundage, M., Clark, J., Askell, A., Herbert-Voss, A., Wu, J., ... & Krueger, G. (2019). Release Strategies and the Social Impacts of Language Models. arXiv preprint arXiv:1908.09203.

Sportier, D. et al. (2013) An Introduction to Syntactic Analysis and Theory. Wiley-Blackwell.

Tuchong (2020a) The Turing Test. https://stock.tuchong.com/image/detail?imag eId=921224657742331926. Accessed 17 Dec 2024.

Tuchong (2020b) NLP and AI. https://stock.tuchong.com/image/detail?imag eId=1069700818174345308. Accessed 17 Dec 2024.

Turing, A. (1936) On computable numbers, with an application to the Entscheidungsproblem. In: Proc. London Mathematical Society, Series 2, 42:230–26

Turing, A. (1950) Computing Machinery and Intelligence. Mind, LIX (236): 433–460.

Chapter 2
N-Gram Language Model

2.1 Introduction

NLP entities like *word-to-word tokenization* using NTLK and spaCy technologies in Workshop 1 (Chap. 11) analyzed words in insolation, but the relationship between words is important in NLP. This chapter will focus on *word sequences*, their modeling, and analyses.

In many NLP applications, there are noises and disruptions affecting incorrect word pronunciation regularly in applications like speech recognition, text classification, text generation, machine translation, Q&A chatbots, Q&A conversation machines, or agents being used in auto-driving.

Humans experience mental confusion about spelling errors as in Fig. 2.1 often caused by pronunciations, typing speeds, and keystroke's location. These errors can be corrected by looking up a dictionary, a spell checker, and grammar usage.

Word prediction in a word sequence can provide automatic spell-check corrections, and its corresponding concept terminology can model words relationships, estimate occurrence frequency to generate new texts with classification, and apply in machine translation to correct errors.

Probability or *word counting* method can work on a large databank called a *corpus* (Pustejovsky and Stubbs 2012), which can be the collection of text documents, literatures, public speeches, conversations, and other online comments or opinions.

Figure 2.2 shows the text with spelling and grammatical errors highlighted in yellow and blue. This method can calculate the probability of word occurrence and provide alternatives with higher frequency probability, but it cannot always provide accurate options.

Figure 2.3 illustrates a simple scenario of *next-word prediction* in sample utterances *I like photography, I like science,* and *I love mathematics.* The probability of *I like* is *0.67 (2/3)* compared with *I love* is *0.33 (1/3)*, the probability of *like photography* and *like science* are similar at *0.5 (1/2)*. Assigning probability to scenarios, *I*

R. Lee, *Natural Language Processing*,
https://doi.org/10.1007/978-981-96-3208-4_2

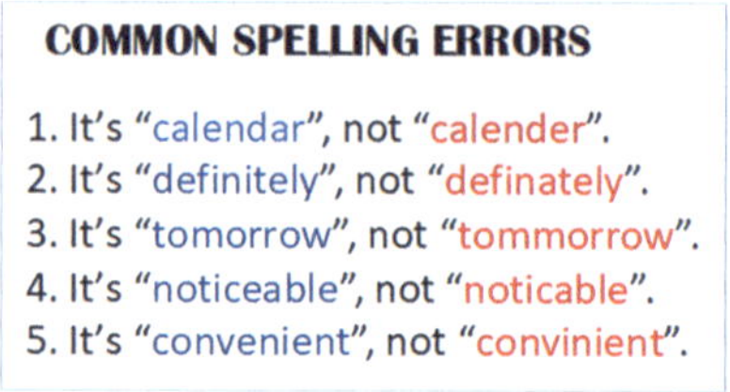

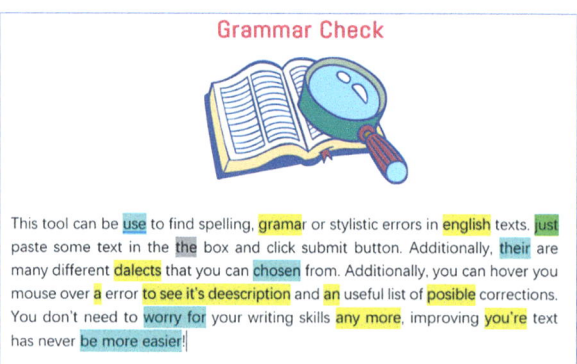

COMMON SPELLING ERRORS
1. It's "calendar", not "calender".
2. It's "definitely", not "definately".
3. It's "tomorrow", not "tommorrow".
4. It's "noticeable", not "noticable".
5. It's "convenient", not "convinient".

Fig. 2.1 Common spelling errors

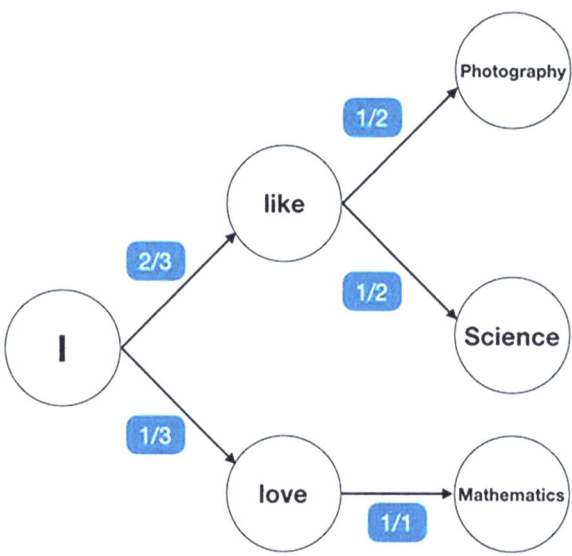

Grammar Check
This tool can be use to find spelling, gramar or stylistic errors in english texts. just paste some text in the the box and click submit button. Additionally, their are many different dalects that you can chosen from. Additionally, you can hover you mouse over a error to see it's deescription and an useful list of posible corrections. You don't need to worry for your writing skills any more, improving you're text has never be more easier!

Fig. 2.2 Spelling and grammar checking tools

I
like
love
Photography
Science
Mathematics
2/3
1/3
1/2
1/2
1/1

Fig. 2.3 Next word prediction in simple utterances

like photography and *I like science* are both *0.67 × 0.5 = 0.335*, and *I love mathematics* is *0.33 × 1 = 0.33*.

When applying probability to language models, it must always note (1) domain-specific verity of keywords togetherness and terminology knowledge varies according to domains, for example, medical science, AI, etc., (2) syntactic knowledge attributes to syntax and lexical knowledge, (3) common sense or world knowledge attributes to the collection of habitual behaviors from past experiences, and (4) languages usage significance in high-level NLP.

When applying probability to word prediction in an utterance, there are words often proposed by *rank* and *frequency* to provide a sequential optimum estimation.

For example:

[2.1] I notice three children standing on the??? (ground, bench …).
[2.2] I just bought some oranges from the??? (supermarket, shop …)
[2.3] She stopped the car and then opened the??? (door, window …)

The structure of [2.3] is perplexed because the word counting method with a sizeable knowledge domain is adequate but *common sense*, *world knowledge* or specific *domain knowledge* are among the sources. It involves scenario syntactic knowledge that attributes to do something with the superior level at the scene such as descriptive knowledge to help the guesswork. Although it is plain and mundane to study preceding and word tracking, it is one the most useful techniques for word prediction. Let's begin with some simple word-counting methods in NLP, the *N-gram language model*.

2.2 N-Gram Language Model

It was learned that the motivations for word prediction can apply to voice recognition, text generation, and Q&A chatbot. The *N-gram language model*, also called *N-gram model* or *N-gram* (Sidorov 2019; Liu et al. 2020), is a fundamental method to formalize word prediction using probability calculation. An *N*-gram is a statistical model consisting of a word sequence in N number, commonly used N-grams include:

- *Unigram* refers to a single word, that is, $N = 1$. It is seldomly used in practice because it contains only one word in N-gram. However, it is important to serve as the base for higher-order N-gram probability normalization.
- *Bigram* refers to a collection of two words, that is, $N = 2$. For example: *I have, I do, he thinks, she knows, etc*. It is used in many applications because its occurrence frequency is high and easy to count.
- *Trigram* refers to a collection of three words, that is, $N = 3$. For example: *I noticed that, noticed three children, children standing on, standing on the*. It is useful because it contains more meaning and is not lengthy. Given a count knowl-

edge of the first three words can easily guess the next word in a sequence. However, its occurrence frequency is low in a moderate corpus.

- *Quadrigram* refers to a collection of four words, that is, $N = 4$. For example: *I noticed that three, noticed that three children, three children standing on*. It is useful with literature or large corpus like Brown Corpus because of their extensive words' combinations.

A sizeable N-gram can present more central knowledge but can pose a problem. If it is too large, it means that the probability and occurrence of a word sequence are infrequent and even 0 in terms of probability counts.

Corpus volume and other factors also affect performance. N-gram model training is based on an extensive *knowledgebase (KB)* or *databank* from specific domains such as public speeches, literature, and topic articles like news, finance, medical, science, or chat messages from social media platforms. Hence, a moderate N-gram is the balance by *frequency* and *proportions*.

The knowledge of counts acquired by a N-gram can assess to conditional probability of candidate words as the next word in a sequence, e.g., *It is not difficult. It is* a bigram which means to count the occurrence of *is* given that *it* has already been mentioned from a large corpus, or the conditional probability of *it is* given that *it* has already been mentioned or can be applied one by one to calculate the conditional probability of an entire words sequence. It is like words and sentences formation of day-to-day conversations which is a psychological interpretation in logical thinking. N-gram progresses in this orderly fashion.

It serves to rank the likelihood of a sequence consisting of various alternative hypotheses in a sentence/utterance for an application like automatic speech recognition (ASR), for example, *[2.4] The cinema staff told me that popcorn/amelcorn sales have doubled*. It is understood that it refers to *popcorn* and not *amelcorn* because the concept of *popcorn* is always attributed to conversations about cinema. Since the occurrence of *popcorn* in a sentence/utterance has a higher rank than *amelcorn*, it is natural to select *popcorn* as the best answer.

Another purpose is to assess the likelihood of a sentence/utterance for text generation or machine translation, for example, *[2.5] The doctor recommended a <u>cat scan</u> to the patient*. It may be difficult to understand what a *cat scan* is or how can a scan be related to a *cat* without any domain knowledge. Since the word "*doctor*" is attributed to the medical field, it is natural that by searching articles, literature, and websites about medical knowledge, we will learn that *cat scan* refers to a computer axial tomography scanner as shown in Fig. 2.4, not a *cat*. This type of word prediction is usually domain-specific and works together with previous words as a guide to choosing an appropriate expression.

Fig. 2.4 Computerized axial tomography scanner (aka. *CAT scan*) (Tuchong 2022)

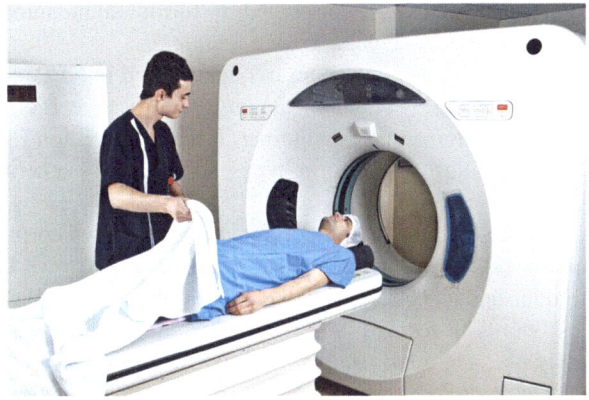

Table 2.1 Stemming vs. lemmatization

Word	Stemming	Lemmatization
information	inform	information
informative	inform	information
computers	comput	computer
feet	feet	foot

2.2.1 Basic NLP Terminology

Here is a list of common terminologies in NLP (Jurafsky et al. 1999; Eisenstein 2019):

- *Sentence* is a unit of written language. It is a basic entity in a conversation or utterance.
- *Utterance* is a unit of spoken language. Different from the concept of *sentence*, *utterance* is usually domain and culture specific, which means it varies according to countries and even within country.
- *Word Form* is an inflected form that occurs in a corpus. It is another basic entity in a corpus.
- *Types/Word Types* are distinct words in a corpus.
- *Tokens* are generic entities or objects of a passage. It is different from word form as tokens can be meaningful words or symbols, punctuations, or simple and distinct character(s).
- *Stem* is a root form of words. *Stemming* is the process of reducing inflected, or derived words from their word stem.
- *Lemma* is an abstract form shared by word forms in the same stem, part of speech, and word sense. *Lemmatization* is the process of grouping together the inflected forms of a word so that they can be analyzed as a single item, which can be identified by the word's lemma or dictionary form.

An example to demonstrate meaning representations between *lemma* and *stem* is shown in Table 2.1. *Lemmatization* is the abstract form to generate a concept. It indicated that *stem* or *root word* can be a meaningful word, or meaningless, or a symbol such as *inform* or *comput* to formulate meaningful words such as *informa-tion*, *informative*, *computer*, or *computers*.

There are several corpora frequently used in NLP applications.

Google (2022) is one of the largest corpora as it contains words and texts from its search engine and the internet. It has over trillion English tokens with over million meaningful wordform types sufficient to generate sentences/utterances for daily use.

Brown Corpus is an important and well-known corpus because it is the first well-organized corpus in human history founded by Brown University in 1961 with continuous updates. At present, it has over 583 million tokens, 293,181 wordform types and words in foreign languages. It is one of the most comprehensive corpora for daily use, and a KB used in many N-grams-related NLP models and applications.

The *Wall Street Journal* is one of the earliest domain-specific corpora to discover knowledge from financial news, the *Associated Press* focuses on news and international events, *Hansard* is a famous *corpus of British parliamentary speeches*; *Boston University Broadcast News Corpus*; *NLTK Corpus*, etc. (Bird et al. 2009; Eisenstein 2019; Pustejovsky and Stubbs 2012).

A *language model* is a traditional word counting model to count and calculate conditional probability to predict the probability based on a word sequence, e.g., applying utterance *it is difficult to… that* with a sizeable corpus like *Brown Corpus*. This traditional word counting method may suggest either *say/tell/guess* based on occurrence frequency for predictions and forecasts at advanced computer systems and research in specialized deep networks and AI models. Although there has been a technology shift, a statistical model is always the fundamental model in many cases (Jurafsky et al. 1999; Eisenstein 2019).

2.2.2   Language Modeling and Chain Rule

Conditional probability calculation is to study the definition of conditional probabilities and look for counts, given by:

$$P(A|B) = \frac{P(A \cap B)}{P(B)} \tag{2.1}$$

For example, to evaluate conditional probability: *The garden is so beautiful that* given by the word sequence "*The garden is so beautiful*" will be:

$$
\begin{aligned}
P(\text{that}|\text{The garden is so beautiful}) &= \frac{P(\text{The garden is so beautiful that})}{P(\text{The garden is so beautiful})} \\
&= \frac{\text{Count}(\text{The garden is so beautiful that})}{\text{Count}(\text{The garden is so beautiful})}
\end{aligned} \tag{2.2}
$$

Although the calculation is straightforward but if the corpus or text collection is moderate, this conditional probability (counts) will probably be zero.

The *Chain Rule* of probability can be used as an independent hypothesis to correct this problem.

By rewriting the conditional probability equation (2.1), it will be:

$$P(A \cap B) = P(A|B)P(B) \tag{2.3}$$

For a sequence of events, A, B, C, and D, the Chain Rule formulation will become:

$$P(A,B,C,D) = P(A)P(B|A)P(C|A,B)P(D|A,B,C) \tag{2.4}$$

In general:

$$P(x_1,x_2,x_3,\ldots x_n) = P(x_1)P(x_2|x_1)P(x_3|x_1,x_2)\ldots P(x_n|x_1\ldots x_{n-1}) \tag{2.5}$$

If the sequence of words from position 1 to n is defined as w_1^n, then the *Chain Rule* applied to word sequences becomes:

$$\begin{aligned} P(w_1^n) &= P(w_1)P(w_2|w_1)P(w_3|w_1^2)\ldots P(w_n|w_1^{n-1}) \\ &= \prod_{k=1}^{n} P(w_k|w_1^{k-1}) \end{aligned} \tag{2.6}$$

So, the conditional probability for the previous example will be:

$$\begin{aligned} P(\text{the garden is so beautiful that}) &= P(\text{the}) * P(\text{garden|the}) \\ &{}^* P(\text{is|the garden}) * P(\text{so|the garden is}) * P(\text{beautiful|the garden is so}) \\ &{}^* P(\text{that|the garden is so beautiful}) \end{aligned} \tag{2.7}$$

Note: Normally, <s> and </s> are used to denote the start and end of a sentence/utterance for better formulation.

This approach seems fair and easy to understand, but there are two major problems. First, it is unlikely to collect correct prefix statistics, which means that the starting point of the sentence is unknown. Second, the calculation of word order probabilities is mundane. If the sentence is long, the conditional probability at the end of this equation is complicated to calculate.

Let's explore how the genius *Markov Chain* is applied to solve this problem.

2.3   Markov Chain in N-Gram Model

Prof. Andrey Andrevevich Markov (1856–1922) was a renowned Russian mathematician and academician who made a significant contribution to science by studying the theory of probability, his primary contribution called *Markov chains* or *Markov process* had applied to biology, chemistry, computer science, and statistics (Ching et al. 2013). *Markov chains theory* can be applied to speech recognitions, N-gram language model, internet ranking, information, and queueing theories (Eisenstein 2019). There is a single-dimension domain *Markov chain* modeling called *Hidden Markov Chain* in handwritten characters and human voice recognitions. This model has an important concept called *Markov Assumption* which assumes the entire prefix history is not necessary, in other words, an event doesn't depend on its whole history; it is only a fixed length nearby history is the essence of *Markov chain theory*.

A Markov chain is a stochastic process that describes a sequence of possible events where the probability of each event depends only on the state reached by the previous event. There are many kinds of *Markov chain* conventions. An important convention called *descriptive Markov chain* is shown in Fig. 2.5. It revealed that an event of Markov chain can be a list of relationships of every single event. Another concept is that the previous state is important but not all previous sequences. Hence, this model can apply in thermodynamics, statistical mechanics, physics, chemistry, economy, finance, information theory and NLP. A complete *Markov Chain event* is like a conversation in a sentence/utterance, each word is equivalent to an object in *Markov chain*. Although the whole chain of conditional probability can be calculated, the last event is the most important one.

By applying the *Markov chain model*, the conditional probability for N-gram probability of a word sequence w_1^n will be approximated by (assuming a prefix of N words):

$$P\left(w_n | w_1^{n-1}\right) \approx P\left(w_n | w_{n-N+1}^{n-1}\right) \tag{2.8}$$

In general:

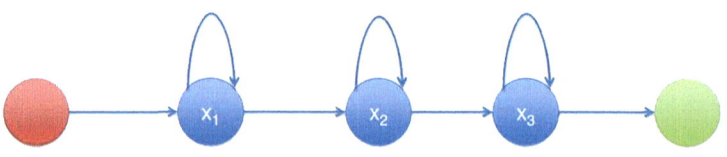

$$p(x_1, x_2, x_3, \ldots, x_n) = p(x_1)p(x_2|x_1)p(x_3|x_2)\ldots p(x_n|x_{n-1})$$

Fig. 2.5   Markov chain model

$$P\left(w_1^n\right) \approx \prod_{k=1}^{n} P\left(w_k | w_{k-1}\right) \tag{2.9}$$

In other words, the original complex conditional probability of a word sequence stated in (2.6) can be easily evaluated by a sequence of bigram probability calculations.

Let's look at an N-gram example *The big white cat.* Unigram probability is *P(cat).* Bigram probability is the *P(cat | white).* Trigram probability is *P(cat | big white)* = *P(white | big)*P(cat | white)* and quadrigram probability is *P(cat | the big white)* = *P(big | the)*P(white | big)*P(cat | white).* All can be easily evaluated by a simple sequence of bigram probability multiplications applying (2.9).

However, it is cautious to note that the probability of a word formulation given fixed prefixes may not always be appropriate in many cases. They may be verifiable events in real-time speeches as words uttered are often correlated to the previous but in cases with pragmatic or embedded meanings at both right and left contexts, there is no priori reason adhered to left contexts.

2.4   Example: The Adventures of Sherlock Holmes

N-gram probability calculation usually come from a training *corpus* or *knowledge-base (KB)* in two extremes. One is an overly narrow corpus, and the other one is an overly general corpus. An overly narrow corpus is a restricted, specific corpus, can be domain specific on a particular knowledge with significant counts to be found during conditional probability counting. An overly general corpus cannot reflect a specific domain but counting can always be found. Hence, a balance between the two dimensions is required. Another consideration is a separate text corpus applied to evaluate standard metrics called *held out* test set, or *development* test set. Further, cross-validations and results tested for statistical significance are also required.

Let's begin with a corpus came frthe om *Project Gutenberg* website (Gutenberg 2022) on *The Adventures of Sherlock Holmes* (Doyle 2019), a famous literature by writer and physician Sir Arthur Conan Doyle (1859–1930). Gutenberg is a website consisting of primarily copyright clearance, free access, and download western cultural tradition literature available to the public. This literature has 12 outstanding detective stories of *Sherlock Holmes* ranging from *A Scandal in Bohemia* to *The Adventure of the Copper Beeches* with other statistics below. It is a domain-specific corpus with comprehensive detective knowledge to form a meaningful knowledge base to perform N-gram modeling in NLP.

No. of pages	:	424
No. of characters (exclude spaces)	:	470,119
No. of words	:	110,087
No. of tokens	:	113,749

No. of sentences	:	6830
No. of word types (V)	:	9886

N-gram modeling in this example is to analyze an influential quote of Sherlock Holmes: *I have no doubt that I* …. This quote does not occur often in other literature but because it is a detective story, the character has a unique aptitude for deducing hypotheses and notions to solve cases. Applying the *Markov chain model* can avoid mundane conditional probability, the N-gram probability is given by:

$$P(\text{I have no doubt that I}) = P(I|\langle s \rangle) * P(\text{have | I}) * P(\text{no | have})$$
$$^* P(\text{doubt | no}) * P(\text{that | doubt}) * P(\text{I | that}) \tag{2.10}$$

Unigram checking on word counting for *I have no doubt* is necessary as a basis to calculate the conditional probability for all bigrams as shown in Table 2.2. So, given the unigram count of *I* is 2755, the bigram probability of *I have* applying Markov chain method will be 288/2755, which is 0.105 as shown in Table 2.3. It is a list of all related bigram counts and probabilities for a given bigram such as *I have, I had, I am, I was, I knew, I hear, I don't* up to *I should* which are common words found in many literatures. The probability also showed *I have* is the most frequent with 0.105 which means that *I have no doubt that* is quoted by the character regularly. The occurrence of *I think* is high and general phrases such as *I have, I had*.

A bigram grammar fragment related to *I have no doubt that* … is shown in Table 2.4 for the counting and probability occurrence frequency beginning with *<s>I, <s>I'd, <s>The, <s>It, I have, I had, I can, have no, have to, have been* to compare with several versions or combinations related to, *I have no doubt that* means to compare occurrence frequency of *I have* with *I had* or *I can,* which is similar to compare the occurrence of *no doubt, no sign* and *no harm* or *that I, that he, that she, that it*. It is noted that the occurrence of *I have no doubt that* is high and distinct in this literature.

Counting all conditional bigram probabilities based on unigram count in Table 2.2 showed *I have no doubt that* for *I* is at 0.138 which is very high, but it is interesting to note that *no doubt* is even higher at 0.167 but again since it is a detective story with a restricted domain, *doubt that* is very high at 0.202 because the character always involves guesswork and frequent grammar usage. Further, the probability of bigram *that I* is much higher than other combinations like *that he, that she* and *that it*. The occurrence frequency in other literature is much lower but because the character is a self-assured and intelligent expert, he said *that I* is more

Table 2.2 Unigram counts for words "*I have no doubt that*" from *The Adventures of Sherlock Holmes*

I	have	no	doubt	that
2755	867	276	84	1767

Table 2.3 Bigram grammar fragment from *The Adventures of Sherlock Holmes*

Bigram with "I" (by counts)				Bigram with "I" (by probability)			
I have	288	I observe	8	I have	0.105	I observe	0.003
I had	161	I deduce	3	I had	0.058	I deduce	0.001
I am	159	I can	37	I am	0.058	I can	0.013
I was	147	I can't	6	I was	0.053	I can't	0.002
I knew	34	I may	23	I knew	0.012	I may	0.008
I hear	33	I must	32	I hear	0.012	I must	0.012
I don't	14	I could	77	I don't	0.005	I could	0.028
I saw	42	I passed	8	I saw	0.015	I passed	0.003
I think	72	I take	4	I think	0.026	I take	0.001
I should	90	I see	32	I should	0.033	I merely	0.012

Table 2.4 Bigram grammar fragment related to utterance "*I have no doubt that I*" from *The Adventures of Sherlock Holmes*

Bigram related to "I have no doubt that I" (by counts)				Bigram related to "I have no doubt that I" (by probability)			
<s>I	883	no doubt	46	<s>I	0.138	no doubt	0.167
<s>I'd	4	no sign	9	<s>I'd	0.001	no sign	0.033
<s>The	164	no harm	4	<s>The	0.026	no harm	0.014
<s>It	229	doubt that	17	<s>It	0.036	doubt that	0.202
I have	288	doubt as	3	I have	0.105	doubt as	0.036
I had	161	doubt upon	2	I had	0.058	doubt upon	0.024
I can	37	that I	228	I can	0.013	that I	0.129
have no	35	that he	139	have no	0.040	that he	0.079
have to	12	that she	61	have to	0.014	that she	0.035
have been	122	that it	109	have been	0.141	that it	0.062

often than *that he* or *that she*. That is the significance of a domain-specific corpus to check for N-gram probability.

So, let's look at some N-gram probabilities calculation, e.g., the probability of *P(I have no doubt that I)* given by Eq. 2.10:

$$P\left(\text{I have no doubt that I}\right) = 0.138 \times 0.105 \times 0.040 \times 0.167 \times 0.202 \times 0.129$$
$$= 0.000002526$$

It is compared with *P(I have no doubt that he)*:

$$P\left(\text{I have no doubt that he}\right) = 0.138 \times 0.105 \times 0.040 \times 0.167 \times 0.202 \times 0.079$$
$$= 0.000001540$$

This example test results led to several observations. It is noted that all these probabilities are limited in general. Conditional probability is limited in a long sentence and required for the *Markov chain*. If applying the traditional method on conditional probability with complex calculation, most of the time the probability is diminished. Further, the probability seems to capture both *syntactic facts* and *world knowledge*. Although *that I* or *that he* is often used in English grammar, the probability in this literature *that I* is more frequent. Hence, it is related to both *syntactic usage*, *common sense*, and *specific domain knowledge*. It depends on knowledge domains leading to diverse probability calculation results.

It is also noted that most of the conditional probabilities are limited because the multiplication of all probability calculations in a long sentence becomes diminished, so it is important to apply the Markov chain and convert complex conditional probabilities into bigram probabilities. Although the occurrence of bigram is infrequent but still exists. Nevertheless, if it is not a sizeable knowledge base or corpus, most of the bigrams will be 0. Hence, the selection of a corpus knowledge base is important. An effective N-gram is related to word counting, conditional probabilities calculation, and normalization.

Another observation is that it showed all these conditional probabilities are limited and underflows as mentioned. A method is to convert them into *natural log*. Applying a *natural log* will become additions to calculate conditional probability with *Markov chain* operation.

Maximum Likelihood Estimates (MLE) is another principal method to calculate the N-gram model. They are parameters of a model M from a training set T. It is the estimate that maximizes the likelihood of training set T given the model M. Suppose the word *language* occurred 380 times in a corpus with a million words, for example, Brown corpus, the probability of a random word from other text forms with the same distribution will be *language*, which it will be 380/1,000,000 = 0.00038. This may be a poor estimate for other corpora, but this type of calculation is domain specific as mentioned meaning that the calculation varies according to different corpora.

Let's return to *The Adventures of Sherlock Holmes'* famous quote *I have no doubt that* example. This time the counting and probability calculation of these words are tabulated as shown in Tables 2.5 and 2.6, respectively. It showed that *I have* has the most occurrence frequency with 288, *that I* is the next with 228 occurrences, *no doubt* with surprising high 46 occurrences, *doubt that* is 17 followed by *that no* and so on. Another discovery is that most of the other combinations is 0. It is intuitive because they are not grammatically or syntactically possible, e.g., *no I* or *I I* and many are infrequent in English usage.

Table 2.5 Bigram counts for "*I have no doubt that I*" in *The Adventures of Sherlock Holmes*

	I	have	no	doubt	that	I
I	0	288	0	1	0	0
have	5	0	35	0	2	5
no	0	0	0	46	0	0
doubt	0	0	0	0	17	0
that	228	1	10	0	7	228

Table 2.6 Bigram probability (normalized) for "*I have no doubt that I*" in *The Adventures of Sherlock Holmes*

	I	have	no	doubt	that	I
I	0.000	0.105	0.000	0.000	0.000	0.000
have	0.006	0.000	0.040	0.000	0.002	0.006
no	0.000	0.000	0.000	0.167	0.000	0.000
doubt	0.000	0.000	0.000	0.000	0.202	0.000
that	0.129	0.001	0.006	0.000	0.004	0.129

Bigram normalization is achieved by the division of each bigram counts by appropriate unigram counts for w_{n-1}. Here is the bigram normalization result for *I have no doubt that*, for example, computing bigram probability of *no doubt* is the counting of *no doubt* which is *46* as shown in Table 2.5 against the counting of *no* which is *276* as in Table 2.2 which becomes *46/276 = 0.167*. In fact, such bigram probability *P(no doubt)* is much higher than *P(I have) = 0.105*, which is infrequent in other corpora because not many corpora have a frequency of *no doubt* as compared with *I have* as *I have* is common in English usage. Since it is detective literature and the character is an expert in his field, it is unsurprised to identify the occurrence frequency of *no doubt* is very high.

The overall *bigram probability* (normalized) findings are: *I have* is *0.105*, *no doubt* is *0.167* the highest, *that I* is *0.129* as shown in Table 2.6. This is special because the occurrence frequency of *I* is not high as compared with *I have*. *Doubt that* is *0.202* which is very high, and others are mostly *0*. These findings showed that, first, all conditional probabilities are limited because N-gram calculation characteristics come from an extensive corpus. But it doesn't mean that there is no comparison. It can be compared if they are not *0*. Second, 0 s are meaningful as most of these words' combinations are neither syntactically nor grammatically incorrect. Third, these conditional probabilities and MLE are domain specific, which may not be the same in other situations.

2.5 Shannon's Method in N-Gram Model

Shannon's method (Jurafsky et al. 1999) is another important topic in N-*gram* model. Professor Claude Shannon (1916–2001) was a renowned American mathematician, electrical engineer, cryptographer, also known as the father of *information theory* and a major founder of contemporary *cryptography*. He wrote his famous thesis at age 21, a master's degree student at MIT demonstrating Boolean algebra electrical applications to construct any logical numerical relationship with meaning. One of his most influential papers, *A mathematical theory of communications* (Shannon 1948) published in 1948 had defined a mathematical notion by which information could be quantified and delivered reliably over imperfect communication channels like phone lines or wireless connections nowadays. His

Table 2.7 Algorithm of Shannon's method for language generation

Shannon's method on language generation
1. Choose a random N-gram (<s>, w) according to its probability
2. Now choose a random N-gram (w, x) according to its probability
3. And so on until we choose </s>
4. Then string the words together into a sentence

Table 2.8 Sentence generation using the Shannon's method from *The Complete Works of Shakespeare*

N-gram	Generated sample sentences from *The Complete Works of Shakespeare*
Unigram	• To him swallowed confess hear both which of save on trail for are ay device and rote life have
	• Every enter now severally so
	• Hill he late speaks a more to leg less first you enter
	• Are where exeunt and sighs have rise excellency took of sleep knave we near vile like
Bigram	• What means sir I confess she?
	• Why dost stand forth thy canopy for sooth
	• What we hath got so she I rest and sent to scold and nature bankrupt nor the first gentlemen?
	• Ener Menenius if it so many good direction found thou art a strong upon command of fear not a liberal largess given away
Trigram	• Sweet prince Falstaff shall die
	• This shall forbid it should be branded if renown made it empty
	• Indeed the duke and had a very good friend
	• Fly and will rid me these news of price
Quadri-gram	• King Henry I will go seek the traitor Gloucester
	• Will you not tell me who I am?
	• It cannot be but so
	• Indeed the short and the long

groundbreaking innovation had provided the tools for network communications and internet technologies. This method showed that assigning probabilities to sentences are well but less informative for language generation in NLP. However, it has a more interesting task to turn it around by applying N-gram and its probabilities to generate random sentences like human sentences by which the model is derived.

There are four steps of *Shannon's Method* for *language generation* as shown in Table 2.7:

An example of four N-gram texts generation methods from *The Complete Works of Shakespeare* by William Shakespeare (1564–1616) (Shakespeare 2021) applying Shannon's Method is shown in Table 2.8.

In summary:

- *Unigram* results showed that the four random sentences are almost meaningless because they used a single word to calculate the probability mostly without relations.
- *Bigram* results showed that the four random sentences have little meaning because they used two words to calculate. It reflected its high occurrence probability frequency but was not grammatically correct.
- *Trigram* results showed that word relations are coherent because it used three words to calculate. It reflected the conditional probability ranking had improved grammar and meanings like human language.
- *Quadrigram* results showed that the language of sentences is almost perfect per original sentences since it used four words co-relation to calculate, but its high occurrence conditional probability frequency are the words encountered with low-ranking options due to copious information to search. It may not be beneficial to text generation.

Although *quadrigrams* can provide realistic language, sentences/utterances lack freedoms to generate new sentences. Hence, *trigrams* are often a suitable option for language generation. Again, if corpus is not sizeable enough to accommodate tokens and words volume like this literature, trigram will be unable to provide the frequent words for N-gram may need to switch using bigram in this case. Hence, quadrigram is unsuitable for text generation because it will be too close to the osriginal words or sentences.

Corpus used in this example is also domain specific from *The Complete Works of Shakespeare*. It consists of 884,647 tokens and 29,066 distinct words that are approximately 10 times more as compared with *The Adventures of Sherlock Holmes*. It has approximately 300,000 bigram types out of all these tokens and the number of bigram combinations will be 844 million possible bigrams. In other words, less than 1% is used and another 99.96% of possible bigrams are never used. It makes sense because most of these random bigram generations are grammatic, syntactic or

Table 2.9 Sample sentence generation using Shannon's method with *Wall Street Journal* articles

N-gram	Generated sample sentences from *Wall Street Journal* articles
Unigram	Months the my and issue of year foreign new exchange's September were recession exchange new endorsed a acquire to six executes
Bigram	Last December through the way to preserve the Hudson corporation N. B. E. C. Taylor would seem to complete the major central planners one point five percent of U. S. E. has already old M. X. corporation of living on information such as more frequently fishing to keep her
Trigram	They also point to ninety nine point six billion dollars from two hundred four oh six three percent of the rates of interest stores as Mexico and Brazil on market conditions
Quadri-gram	Executives from some of the biggest U.S. news organizations check with a British economist last year at Washington's exclusive Metropolitan Club to strategize with a mutual obsession of getting their industry out from under the thumb of Google and Facebook

even pragmatic meaningless, but pose a problem in N-gram calculations for text generation.

For illustration purposes on how domain knowledge affects N-gram generation, Table 2.9 shows some sample sentences generated by Wall Street Journal (WSJ) articles as the corpus (Jurafsky et al. 1999). It showed that trigram has the best performance in terms of sentence structure and meaningfulness on text generation.

2.6   Language Model Evaluation and Smoothing Techniques

Language Model Evaluation (LME) (Jurafsky et al. 1999) is a standard method to train parameters on a training set and to review model performance with new data constantly. That often occurred in real world to learn how the models perform called *training data (training set)* on language model and see whether it works with unseen information called *test data (test set)*. A *test set* is a data set completely different than the training set model but is drawn from the same source, which is a specific domain and applies an evaluation metric, for example, *perplexity* to determine language model effectiveness.

Unknown words are words unseen prior looking at test data regardless of how much training data is available. It can be managed by an open vocabulary task with steps below:

1. Create an unknown word token <UNK>.
2. Train <UNK> probabilities.

    (a) create a fixed lexicon L, of size V from a dictionary or a subset of terms from the training set.
    (b) a subnet of terms from the training set.
    (c) at text normalization phase, any training word not in L changed o <UNK>.
    (d) now can count that like a normal word.

3. Test.

    (a) use <UNK> counts for any word not in training.

2.6.1   Perplexity

Perplexity (PP) is the probability of the test set assigned by the language model, normalized by the number of words as given by:

$$PP(W) = \sqrt[N]{\frac{1}{P(w_1 w_2 \ldots w_N)}} \qquad (2.11)$$

By applying the Chain rule, it will become:

$$\text{PP}(W) = \sqrt[N]{\prod_{k=1}^{N} \frac{1}{P(w_k | w_1 w_2 \dots w_{k-1})}} \tag{2.12}$$

For bigrams, it will be given by:

$$\text{PP}(W) = \sqrt[N]{\prod_{k=1}^{N} \frac{1}{P(w_k | w_{k-1})}} \tag{2.13}$$

In general, minimizing perplexity is the same as maximizing probability for model performance, which means the best language model is the one that can best predict an unseen test set with minimized perplexity rate.

An example of perplexity values for WSJ is shown in Table 2.10 indicating that trigram with minimized perplexity has performed better than bigram and unigram supported this principle for text generation (Jurafsky et al. 1999).

2.6.2 Extrinsic Evaluation Scheme

An *extrinsic evaluation* is a popular method for N-gram evaluation, its theory is straightforward as follows:

1. Put model A into an application, for example, a speech recognizer or even a QA chatbot.
2. Evaluate application performance with model A.
3. Put model B into the application and evaluate.
4. Compare two models' application performance.

The good thing about extrinsic evaluation is that it can perform exact testing at two models which is fair and objective, but it is time consuming for system testing and implementations, i.e., take days to perform experiments if is a sophisticated system. So, a temporary solution is to use *intrinsic evaluation* with an approximation called perplexity to evaluate N-gram. It is easier to implement if the same system is used but perplexity is a poor approximation unless the test data looks identical to the training data. Hence, it is generally useful in pilot experiments.

Table 2.10 Perplexity values for WSJ from unigram to trigram

N-gram order	Unigram	Bigram	Trigram
Perplexity	962	170	109

2.6.3   Zero Counts Problems

Next step is to manage *zero counts problems*. Let's return to *The Adventures of Sherlock Holmes* example, this literature had produced 109,139 bigram types over 100 million possible bigrams as recalled, so there are approximately 99.89% of possible bigrams never seen that have zero entries in the bigram table. In other words, most of these zeros conditional probabilities are bigrams that need to be managed, especially in different NLP applications such as text generation and speech recognition.

There is a brief synopsis of such *zero-count dilemma*. Some of these zeros are truly zeros which means that can't and shouldn't occur because they won't make grammatical or syntactic sense, on the other hand, some are only rare events which means they occurred infrequently, for example, with an extensive training corpus.

Further, *Zipf's law* (Saichev et al. 2010) states that a long tail phenomenon is a rare event that occurs in a very high frequency, and large events numbers occur in a low frequency constantly. These are two extremes, which means some popular words always occur in a high frequency, and most are bigrams in a low frequency. Hence, it is clear to collect statistics on high-frequency events and may have to wait for a long time until a rare event occurs, e.g., a bigram to take a count on this low occurrence frequency event. In other words, high occurrence frequency events always dominate the whole corpus. This phenomenon is essential because it always occurs in website statistics or website counting. These high-frequency websites and N-grams are usually the top 100 and others with limited visit counts and occurrence, so the estimate results are sparse as there are neither counts nor rare events that are required to estimate the likelihood of unseen or 0 count N-grams.

2.6.4   Smoothing Techniques

Every N-gram training matrix is sparse even with large corpora because of *Zipf's law* phenomenon. The solution is to use likelihood estimation for figures on unseen N-grams or *0 count* N-grams to judge the rest of corpus accommodated with these phantom/shadow N-grams. It will affect the rest of the corpus.

Let's assume that an N-gram is used, all the words are known and seen beforehand. When assigning a probability to a sequence where one of these components is 0, the initial process is to search for a low N-gram order and *back off* from a bigram to unigram and replace 0 with something else, or a value with several methods to resolve zero count problems based on this concept; these collective methods are called *smoothing techniques*.

This section explores four commonly used *smoothing techniques*: (1) *Laplace (Add-one) Smoothing*, (2) *Add-k Smoothing*, (3) *Backoff and Interpolation Smoot,hing* and (4) *Good Turing Smoothing* (Chen and Goodman 1999; Eisenstein 2019; Jurafsky et al. 1999).

2.6.5 Laplace (Add-One) Smoothing

Laplace (Add-one) Smoothing (Chen and Goodman 1999; Jurafsky et al. 1999) logic is to consider all zero counts are rare events and add 1 into them. These rare events are neither occurred nor sampled during corpus training.

For unigram:

1. Add 1 to every single word (type) count.
2. Normalize *N token/(N (tokens) + V (types))*.
3. Smooth count c_i^* (adjusted for additions to N) given by:

$$c_i^* = \left(c_i + 1\right)\frac{N}{N+V} \tag{2.14}$$

4. Normalize *N* to obtain a new unigram probability *p**given by:

$$p^* = \frac{c_i + 1}{N+V} \tag{2.15}$$

For bigram:

1. Add 1 into every bigram $c(w_{n-1}w_n) + 1$.
2. Increase unigram count by vocabulary size $c(w_{n-1}) + V$.

Table 2.11 shows a bigram count with and without Laplace Method for the previous example *I have no doubt that I* from *The Adventures of Sherlock Holmes*. It indicated that all zeros become 1 so that *no I* becomes 1, others like *I have* will come from 288 to 289, the calculation is simple but effective.

For bigram probability calculation is given by:

$$P\left(w_n|w_{n-1}\right) = \frac{C\left(w_{n-1}w_n\right)}{C(w_{n-1})} \tag{2.16}$$

So, the bigram probability with the *Laplace method* will be given by:

Table 2.11 Bigram counts with and without Laplace method

Original bigram table of "I have no doubt that I" (by bigram count)							Bigram table of "I have no doubt that I" with Laplace method (by bigram count)						
	I	have	no	doubt	that	I		I	have	no	doubt	that	I
I	0	288	0	1	0	0	I	1	289	1	2	1	1
have	5	0	35	0	2	5	have	6	1	36	1	3	6
no	0	0	0	46	0	0	no	1	1	1	47	1	1
doubt	0	0	0	0	17	0	doubt	1	1	1	1	18	1
that	228	1	10	0	7	228	that	229	2	11	1	8	229

$$P_{\text{Lap}}\left(w_n|w_{n-1}\right) = \frac{C\left(w_{n-1}w_n\right)+1}{\sum_w\left(C\left(w_{n-1}w\right)+1\right)} = \frac{C\left(w_{n-1}w_n\right)+1}{C\left(w_{n-1}\right)+V} \tag{2.17}$$

Table 2.12 shows the bigram probabilities with and without *Laplace Method* for the previous example *I have no doubt that I* from *The Adventures of Sherlock Holmes*.

Note: The bigram probability is calculated by the division of unigram originally but now it will be the division by *the count of unigram + total number of word type (V)* which is equal to 9886, e.g., *P(have | I) = 288/2755 = 0.105*. Applying Laplace method, it becomes *289/(2755 + 9886) = 0.023*. It showed that all zero cases will become 1 which is simple for text generation, but the problem is, some probabilities have changed notably such as *I have* from *0.105* to *0.023*, and *no doubt* has the highest change from *0.1667* to only *0.00463*.

Although it is adequate to assign a number to all zero events but the one with high frequency becomes insignificant because of copious word types in corpus base, indicating that the performance of *Laplace Add-one smoothing* may not be effective in many cases and required to look for alternatives.

2.6.6   Add-k Smoothing

Add-k Smoothing (Chen and Goodman 1999; Jurafsky et al. 1999) logic is to assume that each N-gram is seen in k times, but the occurrence is rare. These zeros are rare events that are less than 1 and unnoticeable meaning that there is a line between 0 and 1, it can be 0.1, 0.01, 0.2 or even smaller; so, a non-integer count is added instead of 1 to each count, for example, 0.05, 0.1, 0.2, typically, $0 < k < 1$ provided that k must be a small number less than 1 in practical applications. It is because if k is too large, it will cause similar problem occurred in *Laplace method*.

By using the same logical as *Add-1 method*, *Add-k Smoothing* is given by:

$$P_{\text{Add}-k}^*\left(w_n|w_{n-1}\right) = \frac{C\left(w_{n-1}w_n\right)+k}{C\left(w_n\right)+kV} \tag{2.18}$$

where $0 < k < 1$

It is adequate to compare with the *Laplace method* in that the whole V is not used if V is very large such as 9886 in *The Adventure of Sherlock Holmes*. When the event is, say 0.05, means that it will be even smaller, but the new number won't be too small. Aslthough *add-k* is useful for many tasks including text classification and generation, but not for all language modeling, generating counts with poor variance and often inappropriate discounts (Gale and Church 1994). Another add-k model consideration is to select an appropriate k number through trial and error but that will lead to problems in practical applications. Nevertheless, *Add-k smoothing* usually provides a better and more viable solution as compared with the Add-1 method.

Table 2.12 Bigram probabilities with and without Laplace method

Original bigram table of "I have no doubt that I"
(by bigram probability)

	I	have	no	doubt	that	I
I	0.00000	0.10454	0.00000	0.00036	0.00000	0.00000
have	0.00577	0.00000	0.04037	0.00000	0.00231	0.00577
no	0.00000	0.00000	0.00000	0.16667	0.00000	0.00000
doubt	0.00000	0.00000	0.00000	0.00000	0.20238	0.00000
that	0.12903	0.00057	0.00566	0.00000	0.00396	0.12903

Bigram table of "I have no doubt that I" with Laplace method
(by bigram probability)

	I	have	no	doubt	that	I
I	0.00008	0.02286	0.00008	0.00016	0.00008	0.00008
have	0.00056	0.00009	0.00335	0.00009	0.00028	0.00056
no	0.00010	0.00010	0.00010	0.00463	0.00010	0.00010
doubt	0.00010	0.00010	0.00010	0.00010	0.00181	0.00010
that	0.01965	0.00017	0.00094	0.00009	0.00069	0.01965

2.6.7   *Backoff and Interpolation Smoothing*

Backoff and Interpolation (B&I) Smoothing (Chen and Goodman 1999; Suyanto 2020) logic is to look for a lower dimension N-gram if there is no example of a particular N-gram. If $N-1$ gram has an insufficient number count (or doesn't exist), then will switch to $N-2$ gram, and so on. Although it is not the perfect option, at least it can produce some viable counting for word prediction. That is to estimate a probability with a bigram instead of a trigram if there is none to be found. Furthermore, it can look up to unigram if no bigram either. This is a kind of *backoff method* and by *interpolation*, can always weigh and combine with quadrigram, trigram, bigram, and unigram probabilities counts, for example, when calculating trigram probability with unigram, bigram, and trigram, each weighted by some λ values. Note the sum of all λs must be 1 given by these equations:

$$
\begin{aligned}
P_{B\&I}\left(w_n | w_{n-2} w_{n-1}\right) &= \lambda_1 P\left(w_n\right) \\
&\quad + \lambda_2 P\left(w_n | w_{n-1}\right) \\
&\quad + \lambda_3 P\left(w_n | w_{n-2} w_{n-1}\right)
\end{aligned}
\tag{2.19}
$$

For a sophisticated version of linear interpolation, each λ value can be calculated by conditioning on the context which means it can be done by using conditional probabilities as well. In this way, if a particular bigram has accurate numbers, it can assume that the trigrams numbers are based on this bigram, which will be a robust method to implement given by the following equation:

$$
\begin{aligned}
P_{B\&I}\left(w_n | w_{n-2} w_{n-1}\right) &= \lambda_1 \left(w_{n-2:n-1}\right) P\left(w_n\right) \\
&\quad + \lambda_2 \left(w_{n-2:n-1}\right) P\left(w_n | w_{n-1}\right) \\
&\quad + \lambda_3 \left(w_{n-2:n-1}\right) P\left(w_n | w_{n-2} w_{n-1}\right)
\end{aligned}
\tag{2.20}
$$

It is noted that by comparing with the previous equation (2.19), this equation also considers conditional probability in all N-gram levels. Hence, both simple interpolation and conditional interpolation methods are learned from a *held-out corpus*. A *held-out corpus* is an additional training corpus to set hyperparameters like λ values by choosing λ values that can maximize the likelihood of held-out corpus. By adjusting N-gram probabilities and search for λ value is to provide the highest probability of *held-out set*. In fact, there are numerous approaches to find this optimal set of λ, a simple way is applying EM algorithm which is an interactive learning algorithm to converge locally optimal λ (Jelinek and Mercer, 1980).

2.6.8   Good Turing Smoothing

Good Turing (GT) Smoothing (Chen and Goodman 1999; Gale and Sampson 1995) logic is to use the total frequency of events that occurred only once to estimate how much mass shift to unseen events, e.g., using a bag of green color beans to estimate the probability of an unseen red color bean.

This technique uses the frequency of N-gram occurrence to reallocate probability distribution in two criteria, for example, N-gram statistics of *The Adventures of Sherlock Holmes* in Table 2.12. It showed that the probability of *have doubt* = 0 without smoothing, so by using bigrams frequency that occurred once, i.e., probability of *I doubt* to represent the total number of bigrams for unknown bigrams given by:

$$P_{\text{unknown}}\left(w_i|w_{i-1}\right) = \frac{\text{Count of bigrams that appeared once}}{\text{Count of total bigrams}} \tag{2.21}$$

It is an intuitive method because it only considers the conditional probability of bigrams that occurred once to represent unknown probabilities instead of adding 1 to them. In other words, the conditional probability of an unknown bigram of the word will be the count for the bigram that occurred once over the count of total bigrams.

For known bigrams such as *no doubt*, the frequency of bigrams that occurred more than one of the current bigram frequency N_{c+1}, frequency of bigrams that occurred the same as the current bigram frequency N_c, and the total number of bigram N are given by:

$$P_{\text{known}}\left(w_i|w_{i-1}\right) = \frac{c^*}{N}$$

$$\text{where } c^* = \left(c+1\right)*\frac{N_{c+1}}{N_c} \text{ and } c = \text{count of input bigram.} \tag{2.22}$$

Exercise: Try to calculate these probabilities from data provided by Table 2.12.

Exercises

2.1. What is a *Language Model (LM)*? Discuss the roles and importance of language models in NLP.

2.2. What is an N-gram? Discuss and explain the importance of N-gram in NLP and text analysis.

2.3. State the *Chain Rule* and explain how it works for the formulation of N-gram probabilities. Use trigram as an example to illustrate.

2.4. What is a *Markov Chain*? State and explain how it works for the formulation of N-gram probabilities.

2.5. Use *The Adventures of Sherlock Holmes* as corpus, calculate N-gram probability the for sentence *"I don't believe in that"* with *Markov Chain* and evaluate all related bigram probabilities.

2.6. Repeat Exercise 2.5 by using another famous literature *Little Women* by Louisa May Alcott (1832–1888) *(*Alcott 2017*)* to calculate N-gram probability of the sentence *"I don't believe in that"* and compare it with results in 2.5. What are the findings?

2.7. Use *Shannon's text generation* scheme on *The Adventures of Sherlock Holmes* as corpus, generate sample sentences like Table 2.9 using unigram, bigram, trigram, and quadrigram text generation methods.

2.8. Repeat Exercise 2.7 using the literature *Little Women (*Alcott 2017*)* to generate corresponding sample sentences and compare them with results in 2.7. What are the findings?

2.9. What is *Perplexity (PP)* in N-gram model evaluation? Use *The Adventures of Sherlock Holmes* as corpus with sample test set, evaluate PP values from unigram to trigram, and compare it with Table 2.10. What are the findings?

2.10. Use *Little Women (*Alcott 2017*)* as a corpus and some sample test sets. Compare the performance of *Add-1 smoothing* against Add-k ($k = 0.5$). Which one is better? Why?

2.11. What is *Backoff and Interpolation (B&I) method* in N-gram smoothing? Repeat 2.10 using *B&I smoothing method* with $\lambda_1 = 0.4$, $\lambda_2 = 0.3$ and $\lambda_3 = 0.3$. Compare the performance with results obtained in 2.10.

2.12. What is *Good Turing (GT) Smoothing* in N-gram smoothing? Repeat Exercise 2.10 using *GT Smoothing* and compare performance results obtained in 2.10 and 2.11. Which one is better? Why?

References

Alcott, L. M. (2017) Little Women (AmazonClassics Edition). AmazonClassics.

Bird, S. (2009). Natural language processing with python. O'Reilly Media.

Chen, S. F. and J. Goodman. 1999. An empirical study of smoothing techniques for language modeling. Computer Speech and Language, 13:359–394.

Ching, W. K., Huang, X., Ng, M. K. and Siu, T. K. (2013) Markov Chains: Models, Algorithms and Applications. Springer.

Doyle, A. C. (2019) The Adventures of Sherlock Holmes (AmazonClassics Edition). AmazonClassics.

Eisenstein, J. (2019) Introduction to Natural Language Processing (Adaptive Computation and Machine Learning series). The MIT Press.

Gale, W. A. and Church, K. W. (1994) What is wrong with adding one? In N. Oostdijk and P. de Haan (eds), Corpus-Based Research into Language, pp. 189–198. Rodopi.

Gale, W. A. and Sampson, G. (1995). Good-Turing frequency estimation without tears. Journal of Quantitative Linguistics, 2(3), 217-237.

Google (2022) Google official site. http://google.com. Accessed 12 July 2022.

Gutenberg (2022) Project Gutenberg official site. https://www.gutenberg.org/ Accessed 13 July 2022.

Jelinek, F. and Mercer, R. L. (1980) Interpolated estimation of Markov source parameters from sparse data. In Proceedings of the Workshop on Programming in Natural Language, pp. 1–11.

Jurafsky, D., Marin, J., Kehler, A., Linden, K., Ward, N. (1999). Speech and Language Processing: An Introduction to Natural Language Processing, Computational Linguistics and Speech Recognition. Prentice Hall.

Liu, Z., Lin, Y. and Sun, M. (2020) Representation Learning for Natural Language Processing. Springer.

Pustejovsky, J. and Stubbs, A. (2012) Natural Language Annotation for Machine Learning: A Guide to Corpus-Building for Applications. O'Reilly Media.

Saichev, A. I., Malevergne, Y. and Sornette, D. (2010) Theory of Zipf's Law and Beyond (Lecture Notes in Economics and Mathematical Systems, 632). Springer.

Shakespeare, W. (2021) The Complete Works of Shakespeare (AmazonClassics Edition). AmazonClassics.

Shannon, C. (1948). A Mathematical Theory of Communication. Bell System Technical Journal. 27 (3): 379–423.

Sidorov, G. (2019) Syntactic n-grams in Computational Linguistics. Springer.

Suyanto, S. (2020). Phonological similarity-based backoff smoothing to boost a bigram syllable boundary detection. International Journal of Speech Technology, 23(1), 191-204.

Tuchong (2022) Computerized Axial Tomography Scanner ("Cat scan"). https://stock.tuchong.com/image/detail?imageId=902001913134579722. Accessed 17 Dec 2024.

Chapter 3
Part-of-Speech (POS) Tagging

3.1 What Is Part of Speech (POS)?

Part of Speech (*PoS* or *POS*) is a category of words normally in lexical terms that have similar grammatic behaviors or properties (Bender 2013; Jurafsky et al. 1999). These are words assigned to the same POS exhibited in syntactic or functional behaviors and roles in grammatic structure sentences, for example, English *verbs* and *nouns*. They sometimes have similar morphology and can be *inflected* to produce similar properties and semantic behavior. To explore how POS works, it is important to understand the concept of *inflection*.

Inflection can be considered as the process of word formation in which items are added to the base form of a word to convey grammatical meanings. The word *inflection* comes from the Latin word *inflectere*, which means *to bend*, for example, (1) inflection *-s* of *cats* signifies the noun is plural, (2) the same *-s* inflection of *gets* signifies the subject is a third-person singular (e.g., [3.1] *He gets the book*), and (3) inflection of *-ed* often signifies past tense (e.g., *arrive → arrived, close → closed*, etc.). Thus, *inflections* are to express grammatical types such as persons, quantities, and tenses. There are several types of POS to define inflection characteristics.

3.1.1 Nine Major POS in the English Language

Every word in English sentences falls into nine major POS types. They are (1) *adjectives*, (2) *verbs*, (3) *pronouns*, (4) *conjunctions*, (5) *prepositions*, (6) *articles* (*determiners*), (7) *adverbs*, (8) *nouns*, and (9) *interjections* as shown in Fig. 3.1. Some linguists include only first eight as major POS and leave *interjections* as an individual category.

© The Author(s), under exclusive license to Springer Nature Singapore Pte Ltd. 2025
R. Lee, *Natural Language Processing*,
https://doi.org/10.1007/978-981-96-3208-4_3

Fig. 3.1 Major POS in the English language

POS is important to study:

1. Word class categorization and usage in linguistics
2. Grammars in English usage
3. Word functions categorization in NLP
4. POS tagging

3.2 POS Tagging

3.2.1 What Is POS Tagging in Linguistics?

Part-of-Speech Tagging (Khanam 2022; Sree and Thottempudi 2011), also called *POS tagging, POST,* or *grammatical tagging*, is the operation of labelling a word in a text, or corpus according to a particular POS based on definition and contexts in linguistics. A simplified format is usually learned by students to identify word types such as *adjectives, adverbs, nouns, verbs*, etc. Grammars vary in foreign languages leading to several POS tagging categorizations.

3.2.2 What Is POS Tagging in NLP?

Tagging is a kind of classification process that may be defined as an automatic description assignment to words or tokens in NLP (Eisenstein 2019). They are called *POS tags* or *tags* to represent one of the POS, semantic information in a

Fig. 3.2 POS example for utterance *"She sells seashells on the seashore"*

sentence or utterance. Some words may have different meanings and roles in POS, for example, *book* can be used as a noun or *booking a table* as a verb.

In NLP, *POS tagging* is the operation of converting a sentence/utterance to forms, or a list of words and list of *tuples*, where each *tuple* has a word or tag form to signify *noun, verb, adjective, pronoun, conjunction*, and their subcategories. Figure 3.2 shows how tagging is applied to sample sentence/utterance: [3.2] *She sells seashells on the seashore.*

Machine learning and rule-based models can produce POS tags in NLP. They generally fall into (1) *Rule-based POS tagging*, (2) *Stochastic POS tagging*, and (3) *Hybrid POS tagging* using advanced technology like *Transformation-based tagging* (Jurafsky et al. 1999; Khanam 2022; Pustejovsky and Stubbs 2012). We will study how they work with NLTK and spaCy technologies at workshops in Part II. First, let's look at some realistic POS databanks.

3.2.3   POS Tags Used in the PENN Treebank Project

PENN Treebank is a frequently used POS tag databank provided by the *PENN Treebank corpus* (Marcus et al. 1993). It is an English corpus marked by a *TreeTagger* tool developed by Professor Helmut Schmid at the University of Stuttgart in Germany. It classifies nine major POS into subclasses that have a total of 45 POS tags with punctuation and examples as shown in Table 3.1, and its English *Penn Treebank* (PTB) corpus has a comprehensive section of *Wall Street Journal (WSJ)* articles to be used on sequential labeling models' evaluation as well as characters and word levels language modeling.

A POS tagging table for sentence [3.3] *David has purchased a new laptop from Apple store* in Fig. 3.3 showed that *Apple* is a proper noun because it can be differentiated by capital letter A as a product brand name.

Table 3.1 Penn Treebank POS tags (with punctuation)

No	POS tag	Description	Example	No	POS tag	Description	Example
1	CC	coordinating conjunction	and, but, or	24	SYM	Symbol	$ / [= *
2	CD	cardinal number	1, third	25	TO	infinitive 'to'	to
3	DT	determiner	a, the	26	UH	interjection	haha, oops
4	EX	existential there	there is	27	VB	verb—base form	drink
5	FW	foreign word	les	28	VBD	verb—past tense	drank
6	IN	preposition, sub-conj	in, of, by, like	29	VBG	verb—gerund	drinking
7	JJ	adjective	big, wide, green	30	VBN	verb—past participle	drunk
8	JJR	adjective, comparative	bigger, wider, greener	31	VBP	verb—non-3sg pres	drink
9	JJS	adjective, superlative	biggest, wildest, greenest	32	VBZ	verb—3sg pres	drinks
10	LS	list marker	1), One, i	33	WDT	wh-determiner	which, that
11	MD	modal	can, could, shall, will	34	WP	wh-pronoun	who, what
12	NN	noun, singular or mass	table, shop	35	WP$	possessive wh-pronoun	whose, those
13	NNS	noun plural	tables, shops	36	WRB	wh-abverb	where, when, how
14	NNP	proper noun, singular	Samsung	37	#	#	#
15	NNPS	proper noun, plural	Vikings	38	$	$	$
16	PDT	predeterminer	all/both the students	39	"	Left quotation	' "
17	POS	possessive ending	friend's	40	"	Right quotation	' "
18	PP	personal pronoun	I, he, it, you	41	(	Opening brackets	({
19	PPZ	possessive pronoun	my, his, your, one's	42	)	Closing brackets	) }
20	RB	adverb	however, quickly, here	43	,	Comma	,
21	RBR	adverb, comparative	better, quicker	44	:	Sent-final punc	. ! ?
22	RBS	adverb, superlative	best, quickest	45	:	Mid-sentence punc	: ; ... -
23	RP	particle	of, up (e.g. give up)				

David has purchased a new laptop from Apple store

Fig. 3.3 Penn Treebank POS tags of sample sentence "*David has purchased a new laptop from Apple store*"

3.2.4 Why Do We Care About POS in NLP?

POS is a fundamental concept to understand the proper use of language, for example, English. Without this, we cannot differentiate the usages or roles of different words in a sentence whether it is a *noun, verb, adjective,* and *determiner.* The major concerns include:

1. Pronunciation often differs from the same word with different roles.
   For example [3.4] *Here are the students' records* versus [3.5] *The teacher records his lecture.*
2. Prediction of the following word, for example, (a) *they* should use *will* instead of shall and (b) the word after *to* is not past tense. It is natural in grammar rules as compared with *N*-gram solely relied on counting words relationship.
3. Stemming is within a restricted tag set, for example, *comput* for *computer.*
4. Syntactic parsing base and then meaning extraction.
   For example [3.6] *Better get going or you will be late.*
5. Machine translation for the same word with different POS classes most likely has a different translation in other languages, for example, translation from English to French.
   (E) book + N → (F) acheter + N (Buy a book → Achète un livre)
   (E) book + VB → (F) réserver + VB (Book a room → Réserver une chambre)

A proper POS tagging can provide correct translation between foreign languages. Further, it is to stress different accents and avoid confusion of the same word (word type) with different POS in a sentence/utterance. There are three types:-

1. Noun vs verb confusion, for example, *ABstract (noun) vs. abstRACT (verb).*
2. Adjective vs Verb confusion, for example, *PERfect (adjective) vs. perFECT (verb).*
3. Adjective vs Noun confusion, for example, *miNUTE (adjective) vs MInute (noun).*

Table 3.2 shows some examples of common English words from the CELEX online dictionary, which have different stresses and meanings to distinguish the role of each word in the sentence/utterance when dealing with noisy channels. These problems can be solved by applying statistical probability *N*-gram methods or stochastic techniques and corpus for fact analysis. Nevertheless, part-of-speech tagging is the first step to solve the problem.

Table 3.2 Common example of same English word with different stress accents

Noun	Verb	Noun	Verb	Noun	Verb
ABstract	abstRACT	ENvelope	enVELope	REBel	reBEL
ACcent	acCENT	EScort	esCORT	REcap	reCAP
ADdict	adDICT	EXploit	exPLOIT	REcall	reCALL
ADdress	adDRESS	EXport	exPORT	REcord	reCORD
ANnex	anNEX	EXtract	exTRACT	REfill	reFILL
ALly	alLY	FInance	fiNANCE	REfund	reFUND
ATtribute	atTRIBute	FRAgment	fragMENT	REfuse	refUSE
COMbat	comBAT	IMpact	imPACT	REject	reJECT
COMmune	comMUNE	IMprint	imPRINT	REplay	rePLAY
COMpact	comPACT	INcrease	inCREASE	SUBject	subJECT
COMpound	comPOUND	INsert	inSERT	SURvey	surVEY
COMpress	comPRESS	INsult	inSULT	SUSpect	susPECT
CONduct	conDUCT	MANdate	manDATE	TORment	torMENT
CONfines	conFINES	OBject	obJECT	TRANSfer	transFER
CONflict	conFLICT	OVERcharge	overCHARGE	TRANSplant	transPLANT
CONscript	conSCRIPT	OVERwork	overWORK	TRANSport	transPORT
CONsort	conSORT	PERmit	perMIT	UPset	upSET
CONtract	conTRACT	PERvert	perVERT		
CONtrast	conTRAST	PREfix	preFIX	**Adjective**	**Verb**
CONverse	conVERSE	PREsent	preSENT	ABsent	abSENT
CONvert	conVERT	PROceeds	proCEEDS	FREquent	freQUENT
CONvict	conVICT	PROcess	proCESS	PERfect	perFECT
DEcrease	deCREASE	PROduce	proDUCE		
DEsert	deSERT	PROgress	proGRESS	**Adjective**	**Noun**
DEtail	deTAIL	PROject	proJECT	inVALid	INvalid
DIScard	disCARD	PROtest	proTEST	miNUTE (my noot)	MInute (min it)
DIScharge	disCHARGE	RAMpage	ramPAGE	comPLEX	COMplex

3.3   Major Components in NLU

Natural Language Understanding (NLU) (Allen 1994; Mitkov 2005) is a critical component in various NLP applications including text summarization, sentiment analysis, information retrievals to Q&A chatbot systems. It composes of five basic modules: (1) *morphology*, (2) *POS tagging*, (3) *syntax*, (4) *semantics*, and (5) *discourse integration* as shown in Fig. 3.4.

Morphology is the understandings of shapes and patterns for every word of a sentence/utterance.

POS tagging is key process to provide functions and categories of words.

Syntax is syntactic analysis to understand the syntactic role and usage of every word or word pattern.

Fig. 3.4 Major
components in NLU

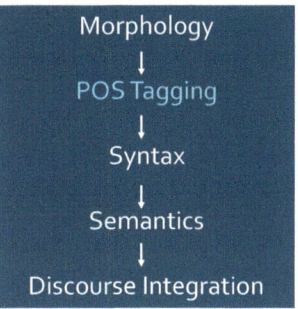

Semantics is an analysis to understand semantic meaning of a sentence/utterance
and its overall meaning.
Discourse integration is to understand the relationship between different sentences
and its contents.

3.3.1 Computational Linguistics and POS

Computational linguistics (CL) (Bender 2013; Clark et al. 2012; Mitkov 2005) can
be regarded as understanding written or spoken language from a computational and
scientific perspective. It focuses on building *artifacts* that process and analyze language. Language is like a *mirror of the mind* that reflects *human thoughts*.
Computational interpretations of language provide new insights into how human
thinking and intelligence work.

As human language is natural and the most polytropic means of communication
either person-to-person or person-to-machine, linguistically enabled computer systems provide a new era of NLP applications. There are two major issues to address
in computational linguistics: (1) linguistic itself refers to facts about language and
(2) algorithmic refers to effective computational procedures dealing with these facts.

The major goals of computational linguists include:

1. Construction of grammatical and semantic frameworks/models for language
   characterization.
2. Realization of learning models for the exploration of both structural and distributional properties of language, and,
3. Exploration of neuroscience and cognitive-oriented computational models of
   how language processing and learning work in our brains.

Thus, POS and POS tagging can be considered as the fundamental process in
computational linguistics to understand and model human languages.

3.3.2   POS and Semantic Meaning

The elementary level of language *semantics* (Goddard 1998) is to describe actual meaning of *word forms*. For example, a *noun* may be a category of words for people, locations, and things. *Adjectives* may be the category of words for properties of nouns.

Consider: [3.7] *green book* in which *green* is an *adjective* while *book* is a *noun*.

In fact, the word *book* can have two meanings: (1) description of word *book* from the dictionary and (2) noun in a sentence, which is an object. For the word *green*, it has an in-depth interpretation of (1) an adjective to describe the book in green color and (2) a semantic meaning to describe the book in green.

Now consider: [3.8] *book worm???*

[3.9] *This green is very smoothing???*

Here the word *book* has the same spelling and pronunciation as [3.8] but it becomes an adjective instead of a noun because of the *semantic meaning* of *book worm*. In [3.9], *green* becomes a noun instead of an adjective because of semantic meaning consideration in the whole sentence/utterance. So, the POS of every word/word pattern can be varied when considering the role in the overall semantic meaning of a sentence/utterance.

3.3.3   Morphological and Syntactic Definition of POS

According to the structure of morphologically clear grammatical rules, when there is an adjective to fill in the blank, for example [3.10] *It's so* _____, it can be *difficult, expensive, small*, etc. This rules' structure gives shape to appropriate POS tags for description, for example, when a noun is a word that can be labeled as plural means, it can be defined in either singular or plural form with *s*, or the other way round which is a two-way process. Thus, when a tagger tags a word with *s*, it gives hints that the word may contain *s* or a noun in plural, for example, cat or cats.

On the other hand, when there is a noun that can fill in the blank. For example, [3.11] *the* _____ *is so pretty*, it can be *decoration, house, painting*, etc. and conscious of not using a proper noun, for example, the *Tesla*.

Consider the following situations, what is the POS for the word *purple*:

[3.12] *It's so purple.*
[3.13] *Both purples should be okay for the room.*
[3.14] *The purple is a bit odd for the white carpet.*

In [3.12], it is an adjective. However, in [3.13] it is a particular noun in plural forms. In [3.14], it is also an indifferent noun to classify as a group against uncountable objects in purple.

3.4 Nine Key POS in English

There are nine key POS in English: (1) *pronoun*, (2) *verb*, (3) *adjective*, (4) *interjection*, (5) *noun*, (6) *adverb*, (7) *conjunction*, (8) *preposition*, and (9) *article* as shown in Fig. 3.5. Some linguists consider *interjections* as separate POS category to express strong feeling or emotion in a single word or a phrase, for example, [3.15] *Hooray! It's the last day of school.* It is distinct compared with other POS.

3.4.1 English Word Classes

There are two types of English *word classes*: (1) *closed class* and (2) *open class*. Both classes are important to understand proper sentences in different languages.

Closed-class words are also known as *functional/grammar words*. They are *closed* since new words are seldom created in the class. For example, *conjunctions*, *determiners*, *pronouns*, and *prepositions* are *closed class*. On the other hand, new items are added to *open classes* regularly. As *closed-class* words are usually used with a particular grammatical structure, it cannot be interpreted in isolation, for example, [3.16] *the style of this painting*, both *the* and *this* have no special meaning as compared with *painting* that has a specific meaning in usual knowledge.

Open-class words are also known as *lexical/content* words. They are *open* because the meaning of *open-class* words can be found in dictionaries and therefore their meaning can be interpreted individually. For example, *nouns*, *verbs*, *adjectives*, and *adverbs* are *open class* made up of the entire sub-class of words. These connective words are restrictive and used frequently to describe different *scenarios*

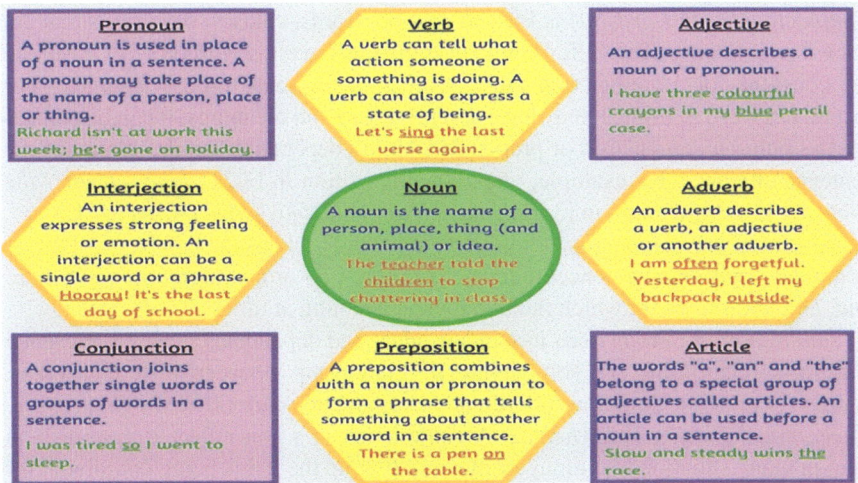

Fig. 3.5 Nine major POS in the English language with description

or *meanings* about spatial positions of two object nouns, for example, [3.17] *The cat sits by/under/above the piano*. Further, there are new types of *open-class* objects created from scratch or a combination of the existing words according to contemporary times, for example, *fax, telex, internet, iPhone, hub, bitcoin, metaverse*, etc.

3.4.2   What Is a Preposition?

Preposition (PP) is POS with a word (group of words) being used before a noun, pronoun, or noun phrase to indicate direction, location, spatial relationships, time or to describe an object or information to the recipient. There are approximately 80–100 prepositions in English to generate functional sentences/utterances.

This information can include where something takes place, for example, [3.18] *before dinner*, or general descriptive information, for example, [3.19] *the girl with ponytail*. The target of the preposition is the noun that followed the preposition. It is also the ending point for each preposition phrase. For instance, [3.20] *to the supermarket*. The word *to* is a preposition and *supermarket* is the target of the preposition, and [3.21] *over the rainbow*, the word *over* is the preposition and *rainbow* is the target of the preposition. A list of the top 40 prepositions from the CELEX online dictionary (CELEX 2022) of the COBUILD 16-million-word corpus is shown in Table 3.3. It showed that *of, in, for, to,* and *with* are the top five prepositions to correlate with ideas and additional information of a sentence/utterance.

3.4.3   What Is a Conjunction?

Conjunction (CONJ or CNJ) is POS to connect words, clauses, or phrases that are known as *conjuncts*. This definition may sometimes overlap with other POS so that the constitute of a conjunction must be defined for each foreign language. For instance, a word in English may have several senses and meanings. It can be considered as either a conjunction or preposition highly dependable on the syntax of the sentence/utterance, for example, *after* is a preposition in [3.22] *Jane left after the show* but is a conjunction in [3.23] *Jane left after she finished her homework*.

Co-ordinating conjunction allows joining words, clauses, or phrases of equal grammatic rank in a sentence/utterance. Common co-ordinating conjunctions are *and, but, for, nor,* or *yet* which include logical meaning at times.

Subordinating conjunctions join independent and dependent clauses to present a *causation* relationship, or some kind of relationship between different words, clauses, or phrases. Common subordinating conjunctions are *as, although, because, since, though, while,* and *whereas*. A conjunction is a non-inflected grammatical item in many situations as it may or may not link up the items being conjoined, for example, [3.24] *the book is so difficult that is hard for children to read. That* is to describe about the book to connect two ideas and [3.25] *this painting is very*

Table 3.3 Top 40 commonly used prepositions extracted from the CELEX online dictionary

Rank	PP	Freq.	Rank	PP	Freq.
1	of	540,085	21	above	3056
2	in	331,235	22	near	2026
3	for	142,421	23	off	1695
4	to	125,691	24	past	1575
5	with	124,965	25	worth	1563
6	on	109,129	26	toward	1390
7	at	100,169	27	plus	750
8	by	77,794	28	till	686
9	from	74,843	29	amongst	525
10	about	38,428	30	via	351
11	than	20,210	31	amid	222
12	over	18,071	32	underneath	164
13	through	14,964	33	versus	113
14	after	13,670	34	amidst	67
15	between	13,275	35	sans	20
16	under	9525	36	circa	14
17	per	6515	37	pace	12
18	among	5090	38	nigh	9
19	within	5030	39	re	4
20	towards	4700	40	mid	3

beautiful but is expensive. In this case *but* is to explain an initial idea to correlate with second idea. A list of top 50 commonly used co-ordinating and subordinating conjunctions from the CELEX online dictionary is shown in Table 3.4. It showed *and, that, or,* and *as* are used frequently to convey more than one concept at the same time or further explanation.

3.4.4 What Is a Pronoun?

Pronoun (PRN or PN) is POS that can be considered as a word (phrase) to serve as a substitution for a noun or noun phrase. It is also called the pronoun's *antecedent*. Pronouns usually appear as short words to replace a noun (noun phrase) for the construction of a sentence/utterance. Commonly used pronouns are *I, he, she, you, me, we, us, this, them, that*.

A pronoun can be used as a subject, direct (indirect) object, object of preposition and more to substitute any person, location, animal, or thing. It can replace a person's name in a sentence/utterance, for example, [3.26] *Jack is sick today, he cannot attend the evening seminar*. Pronoun is also a powerful tool to simplify the contents of dialogue and conversation by replacing them with simple tokens. A list of the top

Table 3.4 Top 50 commonly used conjunctions extracted from the CELEX online dictionary

Rank	CONJ.	Freq.	Rank	CONJ.	Freq.
1	and	514,946	26	now	1290
2	that	134,773	27	neither	1120
3	but	96,889	28	whenever	913
4	or	76,563	29	whereas	867
5	as	54,608	30	except	864
6	if	53,917	31	till	686
7	when	37,975	32	provided	594
8	because	23,626	33	whilst	351
9	so	12,933	34	suppose	281
10	before	10,720	35	cos	188
11	though	10,329	36	supposing	185
12	than	9511	37	considering	174
13	while	8144	38	lest	131
14	after	7042	39	albeit	104
15	whether	5978	40	providing	96
16	for	5935	41	whereupon	85
17	although	5424	42	seeing	63
18	until	5072	43	directly	26
19	yet	5040	44	ere	12
20	since	4843	45	notwithstanding	3
21	where	3952	46	according as	0
22	nor	3078	47	as if	0
23	once	2826	48	as long as	0
24	unless	2205	49	as though	0
25	why	1333	50	both and	0

50 commonly used pronouns extracted from the CELEX online dictionary is shown in Table 3.5. It showed *it, I, he, you,* and *his* are used frequently.

The truth is without pronouns, nouns become repetitive and cumbersome in speech and writing. However, the pronoun may cause ambiguity, for example, [3.27] *Jack blamed Ivan for losing the car key, he felt sorry for that. He* normally refers to the first person which is *Jack* but makes sense in pragmatic meaning for Ivan to *feel sorry* because *Jack* blamed him for the loss.

3.4.5   What Is a Verb?

Verb (VB) can be considered as a word syntax to conduct an action, process, occurrence, or state of being. In general, verbs are inflected to encode tense, aspect, mood, and voice in many languages, but are interchangeable with nouns of a word in some foreign languages. In English, a verb may also conform with gender, person, or number of arguments such as its subject or object.

Table 3.5 Top 50 commonly used pronouns extracted from the CELEX online dictionary

Rank	PRN	Freq.	Rank	PRN	Freq.
1	it	199,920	26	our	23,029
2	I	198,139	27	these	22,697
3	he	158,366	28	any	22,666
4	you	128,688	29	more	21,873
5	his	99,820	30	many	17,343
6	they	88,416	31	such	16,880
7	this	84,927	32	those	15,819
8	that	82,603	33	own	15,741
9	she	73,966	34	us	15,724
10	her	69,004	35	how	13,137
11	we	64,846	36	another	12,551
12	all	61,767	37	where	11,857
13	which	61,399	38	same	11,841
14	their	51,922	39	something	11,754
15	what	50,116	40	each	11,320
16	my	46,791	41	both	10,930
17	him	45,024	42	last	10,816
18	me	43,071	43	every	9788
19	who	42,881	44	himself	9113
20	them	42,099	45	nothing	9026
21	no	33,458	46	when	8336
22	some	32,863	47	one	7423
23	other	29,391	48	much	7237
24	your	28,923	49	anything	6937
25	its	27,783	50	next	6047

English verbs have tenses consideration: (1) present tense to notify that an action is being carried out, (2) past tense to notify that an action has been completed, (3) future tense to notify that an action to be happened in future, and (4) future perfect tense to notify an action will be completed in future.

A *modal verb* is a category of verb that contextually indicates a modality such as ability, advice, capacity, likelihood, order, obligation, permission, request, or suggestion. It is usually accompanied by the base (infinitive form) of another word with semantic contents. Common modal verbs are *can, could, may, might, shall, should, will, would,* and *must*. A list of the top 25 commonly used verbs from the CELEX online dictionary is shown in Table 3.6. It showed *can, will, may, would,* and *should* are used frequently. They also express significance in the subsequent verb, for example, verb following *can* and *will* must use present tense, not past tense.

Table 3.6 Top 25 commonly used modal verbs extracted from the CELEX online dictionary

Rank	VB	Freq.	Rank	VB	Freq.
1	can	70,930	14	won't	3100
2	will	69,206	15	'd	2299
3	may	25,802	16	ought	1845
4	would	18,448	17	will	862
5	should	17,760	18	shouldn't	858
6	must	16,520	19	mustn't	332
7	need	9955	20	'll	175
8	can't	6375	21	needn't	148
9	have	6320	22	mightn't	68
10	might	5580	23	oughtn't	44
11	couldn't	4265	24	mayn't	3
12	shall	4118	25	dare	3
13	wouldn't	3548			

3.5 Different Types of POS Tagset

3.5.1 What Is Tagset?

There are nine POS in English—pronoun, verb, adjective, interjection, noun, adverb, conjunction, preposition, and article learnt as students but there are clearly more sub-categories that can be further divided. For example, in nouns, the plural, possessive, and singular forms can be distinguished and further classified.

A *Tagset* is a batch of POS tags (*POS tags or POST*) to indicate the part of speech and sometimes other grammatical categories such as case and tense for the classification of each word in a sentence/utterance.

Brown Corpus Tagset (Brown 2022), *PENN Treebank Tagset* (Treebank 2022), and *CLAWS* (CLAWS7 2022) are commonly used. *Brown Corpus* was the first well-organized corpus of English for NLP analysis developed by Profs Emeritus Henry Kučera (1925–2010) and W. Nelson Francis (1910–2002) at Brown University, United States, in mid-1960s. It consists of over one million English words extracted from over 500 samples of randomly chosen publications. Each sample consists of over 2000 words with 87 tags defined (Brown 2022).

The English *PENN Treebank Tagset* originated by English corpora is annotated with the TreeTagger tool. PENN Treebank Tagset developed by Professor Helmud Schmid in the University of Stuttgart, Germany, consists of 45 distinct tags (Abeillé 2003; Treebank 2022).

English *CLAWS part-of-speech Tagset version 7*, also called *C7 Tagset*, is available in English corpora annotated with tools using CLAWS (Constituent Likelihood Automatic Word-tagging System). C7 Tagset developed by the University Centre for Computer Corpus Research on Language at Lancaster University was based on the Hidden Markov model to determine the likelihood of sentences and sequences of words in anticipating each POS label. It consists of 146 distinct tags (CLAWS7 2022).

3.5.2   Ambiguous in POS Tags

It may be the necessity of tagset databank against the dictionary to check out POS. A reason is that there are ambiguities in POS tags for many words:

1. Noun-verb ambiguity.
    For example, record: [3.28] *records the lecture* vs [3.29] *play CD records.*
2. Adjective-verb ambiguity.
    For example, perfect: [3.30] *a perfect plan* vs [3.31] *Jack perfects the invention.*
3. Adjective-noun ambiguity.
    For example, complex: [3.32] *a complex case* vs [3.33] *a shopping complex.*

Table 3.7 shows an ambiguous analysis of words in *Brown corpus* (DeRose 1988). One tag refers to a word tagged with a single POS type; 2–7 tags refer to a word tagged with several POS types. For example, a 3 POS ambiguous tag for *green*: (a) [3.34] *color green* (noun), (b) [3.35] *a green apple* (adjective), and (c) [3.36] *the roof was greening with leaves* (verb). A 7 POS ambiguous tag for *still:* (a) *[3.37] the still status* (adjective), (b) [3.38] *the still of the night* (noun), (c) [3.39] *it was still snowing* (adverb), and (d) [3.40] *Her quiet words stilled the animal* (verb). (Note: As an exercise, find out the other three POS tag usages for *still.*) Overall, there is a total of 10.4% ambiguous word types often used in language in which over 40% of ambiguous words are easy to disambiguate.

3.5.3   POS Tagging Using Knowledge

There are four methods to acquire knowledge from POS tagging: (1) dictionary, (2) morphological rules, (3) *N*-gram frequencies, and (4) structural relationships combination.

Dictionary is the basic method for tag usage, but it may not be fully reliable because there are ambiguous words meaning that the same word can have more than a single POS tagging in diverse scenarios.

Table 3.7 Ambiguous analysis of words in *Brown Corpus*

Unambiguous (1 tag)	35,340
2 tags	3760
3 tags	264
4 tags	61
5 tags	12
6 tags	2
7 tags	1
Ambiguous (2–7 tags)	**4100**
Ambiguous %	**10.40%**

Morphological rules are to identify well-known word shapes and patterns, for example, the inflection *-ed* for *past tense*, verb + *-ing* for *continuous form*, *-tion* for *noun description*, *-ly* for *adjective*, and capitalization such as *New York* for *proper noun*.

N-gram frequencies checking, also called *next word prediction*, for example, grammatic pattern *to ___*. When there is a *to*, if the next word is a *verb*, it must be in *present* and not *past tense*. If it is a *determiner*, the next word must be a *noun*.

Structural relationships combination method means to combine several methods to acquire tag information, for example, [3.41] *She barely heard the foghorns knelling her demise* vs. [3.42] The hunter's horn sounded the final <u>knell</u>. If there is no understanding on what knell means, there is an -ing pattern to indicate that is a verb in continuous tense, and final is an adjective description to indicate that knell is likely a noun.

3.6 Approaches for POS Tagging

There are three basic approaches to POS Tagging: (1) *Rule-based*, (2) *Stochastic-based*, and (3) *Hybrid Tagging*.

3.6.1 Rule-Based Approach POS Tagging

The rule-based approach is a classical approach in linguistics (Sree and Thottempudi 2011). The grammar knowledge learnt in primary schools is in fact grammatic rules, which means that the rule-based approach is the transfer of linguistic rule base usage into POS tagging.

It is a two-stage process: (1) dictionary consists of all possible POS tags for basic concepts of words as abovementioned and (2) words with more than single tag ambiguity applied handwritten or grammatic rules to assign the correct tag(s) according to surrounding words. The obtained rule sets directly affect tagging results accuracy. The lexicon is used initially for basic segmentation and tagging of the corpus, listing all possible lexical properties of the object, and combining rule-base with contextual information to disambiguate and retain the only suitable lexical properties.

The rule generation can be achieved by (1) hand creation and (2) training from a corpus with machine learning. The advantages of hand creation are that it is sensible and explainable to humans, but manual construction of rules is usually labor intensive. Also, if rules are described with too many details, the coverage of rules will be greatly reduced and difficult to adjust according to the actual situation. On the other hand, if rules are not based on contexts but rather on the lexical nature of rules, ambiguity may arise, that is, if the preceding of a word is an article, then the word must be a noun.

For example, consider [3.43] *a book. a* is an article as per possible tags that can be assigned directly, but *a book* can either be a noun or a verb. If considered *a book*, *a* is an article and follows the rules above, *book* should be a *noun* because the *article* is often followed by a *noun,* so a tag of *noun* is assigned to *book*. Word structures are often complex leading to more ambiguities and rules are required for differentiation.

3.6.2 Example of Rule-Based POS Tagging

Step 1: Assign each word with a list of possible tags based on a dictionary.

Step 2: Work out unknown and ambiguous words with two approaches: rules that specify what (1) *to do* and (2) *not to do.*

Figure 3.6 shows a sample adverbial *that* rule (Jurafsky et al. 1999):
It showed that:

- The first two statements of this rule verify the word *that* directly precedes a sentence/utterance's final adjective, adverb, or quantifier.
- For all other cases, the adverb reading is eliminated.
- The last clause eliminates cases that are preceded by verbs like *consider* or *believe* which can take a noun and an adjective.
- The logic behind this is to avoid tagging the following instance of *that* as an adverb such as [3.44] *It isn't that odd.*
- The other rule is used to verify if the previous word is a verb that expects a complement (like *think* or *hope*), and if *that* is followed by the beginning of a noun phrase, and a finite verb such as [3.45] *I consider that a win* or more complex structure such as [3.46] *I hope that she is confident.*

Stochastic-based approach (Dermatas and Kokkinakis 1995) is different from *the rule-based approach* in which it is a supervised model using frequencies or probabilities of tags that appeared in the training corpus to assign a tag to a new word. This tagging method depends on tag occurrence statistics, that is, probability of the tags. Stochastic taggers are further categorized into two parts: (1) *word frequency* and (2) *tag sequence frequency* to determine a tag.

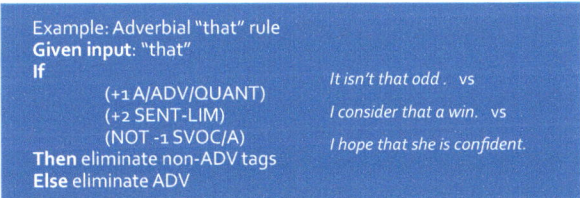

Fig. 3.6 Sample rule for adverbial "*that*" rule

Word frequency is to identify the tag that has a notable occurrence of the word, for example, based on the counting from a corpus, the word *list* occurs ten times in which six times as a *noun* and four times as a *verb*, and the word *cloud* will always be assigned as a *noun* since it has a notable occurrence in the training corpus. Hence, a word frequency approach is not very reliable in certain scenarios.

Tag sequence frequency, also called *N*-gram approach, is assigned the best tag to a word evaluated by the probability of *N* previous words tags. Although it provides better outcomes than word frequency approach, it may be unable to provide accurately for some rare words and phrases.

Stochastic POS tag model allows features to be non-independent and the addition of various granularities features. Hidden Markov Model (HMM) Tagger is a common stochastic-based approach, its Maximum Entropy Markov Model (MEMM) (Huang and Zhang 2009) is a stochastic POS tagging model that determine an exponential algorithm for each state as the conditional probability of the next state given the current state, which has the advantages of a stochastic POS tagging model. However, it also suffers from label bias problems. Unlike MEMM model, the Conditional Random Field (CRF) model adopts only one model as the joint probability of the entire label sequence given the observations sequence. Lafferty et al. (2001) verified that this model can effectively solve the tagging bias problems.

3.6.3   Example of Stochastic-Based POS Tagging

Let's use *HMM Tagger* as example. The rationale of the *HMM tagger* is applying *N*-gram frequencies to determine the best tag for a given word, like the same concept to investigate *N*-gram with *Markov Chain*. Mathematically, all is needed to maximize the conditional probability. The conditional probability w_i is tag t_i in the context given w_i by

$$P\left(t_i \text{ in context}|w_i\right) = \frac{P\left(w_i|t_i \text{ in context}\right)P\left(t_i \text{ in context}\right)}{P\left(w_i\right)} \tag{3.1}$$

In other words, given a sentence/utterance or word sequence, HMM taggers select tag sequence that maximizes the following formula given by:

$$P\left(\text{word}|\text{tag}\right) * P\left(\text{tag}|\text{previous } n \text{ tags}\right) \tag{3.2}$$

For bigram-HMM tagger, select tag t_i for w_i, that is most probable given the previous tag t_{i-1}, and the current word w_i in this equation by:

$$t_i = \mathop{\text{argmax}}_{j} P\left(t_j |, t_{i-1}|, w_i\right) \tag{3.3}$$

By simplifying Markov assumptions, the previous equation is applied to give a basic HMM equation for a single tag as follows:

$$t_i = \overset{\text{argmax}}{\underset{j}{}} P\left(t_j | t_{i-1}\right) P\left(w_i | t_j\right) \tag{3.4}$$

3.6.4 Hybrid Approach for POS Tagging Using Brill's Taggers

Hybrid approach is the integration of rule-based and stochastic with high-level methods including neural networks such as *LSTM* and other machine learning related methods often applied in NLP nowadays. Let's study an important hybrid approach for POS Tagging—*Transformation-based tagging*, also called *Brill's Taggers* invented by Dr. Eric Brill in 1995 (Brill 1995). It is a direct *Transformation-Based Learning (TBL)* implementation based on the integration of these two approaches.

3.6.5 What Is Transformation-Based Learning?

There are five steps in TBL by comparison to analog of oil painting with a *layering-and-refinement* approach.

1. Start with a background *theme* such as sky or household background.
2. Paint the background first, for example, if sky is the background scheme, paint clouds over it.
3. Paint the *main theme* or *object* over the background, for example, landscape, birds.
4. Refine the *main theme* or *object* over the background to make it more precise, for example, paint a landscape, add trees and animals layer-by-layer.
5. Further refine *objects* or the *main theme* until perfect, for example, apply a layering process or refinement for every single tree and animal (Fig. 3.7).

3.6.6 Hybrid POS Tagging: Brill's Tagger

Brill's Tagger is a type of hybrid TBL. Hybrid refers to integrate *rule-based* and *stochastic-based* methods in Brill's algorithm.

Rule 1: Label each word of the tag that is mostly likely given on contextual information, for example.

$$\text{Race}: P\left(\text{NN}|\text{race}\right) = 0.98; \quad P\left(\text{VB}|\text{race}\right) = 0.02$$

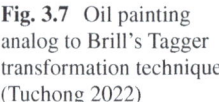

Fig. 3.7 Oil painting analog to Brill's Tagger transformation technique (Tuchong 2022)

Rule 2: Apply the transformation rule based on the context established.

Example:

Race: change NN to VB when the previous tag is TO.
[3.47] *Secretariat is expected to race tomorrow.*—change tag *race* from NN to VB.
[3.48] *The race is already over.* —no change, *race* remains as NN.

For [3.47] *race* has a higher probability of a *noun*, it will be treated as such by applying rule 1 initially. However, when there is a *verb* prior *to*, it should apply rule 2 to change into a *verb* instead of a *noun* according to grammatical rules.

For [3.48] *race* again has a high probability of being a noun but due to the grammatical rule being invalid, it remains as a *noun*. Thus, TBL is often applied to identify stochastic probabilities of tag frequencies for initial guesswork followed by grammatic rules for refinement.

3.6.7   Learning Brill's Tagger Transformations

There are three stages to learn *Brill's tagger* transformations:

1. Label every word with its best tag with the stochastic method,
2. Examine every possible transformation to select one with the most improved tagging and,
3. Retag data according to tagging rules.

These three stages are repetitive until a stopping criterion with no more rules to apply. TBL output is an ordered list of transformations that constitute a POS tagging procedure to a new corpus. The sample rules of Brill's TBL model are shown in Fig. 3.8 (Jurafsky et al. 1999).

Many NLP applications have adopted the *Brill's model* because it is a good combination of *rule-based models* (which provide detailed refinement) and *stochastic*

> The preceding (following) word is tagged **z**.
> The word two before (after) is tagged **z**.
> One of the two preceding (following) words is tagged **z**.
> One of the three preceding (following) words is tagged **z**.
> The preceding word is tagged **z** and the following word is tagged **w**.
> The preceding (following) word is tagged **z** and the word two
>   before (after) is tagged **w**.

Fig. 3.8   Sample rules used in Brill's TBL scheme

models (which provide efficient tagging solutions). In addition, the *Brill's model* can be easily implemented in both the *world domain* and the *knowledge domain* (such as medical knowledge domain), which may have specific rules or terminology for the corpus.

3.7   Taggers Evaluations

There are several considerations when *POS taggers* are implemented (Padro and Marquez 1998):

1. Evaluate algorithm adequacy.
2. Identify error's origin.
3. Repair and solve.

A *confusion matrix* suggests that current taggers face major problems:

1. Noun-single or mass vs. proper noun-singular vs. adjective (NN vs. NNP vs. JJ). These are hard to distinguish as proper nouns is crucial for information extraction, retrieval, and machine translation for different languages that have diverse tagging algorithms or classification schemes.
2. Adverb vs. adverb vs. preposition-sub-conjunction (RP vs. RB vs. IN). All of these can appear in satellite sequences following a verb immediately.
3. Verb-base form vs. verb-past participle vs. adjective (VB vs. VBN vs. JJ). They are crucial to distinguish for partial parsing, i.e., participles to identify passives and to label the edges of noun phrases correctly.

The confusion matrix from HMM error analysis of *The Adventures of Sherlock Holmes* (Doyle 2019) is shown in Table 3.8. For example, the mis-tagging of (1) NN by JJ is 7.56%, (2) NNP by NN is 5.23%, and (3) JJ by NN is 4.35%. Hence, mistaking NN by JJ occurred more often than JJ by NN in English texts but it may vary in other foreign languages.

Table 3.8 Confusion matrix from HMM of *The Adventures of Sherlock Holmes*

	IN	JJ	NN	NNP	RB	VBD	VBN
IN		0.18			0.56		
JJ	0.32		4.35	3.21	2.25	0.31	2.54
NN		7.56					0.35
NNP	0.31	3.12	5.23		0.15		
RB	2.45	3.21	0.43				
VBD		0.56	0.52				4.31
VBN		3.21				2.12	

3.7.1   How Good Is an POS Tagging Algorithm?

A satisfactory POS tagging algorithm depends on the maximum performance it can achieve. It must be realistic and of course the higher the better, but there are limits. For example, a *POS tagging system* has more than 90% accuracy should be considered satisfactory. But how can we define *satisfactory*? For example, (1) it is satisfactory for a voice dialogue system to give the correct meaning to user input 97% of the time, because ambiguity often occurs in noisy backgrounds and incorrect pronunciation or (2) it is satisfactory for an OCR system to correctly determine the word 97% of the time. So, it depends on the scenario, environment, complexity, domain problem, and application to be implemented.

Exercises

3.1 What is Part-of-Speech (POS)? How is it critical for NLP systems/applications implementation?

3.2 State and explain NINE basic types of POS in the English Language. For each POS type, give an example for illustration.

3.3 What is POS Tagging in NLP? How is it important to NLP systems/applications implementation? Give two examples of NLP systems/applications for illustration.

3.4 State and explain THREE types of POS Tagging methods in NLP.

3.5 What is PENN Treebank tagset? Perform POS Tagging for the following sentences/utterances using the PENN Treebank tagset.

[3.47] *POS tagging is a very interesting topic.*

[3.48] *It is not difficult to learn PENN Treebank tagset provided that we have sufficient examples.*

3.6 What is *Natural Language Understanding (NLU)*? State and explain FIVE major components of NLU in NLP.

3.7 Why *semantic meaning* is an important factor in POS tagging? Give two examples to support your answer.

3.8 What is *ambiguous* in POS tags? Give two examples of words with three and four ambiguous POS tags.

3.9 What is the rule-based approach in POS tagging? Give an example of the POS tagging rule to illustrate how it works.

3.10 What is a stochastic-based approach in POS tagging? Give an example to explain how *word frequency* and *tag sequence frequency* are applied for POS tagging.
3.11 State and explain transformation-based learning (TBL). Give an example to support your answer.

References

Abeillé, A. (ed) (2003) Treebanks: Building and Using Parsed Corpora (Text, Speech and Language Technology Book 20). Springer.

Allen, J. (1994) Natural Language Understanding (2nd edition). Pearson

Bender, E. M. (2013) Linguistic Fundamentals for Natural Language Processing: 100 Essentials from Morphology and Syntax (Synthesis Lectures on Human Language Technologies). Morgan & Claypool Publishers

Brill, E. (1995). Transformation-based error-driven learning and natural language processing: A case study in part-of-speech tagging. Computational Linguistics, 21(4), 543–566.

Brown (2022) Brown corpus tagset. https://web.archive.org/web/20080706074336/http://www.scs.leeds.ac.uk/ccalas/tagsets/brown.html. Accessed 15 July 2022.

CELEX (2022) CELEX corpus official site. https://catalog.ldc.upenn.edu/LDC96L14. Accessed 15 July 2022.

Clark, A., Fox, C. and Lappin, S. (2012) The Handbook of Computational Linguistics and Natural Language Processing. Wiley-Blackwell.

CLAWS7 (2022) UCREL CLAWS7 Tagset. https://ucrel.lancs.ac.uk/claws7tags.html. Accessed 15 July 2022.

DeRose, S. J. (1988). Grammatical category disambiguation by statistical optimization. Computational Linguistics, 14, 31–39.

Dermatas, E. and Kokkinakis, G. (1995). Automatic stochastic tagging of natural language texts. Computational Linguistics, 21(2), 137–164.

Doyle, A. C. (2019) The Adventures of Sherlock Holmes (AmazonClassics Edition). AmazonClassics.

Eisenstein, J. (2019) Introduction to Natural Language Processing (Adaptive Computation and Machine Learning series). The MIT Press.

Goddard, C. (1998) Semantic Analysis: A Practical Introduction (Oxford Textbooks in Linguistics). Oxford University Press.

Huang H, Zhang X. (2009) Part-of-speech tagger based on maximum entropy model. 2009 2nd IEEE International Conference on Computer Science and Information Technology. IEEE; pp 26-29. https://doi.org/10.1109/ICCSIT.2009.5234787.

Jurafsky, D., Marin, J., Kehler, A., Linden, K., Ward, N. (1999). Speech and Language Processing: An Introduction to Natural Language Processing, Computational Linguistics and Speech Recognition. Prentice Hall.

Khanam, H. M. (2022) Natural Language Processing Applications: Part of Speech Tagging. Scholars' Press.

Lafferty, J., McCallum, A. and Pereira, F. (2001). Conditional random fields: Probabilistic models for segmenting and labeling sequence data. Proc. 18th International Conf. on Machine Learning. Morgan Kaufmann. pp. 282–289.

Marcus, M., Santorini, B. and Marcinkiewicz, M. A. (1993). Building a large, annotated corpus of English: The Penn Treebank. In Computational Linguistics, volume 19, number 2, pp. 313–330.

Mitkov, R. (2005) The Oxford Handbook of Computational Linguistics. Oxford University Press.

Padro, L. and Marquez, L. (1998). On the evaluation and comparison of taggers: The effect of noise in testing corpora. Cornell University Library, arXiv.org. https://arxiv.org/abs/cs/9809112.

Pustejovsky, J. and Stubbs, A. (2012) Natural Language Annotation for Machine Learning: A Guide to Corpus-Building for Applications. O'Reilly Media.

Sree, R. and Thottempudi, S. G. (2011) Parts-of-Speech Tagging: A hybrid approach with rule based and machine learning techniques. LAP Lambert Academic Publishing.

Treebank (2022) Penn Treebank Tagset. https://www.ling.upenn.edu/courses/Fall_2003/ling001/penn_treebank_pos.html. Accessed 15 July 2022.

Tuchong (2022) Oil painting analog to Brill Tagger transformation technique. https://stock.tuchong.com/image/detail?imageId=965034062997356555. Accessed 17 Dec 2024.

Chapter 4
Syntax and Parsing

4.1 Introduction and Motivation

This chapter will explore *syntax analysis* and introduce different types of *constituents* in the English language followed by the main concept of *context-free grammar (CFG)* and *CFG parsing*. We will also study major parsing techniques including *lexical* and *probabilistic parsing* with examples.

Linguistic and grammatical aspects are addressed in NLP to identify patterns that govern the creation of language sentences like English. They include the investigation of *Part-of-Speech (POS)* mentioned in Chap. 3, and *grammatic rules* to create sentences or utterances with *syntactic rules*. These syntactic rules relied on effective computational procedures such as *rule-based*, *stochastic-based*, techniques and machine learning to deal with language syntax (Bender 2013; Gorrell 2006).

Another motivation is to study syntax and parsing methods or algorithms so that they can fall into an automatic system like forming a parser to understand syntactic structure during the construction process. Figure 4.1 illustrates the relationship between grammar, syntax, and the corresponding parse tree of a sentence/utterance with four tokens: *Tom pushed the car*. Syntax-level analysis is to analyze the structure and the relationship between tokens to create a parse tree accordingly.

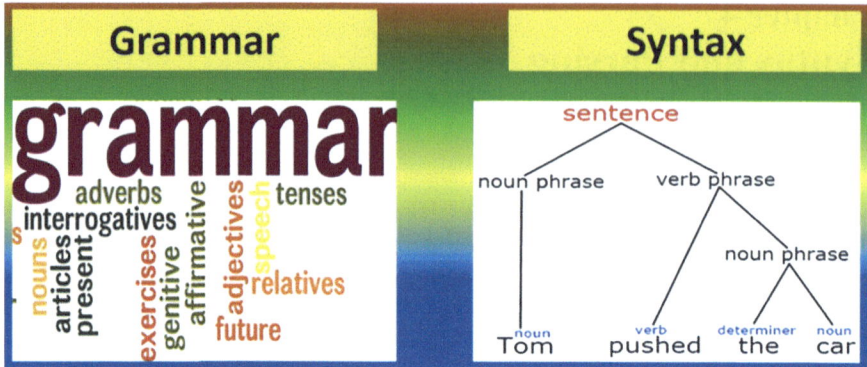

Fig. 4.1 Grammar, syntax, and parse tree

4.2   Syntax Analysis

4.2.1   What Is Syntax

Syntax refers to the set of rules that govern how groups of words are combined to form phrases, clauses, and sentences or utterances in linguistics (Bender 2013; Brown and Miller 2020). The term *syntax* is derived from the Greek word σύνταξη, which means *arrangement of words*.

Syntax provides a structured and organized way to create meaningful phrases and sentences. It is an essential tool in technical writing and sentence construction. The fact is all native speakers learn proper syntax from their mother languages by nature. The complex sentences by a writer or speaker create formal or informal level, or phrases and clauses presentation to audiences. *Syntax* can be defined as the correct arrangement of word tokens in written or spoken sentences and utterances, enabling computer systems to process these tokens without requiring an understanding of their precise meaning from an NLP perspective.

4.2.2   Syntactic Rules

POS in English often follows patterns order in sentences and clauses (Khanam 2022; Jurafsky et al. 1999). For instance, *compound sentences* are combined by conjunctions like *and, or, with* or multiple adjectives transformation of the same noun based on order(s) according to their classes, for example, [4.1] *The big black dog.*

Syntactic rules also described to assist language parts make sense. For example, sentences/utterances in English usually begin with a subject followed by a *predicate* (i.e. a verb in the simplest form) and an object or a complement to show what's acted upon, for example, [4.2] *Jack chased the dog* is a typical sentence with a *subject-verb-object* pattern of syntactic rule in English. However, [4.3] *Jack quickly chased the dog at lush green field* contains adverbs and adjectives to take their places in front of the sentence transformation *(quickly chased, lush green field)* with informative description.

4.2.3   Common Syntactic Patterns

There are seven common syntactic patterns:

1. Subject → Verb

For example, [4.4] *The cat meowed.*

This *syntactic pattern* is a standardized pattern containing only minimum subject and verb requirements. The topic always comes first in usual situations.

2. Subject → Verb → Direct Object.

For example, [4.5] *The cat plays the ball.*

When the verb is *transitive* with a *direct object*, the direct object usually goes after the verb in this syntactic pattern.

3. Subject → Verb → Subject Complement.

For example, [4.6] *The cat is playful.*

Subject complement usually goes after the verb in this syntactic pattern. Linking verbs such as *be, is, like,* or *seem* are usually used with subject complement.

4. Subject → Verb → Adverbial Complement.

For example, [4.7] *The cat paced slowly.*

Adverbial complement usually goes after the verb like the previous case (3).

5. Subject → Verb → Indirect Object → Direct Object

For example, [4.8] *The cat gave me the ball.*

This *syntactic pattern* contains *direct* and *indirect objects*. The direct object usually goes after the indirect object and the indirect object usually goes right after the verb. For example, [4.8] can be rephrased as [4.9] *The cat gave the ball to me.*

6. Subject → Verb → Direct Object → Direct Complement

For example, [4.10] *The cat made the ball dirty.*

Object complement usually goes after the direct object and the direct object is usually followed by a verb in this syntactic pattern.

7. Subject → Verb → Direct Object → Adverbial Complement

For example, [4.11] *The cat perked its ears up.*

up is the *adverbial complement* to describe how the cat behaves. *Direct complement* is replaced by *adverbial complement* like in the previous case (6).

The main purpose of syntactic parsing is a study to formulate rules with POS tags to perform automatic, or semi-automatic *sentence parsing*.

4.2.4 Importance of Syntax and Parsing in NLP

There are five major components in *Natural Language Understanding (NLU)* as shown in Fig. 4.2. Among these, the *syntax* and *parsing* components play central roles in linking natural language with its *syntactic structure* before understanding its *semantic* or *embedded (pragmatic) meanings* in NLP (Allen 1994; Eisenstein 2019). These components form the first layer of analysis to determine whether sentences or utterances are logically sound. In other words, if a sentence or utterance contains a syntactic error, such as *Jack buys (buys what?),* it will be nonsensical, making it impossible to proceed to *semantic analysis.*

Syntax and *parsing* are sole processes beneficial to:

1. Check grammar by word-processing applications such as Microsoft Word.
2. Speech recognizer at human speech real-time syntactic level in noisy environment.

It has significance in high-level NLP applications such as machine translation and Q&A chatbot systems.

4.3 Types of Constituents in Sentences

4.3.1 What Is Constituent?

A *constituent* is considered as the linguistic component of a language (Bender 2013; Brown and Miller 2020). For example, words or phrases that combine into a sentence or utterance are constituents. It can be a word, morpheme, clause, or phrase. *Parsing* is a kind of sentence analysis to identify the subject or predicate with different POS, and parse sentences/utterances into corresponding constituents e.g. There are several ways to describe the cat in Fig. 4.3.

[4.12] *The milky cat with long tail (as a constituent of a clause) is meowing.*

Fig. 4.2 Major
components in NLU

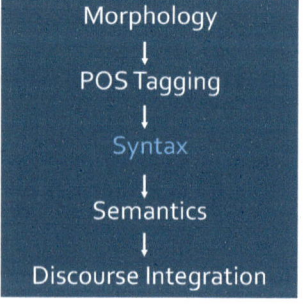

Fig. 4.3 The milky cat
with long tail is meowing
(Tuchong 2022)

A single pronoun *it* to replace the identified constituent. This makes sense as it described the milky cat with long tail is meowing, *it is meowing*, or a name to the cat *Coco*, like:

[4.13] *Coco is meowing* or
[4.14] *Coco with long tail is meowing*

which means a word or phrase form can be replaced by a simple token, or complex constituents with additional description:

[4.15] *The milky cat with long tail is meowing in the late afternoon.*
[4.16] *The milky cat with long tail is meowing the late afternoon while Jack* is asleep.

Constituents can also be a *time unit* with usage variations instead of an object unit in noun phrase (NP), they are syntactically acceptable, but some are not:

[4.17] *Jane wants to go to Greece <u>late this winter</u>.*
[4.18] *<u>Late this winter</u> Jane wants to go to Greece.*
[4.19] *Jane wants <u>late this winter</u> to go to Greece.*

It makes sense wherever the location of *late this winter* as it is a constituent describing a particular time in syntax, but there are syntactic errors as below:

[4.20] *<u>Late</u> Jane wants to go to Greece <u>this winter</u>.*

– Cannot separate time unit into two parts.

[4.21] *Jane wants <u>late</u> to go to Greece <u>this winter</u>.*

– Senseless meaning

[4.22] *The <u>late this winter</u> Jane wants to Greece.*

– Incorrect syntactic pattern

4.3.2   Kinds of Constituents

Constituents are present in every sentence, phrase, and clause. In other words, each sentence is formed by combining these elements into meaningful constructions or utterances (Bender 2013; Brown and Miller 2020). The commonly used constituent types include (1) *noun-phrase*, (2) *verb-phrase*, and (3) *preposition-phrase*. For instance:

[4.23] *My cat Coco scratches the UPS courier on the table.*

– These constituents are made up of noun phrase (*my cat Coco*), predicate, and verb phrase *(scratches the UPS courier on the table)*.

4.3.2.1   Noun Phrase (NP)

A noun phrase (NP) consists of a noun and its modifiers. Modifiers that precede the noun include adjectives, articles, participles, possessive nouns, and possessive pronouns, while those that follow the noun include adjective clauses, participial phrases, and prepositional phrases. For example, in [4.23] *My cat Coco* is an NP consisting of determiner (DT) *My* + noun (NN) *cat* + proper noun (NNP) *Coco*.

There are other NPs appear as objects of prepositions or objects of verbs:

[4.24] <u>*The milky cat with long tail*</u> *is meowing.*

[4.25] <u>*Very few cats*</u> *wore a collar.*

[4.26] <u>*The long tail*</u> *is brought to room.*

[4.27] <u>*Many places*</u> *hear meowing.*

[4.28] <u>*A cat with a long tail and a collar*</u> *is meowing.*

[4.29] <u>*Jane*</u> *saw so many cats in the room.*

4.3.2.2   Verb Phrase (VP)

A *Verb Phrase (VP)* consists of a main verb accompanied by linking verbs or modifiers that function as the verb of the sentence. Modifiers in a VP are words that can alter, specify, limit, or elaborate on the main verb. They typically include auxiliary verbs such as "is," "has," "am," and "are," which work in conjunction with the main verb. The main verb in a VP conveys information about the event or activity being discussed, while the auxiliary verbs provide additional meaning by indicating the tense or aspect of the phrase.

There are nine common VP types:

1. Singular main verb.
[4.30] *Jack catches a deer.*
2. Auxiliary verb (to be) + main verb -ing form
When the main verb is used in *-ing* form, e.g., *walking, talking*, it expresses a continuous aspect to show whether is in the past, present, or future.
[4.31] *Jack is singing.*
3. Auxiliary verb (have) + main verb (past participle form).
When the verb *to have* (i.e., have, has, had) and the main verb in past participle form.
[4.32] *Jack has broken the vase.*
4. Modal verb + main verb.
When a modal verb is combinedly used with a main verb, it includes things such as
   possibility, probability, ability, permission, and obligation. Examples of modal
   words include *must, shall, will, should, would, can, could, may and might.*
[4.33] *Jack will leave.*
5. Auxiliary verb (have + been) + main verb (-ing form).
When both continuous and perfect aspects are expressed, the continuous aspect
   comes from *-ing* verb and the perfect aspect comes from the auxiliary verb
   have been.
[4.34] *Jack has been washing the car.*
6. Auxiliary verb (to be) + main verb (past participle form).
When a verb *to be* is combined with the main verb in past participle form to express
   a passive voice. The passive voice is used to indicate an action is happening to
   the subject of a sentence than the subject performing the action.
[4.35] *The lunch was served.*
7. Negative and interrogative verb phrases.
VP gets separated when sentences have a negative or interrogative nature.
[4.36] *Jack is not answering the exam questions.*
8. Emphasize verb phrases.
Use auxiliary verbs, for example, *do, does, did* to emphasize a sentence.
[4.37] *Jack did enjoy the vacation.*
9. Composite VP.
When it consists of other VP or NP.
[4.38] *My cat Coco scratches the UPS courier on the table.*
   – *scratch* is the main verb in VP to describe an action/event that happens to object
   UPS courier, on the table is auxiliary information to further explain the event. It
   still makes sense with/without it. It includes *scratch* VP + *UPS courier* NP + *on
   the table* PP.

4.3.3 Complexity on Simple Constituents

Single-word constituents are parts of speech (POS) discussed in Chap. 3, where the types of single-word constituents vary based on tagset sizes. Additionally, there are several complex design considerations:

[4.39] *Jane bought the big red handbag* ☑ vs.
[4.40] *Jane bought the red big handbag* ☒

Although there are two parts of speech that can be syntactically correct, such as the placement of "red" before "big," the arrangement can result in incorrect syntax. Additionally, there may be incomplete simple constituent types.

[4.41] *The cat with a long tail meowing a collar.* ☒
   – Doesn't make sense although NP is correct, *collar* is an incorrect description.
[4.42] *Jane imagined a cat with a long tail.* ☑
[4.43] *Jane decided to go.* ☑
   – Both make sense without further description in syntactic structure.
[4.44] *Jane decided a cat with a long tail.* ☒
   – Doesn't make sense again in syntactic correctness.
[4.45] *Jane decided a cat with a long tail should be her next pet.* ☑
   – Syntactic correct although the sentence structure is slightly complex.
[4.46] *Jane gave Lily some food.* ☑
   – Syntactic correct although most of the time it describes food.
[4.47] *Jane decided Lily some food.* ☒
   – Although the syntactic structure is the same, they have different designs to further describe food types and purposes.

4.3.4 Verb Phrase Subcategorization

There is a universal pattern or structure for classifying verbs in verb phrases (VPs). Subcategories reflect the ability of lexical items (typically verbs) to recognize the existence and types of syntactic arguments they can co-occur within linguistics (Brown and Miller 2020; Gorrell 2006). While traditional English grammar classifies verbs into transitive and intransitive subcategories, modern English grammars identify more than 100 subcategories. Subcategorization frames can be viewed as a set of rules that generate syntactic structures from the base form. Five major frame rules are illustrated in Table 4.1.

1. VP with a single verb as member
[4.48] *He talks.* (VP → VB)
[4.49] *I laugh.* (VP → VB)
2. Verbal phrase requires a noun phrase (NP) as a specifier (VS)—intransitive verbs
[4.50] *He finds a clue.* (VP → VB + NP)

Table 4.1 Examples of verbs with different frames of subcategorization in VP syntax

Frame rule	Description	Examples
φ	VP with single verb as member	talk, sleep, eat, laugh, etc.
VS(NP)	The verbal phrase requires a noun phrase (NP) as a specifier (VS)—intransitive verbs	find, see, leave, get, etc.
VS(NP)VC(NP)	The verbal phrase requires a noun phrase (NP) as a specifier (VS) and a noun phrase (NP) as a complement (VC) (direct transitive verbs)—direct transitive verbs	show, make, read, write, etc.
VS(NP) VC(PH([on]))	The verbal phrase requires a noun phrase (NP) as a specifier (VS) and a prepositional phrase (PP) headed by "on" as a complement (VC)—indirect transitive verbs governing "on"	depend, insist, operate, suggest etc.
VS(NP)VC(NP) VC(PH([to]))	The verbal phrase requires a noun phrase as a specifier (VS), a noun phrase as a complement (VC), and a prepositional phrase headed by "to" as a complement (VC)—ditransitive verbs	give, mean, think, etc.

[4.51] *She sees Jack.* (VP → VB + NP)

3. Verbal phrase requires a noun phrase (NP) as a specifier (VS) and a noun phrase (NP) as a complement (VC)—direct transitive verbs

[4.52] *Please show me the map.* (VP → VB + NP + NP)

4. Verbal phrase requires a noun phrase (NP) as a specifier (VS) and a prepositional phrase (PP) headed by *on* as a complement (VC)—indirect transitive verbs governing *on:*

[4.53] *This ingredient can make six muffins depending on size* (VP → VB + PP + NP)

5. Verbal phrase requires a noun phrase (NP) as a specifier (VS), as a complement (VC) and a prepositional phrase headed by *to* as a complement (VC)—ditransitive verbs:

[4.54] *Do you mean that I need to attend the exam?* (VP-VB + S)

4.3.5 The Role of Lexicon in Parsing

A *lexicon* is the vocabulary of a language or a specific field of knowledge, such as medicine or computer science (Bender 2013; Brown and Miller 2020). It serves as an inventory of lexemes in linguistics. The term "lexicon" is derived from the Greek word λεξικόν, which means "of or for words."

Linguists believe that all human languages are composed of two major components: (1) *lexicon* as the list of a language's words and vocabulary and (2) *grammar* as the set of rules to allow word combinations into meaningful sentences.

Items within a lexicon are called *lexemes*, and groups of lexemes are called *lemmas*, often used to describe the size of a lexicon.

Lexical analysis is the process to understand what words mean, intuit contexts, and note the relationship of one word to others. It analyses and converts the sequence of words into a list of lexical tokens. A program that performs such lexical analysis is called *tokenizer, lexer,* or *scanner.* A lexer is combined with a parser generally to analyze the syntax of sentences, texts, or dialogues.

The roles of lexicon in parsing are to:

1. Treat as the starting point for POS tagging.
2. Provide extra information such as subcategorization with frames and syntactic rules.

For verbs, lexicon refers to several types of subcategorizations such as *think* versus *laugh.*

For adjectives:

[4.55] *Jack is angry with Sophia* vs. [4.56] *Jack is angry at Sophia.*
[4.57] *Jack is mad at Sophia* vs. [4.58] *Jack is mad with Sophia.* ☒

There are patterns and rules. Both are correct for [4.55] and [4.56], but for the verb *mad*, [4.57] is correct while incorrect for [4.58] which means subcategorization is acceptable for some pattern but not only on syntax.

For nouns: [4.59] *Janet has a passion for classical music* vs. [4.60] *Janet has an interest in classical music.*

They have different patterns of syntactic rules.

4.3.6   Recursion in Grammar Rules

English sentences can be structurally complex. A concise sentence typically consists of a limited set of constituent types, such as NP (Noun Phrase), VP (Verb Phrase), and PP (Prepositional Phrase), which can recursively combine to form more intricate structures according to specific grammatical rules.

S → NP VP [4.61] *My good friend Jack buys a flat.*
VP → V NP [4.62] *buys a flat.*
NP → NP PP [4.63] *My good friend.*
NP → NP S [4.64] *The boy who come early today won the game.*
PP → prep NP [4.65] *The cupcake with sprinkles is yours.*

4.4 Context-Free Grammar (CFG)

4.4.1 What Is Context Free Language (CFL)?

Context-free language (CFL) is a superset of *Regular Language (RL)* generated by *context-free grammar (CFG)* which means every RL is a CFL but not all CFL is a RL (Eisenstein 2019; Jurafsky et al. 1999). In short, CFL is:

1. Recursively enumerable language as a superset of language model.
2. Context-sensitive language, a subset of recursively enumerable language.
3. Subsets of context-sensitive language.

The four levels of human language are shown in Fig. 4.4.

The set of all Context-Free Languages (CFLs) is identical to the set of languages accepted by pushdown automata, and regular languages (RL) form a subset of CFLs. An input language is accepted by a computational model if it passes through the model and ends in an acceptable final state. Most arithmetic expressions generated by a Context-Free Grammar (CFG) are CFLs.

A CFL is closed under specific operations, meaning that applying these operations to a CFL results in another CFL. These operations include *union, concatenation, Kleene closure, substitution, prefix, cycle, reversal, quotient, intersection, difference with RL, and homomorphism*. CFLs and CFGs play a significant role in both NLP and computer language design in computer science and linguistics.

4.4.2 What Is Context Free Grammar (CFG)?

CFG is to describe CFL as a set of *recursive rules* for generating string patterns, because the application of production rules in grammar is *context-independent*, meaning they do not depend on other symbols with the rules (Bender 2013; Brown and Miller 2020).

CFG is commonly applied in linguists and compiler design to describe programming languages and parsers that can be created automatically.

Fig. 4.4 Level of languages

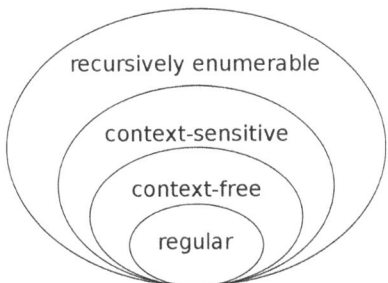

4.4.3 Major Components of CFG

CFG consists of four major components (Bender 2013; Jurafsky et al. 1999):

1. A set of non-terminal symbols N are placeholders for patterns of terminal symbols created by nonterminal symbols. These symbols are usually located at the LHS (left-hand-side) of production rules (P). The strings generated by CFG usually consist of symbols only from nonterminal symbols.
2. A set of terminal symbols Σ (disjoint from N) are characters appear in strings generated by grammar. Terminal symbols usually located only at RHS (right-hand-side) of production rules (P).
3. A set of production rules $P: A \rightarrow \alpha$, where A is a non-terminal symbol and α is a string of symbols from the infinite set of strings ($\Sigma \cup N$).
4. The designated start symbol S is a start symbol of the sentence/utterance.

Σ is a set of POS and N is the set of constituent types, that is, NP, VP, and PP mentioned in Chap. 3 and previous section, respectively.

4.4.4 Derivations Using CFG

The standard formulation of CFG is given by:

Assume L_G generated by grammar G is a set of strings composed of terminal symbols, which is generated from S:

$$L_G = \left\{ w \mid w \text{ is in } \Sigma^* \text{ and } S \Rightarrow w \right\} \tag{4.1}$$

Let Σ be the set of POS, so CFG in (4.1) can create grammar like this:

$$N \, V \, det \, N \tag{4.2}$$

The definition of CFG is given by:

$$L_G = \left\{ s \mid w \text{ is in } \Sigma^* \text{ and } S \Rightarrow w \text{ and } s \text{ can be derived from } w \text{ by substituting words for POS as licensed by the lexicon} \right\} \tag{4.3}$$

Based on this definition can generate numerous productions like this format:

$$S \rightarrow NP \, VP \tag{4.4}$$

Equation 4.4 is the most basic grammar rule where a sentence is generated from an NP and a VP that can be further decomposed recursively as shown in Fig. 4.5.

It shows CFG rules and its corresponding parse tree for sentence/utterance [4.66] *Jane plays the piano*. There are four tokens in this sentence/utterance to form a

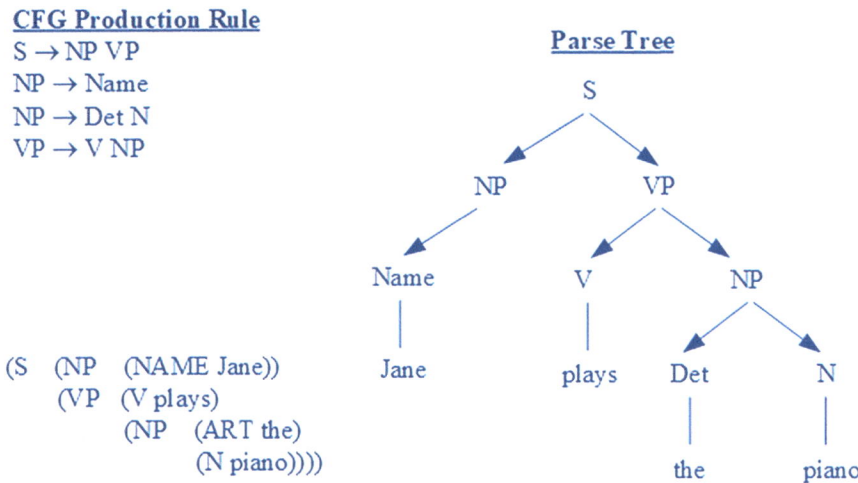

Fig. 4.5 CFG rules and corresponding parse tree for sentence [4.66] *Jane plays the piano*

well-defined syntactic structure generated by NP and VP. NP can be designated to a name pointed to token *Jane*, and for VP, is decomposed into a verb or NP as shown in four production rules shown in the top left corner of Fig. 4.5. In this case, the verb is pointed to *plays,* NP can be decomposed into a determiner and a noun pointed to *the* and *piano*, respectively.

4.5   CFG Parsing

There are three CFG parsing levels: (1) morphological, (2) phonological, and (3) syntactic (Grune and Jacob 2007; Jurafsky et al. 1999).

4.5.1   Morphological Parsing

Morphological parsing is the initial level to determine the morphemes of a word being constructed. For example, a morphological parser can reveal that the word *mice* is the plural form of the noun stem *mouse*, while *cats* are the plural form of the noun stem *cat*. Given the string *cats* as input, the morphological parser will interpret *cats* as *cat* N PL. By using FSA (Finite State Automata), FST (Finite State Transducer), a morphological parser can produce an output with their stems and modifiers.

Originally, FST was generated by algorithmic parsing of word sources such as a complete dictionary with modifier markups but can be realized by recurrent neural networks with training corpus upon advancement in machine learning and artificial neural networks.

4.5.2 Phonological Parsing

Phonological parsing is the second level using the sounds of a language, that is, phonemes to process sentences or utterances (Wagner and Torgesen 1987).

Phonological processing includes (1) *awareness*, (2) *working memory*, and (3) *phonological retrieval*. All three components are important to speech production and written language skills development. Hence, it is necessary to observe children's spoken and written language development with phonological processing difficulties.

Phonological parsing is to interpret sounds into words and phrases to generate parser.

4.5.3 Syntactic Parsing

Syntactic parsing is the third level to identify relevant components and correct grammar of a sentence. Abstract meaning representation is assigned to define legal strings of a language like CFG without recognizing the structure.

Parsing algorithms are applied to analyze sentences or utterances within language and assign appropriate syntactic structures into them. *Parse trees* are useful to study grammar, semantic analysis, machine translation, speech recognition, and Q&A chatbots in NLP.

4.5.4 Parsing as a Kind of Tree Searching

Syntactic parsing can be considered as search within a set of parse trees, its main purpose is to identify the right path and space through automation in an FSA system structure.

CFG is a process to determine the right parse tree among all possible options. If there is more than one possible *parse tree*, *stochastic method* (or other machine learning methods) will be applied to locate a probable one. In other words, it is a process to identify search space defined by grammatical rules so that their constraints can become inputs to perform automatic parsing and study grammar.

4.5.5   CFG for Fragment of English

English grammar and lexicon simplified domains are applied to reveal CFG rules in an example of musical instruments as shown in Table 4.2. It consists of production rules from several categories S → NP VP, S → Aux NP VP, S → VP as well as production rules for NP, Nom, and VP with components Det, N, V, Prep, and PropN.

4.5.6   Parse Tree for "Play the Piano" for Prior CFG

A parse tree of sentence/utterance [4.67] *play the piano* is shown in Fig. 4.6. It has three tokens *play*—Verb, *the*—Det and *piano*—Noun to construct a parse tree from the top node S to generate VP, VP to generate Verb and NP, and NP to decompose into Det Nom, and Nom to generate Noun.

4.5.7   Top-Down Parser

There are (1) *top-down* and (2) *bottom-up parser* approaches to construct a parse tree. *Top-down parser* constructs from *root-node S* down to *leave-nodes* (words in the sentence or utterance). The first step is to identify all trees with root S, the next step is to expand all constituents in these trees based on the given production rules. The whole process is operated level-by-level process until parse trees reach the leaves i.e. POS tokens of the sentence/utterance. For candidate parse trees that cannot match the leave nodes, that is, POS tokens are discarded and considered as failed parse tree(s). Figure 4.7 shows the first three-level construction of all possible parse trees applying a *Top-Down parser*.

It showed that the parse tree construction started from the base level with S tag (root node). The second level has generated an additional layer with three possible production rules: S → NP & VP, S → Aux & NP & VP, and S → VP. The third level is complex because it has decomposed into three levels, S → NP & VP is the first

Table 4.2 A simplified example on English grammar and lexicon	

S → NP VP	VP → V
S → Aux NP VP	Det → this I that I the I a
S → VP	N → play I piano I guitar I flute
NP → Det Nom	V → play I include I prefer
NP → PropN	Aux → does, do
Nom → N Nom	Prep → on I from I to
Nom → N	PropN → Germany I Italy I Yamaha
Nom → Nom PP	
VP → V NP	

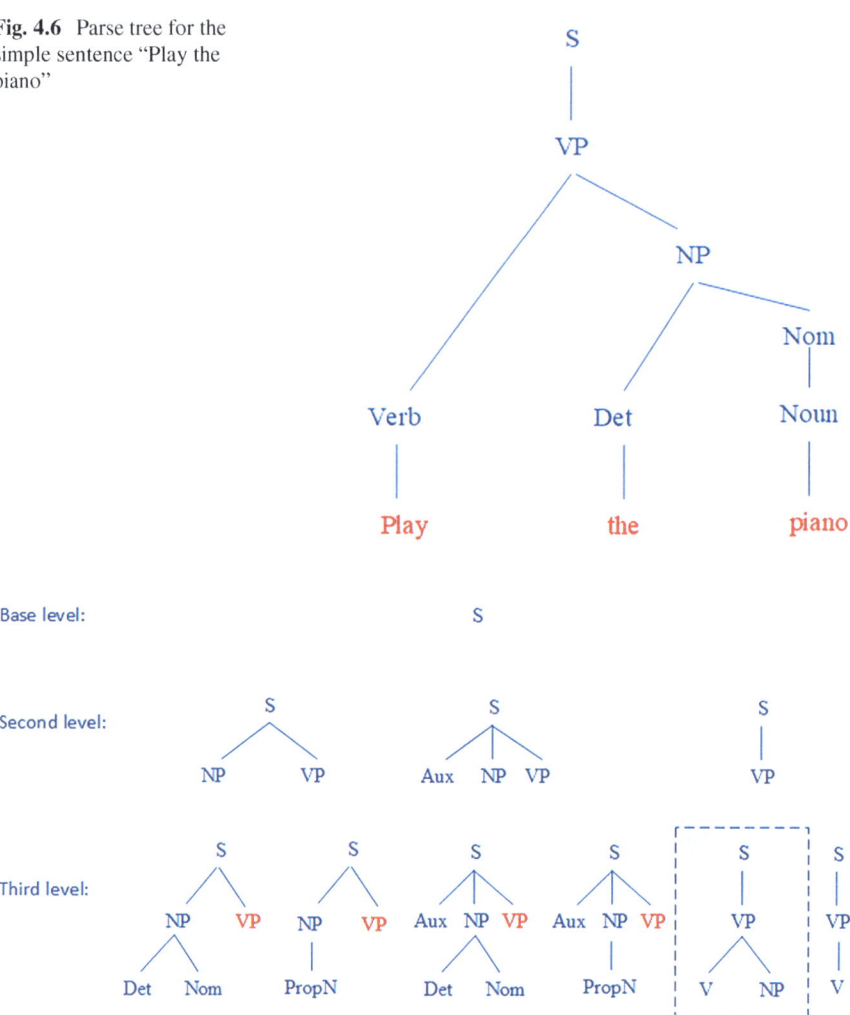

Fig. 4.6 Parse tree for the simple sentence "Play the piano"

Fig. 4.7 A three-level expansion of parse tree generation using a top-down approach

variation to decompose into Det and Nom. NP is the second variation to decompose into PropN. It is noted that LHS is the expanded part for demonstration purposes, but both LHS and RHS require expansion. S → Aux & NP & VP are the second variation where NP decomposes into Det & Nom, and an NP decomposes into PropN. VP decomposition in the first four parse tree is not shown as they all failed to match the leave nodes except only the fifth case is correct to form a complete *play the piano* parse tree.

The top-down approach by CFG on terminals and non-terminals is shown in Table 4.3. It showed rule 3 as the first one to apply and rule 2 for VP decomposed into V NP and V to decompose *play* and then NP to Det and Nom, rule 4 and rule 5

Table 4.3 CFG rules and terminal/non-terminal nodes being used with top-down approach parsing

S → NP VP	VP → V
S → Aux NP VP	Det → this | that | the (5) | a
S → VP (1)	N → play | piano (7) | guitar | flute
NP → Det Nom (4)	V → play (3) | include | prefer
NP → PropN	Aux → does, do
Nom → N Nom	Prep → on | from | to
Nom → N (6)	PropN → Germany | Italy | Yamaha
Nom → Nom PP	
VP → V NP (2)	

are Det points to *the*, and rule 6 Nom points to *an* and final rule points to end, and rule 7 points to *piano*. This will complete a top-down approach parsing with the fifth parse tree end-up as valid solution. Readers can apply these seven-step processes to complete the construction of parse tree for the fifth case as an exercise.

4.5.8 Bottom-Up Parser

Bottom-up parser on the other hand starts from token words of the sentences/utterances to construct a parser tree upward by applying the same set of production rules and try to generate from right-hand-side (RHS) of the production rule in reverse order. In example [4.67] *play the piano* has two variations to start in which the word *play* can be considered either as a noun (N) or a verb (V). So, there are two options: one with the *play the piano* as N Det or as V Det N. Since this approach cannot indicate which one is the correct option so the parsing operation will continue to grow until they can reach the root node S, and if they cannot match the root node, the tree(s) will be discarded.

Figure 4.8 shows the first three-level expansion of a parse tree using *bottom-up approach*. So, in this case *play the piano* has two variations either play is N or V. There are two parts one is *play* consider as N and other as V from base level. So, at second level is to further expand the line pointed to *play* and tried to expand N pointed to play into N in the first case. In second case is to further expand N pointed to Nom in second layer. In the third level, second case is further expanded into two options, one is Nom → V and the other is VP → V & NP, NP → Det & Nom, and further up to S → VP to complete the whole parsing, in which other two parse tree options ended-up with invalid parsing as shown in Fig. 4.8.

Table 4.4 shows CFG rules for terminal and non-terminal nodes applying bottom-up approach parsing. Again, it consists of seven steps. Rule 1 is V pointed to *play*, rule 2 is Det pointed to *the*, rule 3 is N pointed to *piano*, rule 4 is Nom pointed to N, rule 5 is NP pointed to Det & Nom, rule 6 is VP pointed to NP and rule 7 is S point to VP to complete the whole parse tree until it can finally match the root/source node S.

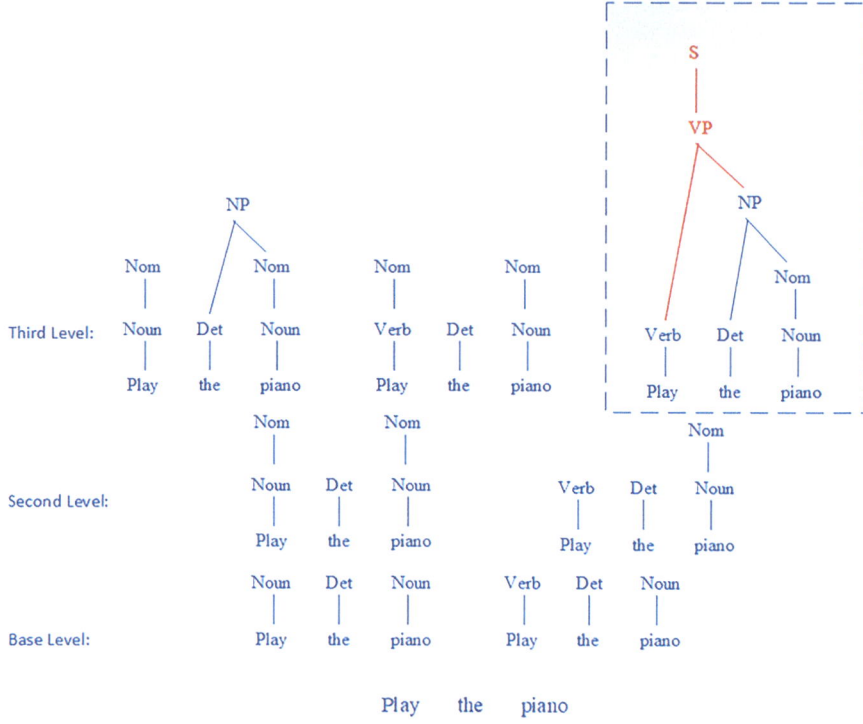

Fig. 4.8 A three-level expansion of parse tree generation using bottom-up approach

Table 4.4 CFG rules and terminal/non-terminal nodes being used with bottom-up approach parsing	S → NP VP	VP → V
	S → Aux NP VP	Det → this I that I the (2) I a
	S → VP (7)	N → play I piano (3) I guitar I flute
	NP → Det Nom (5)	V → play (1) I include I prefer
	NP → PropN	Aux → does, do
	Nom → N Nom	Prep → on I from I to
	Nom → N (4)	PropN → Germany I Italy I Yamaha
	Nom → Nom PP	
	VP → V NP (6)	

4.5.9   Control of Parsing

Although both *top-down* and *bottom-up parsing* are straightforward, the control of parsing is required to consider (1) which node to expand first and (2) select grammatical rules sequence wisely to save time as most of the parse tree generation are dead-end and wastage of resources.

4.5.10   Pros and Cons of Top-Down vs. Bottom-Up Parsing

4.5.10.1   Top-Down Parsing Approach

Pros

Since it starts from root/source node S, it can always generate a correct parse tree unless the sentence has a syntactic error. In other words, it never explores the parse that won't end up in root/source node S, which means it will always find a solution.

Cons

This approach doesn't consider final word/token tags during parsing from the very beginning, it will waste a lot of time to generate tree(s) that may be totally unrelated to the correct result. *Play* should parse as V instead of N as shown in Fig. 4.7, this approach showed that all first and fourth parts of the parse tree using *play* as N are invalid and a waste of time to parse tree generation.

4.5.10.2   Bottom-Up Parsing Approach

Pros

Since it starts from sentence tokens/POS, it can always generate a parse tree with all tokens/POS in the sentence considered and reduced time on rules unrelated to these tokens which means it can sort out problems that occur in the top-down approach for all production rules without POS tags.

Cons

This approach may often end up with broken tree(s) that cannot match the root node S to complete parse tree as it starts from leave node instead of root/source node S. It makes sense because although there are many ways to match production rules, the variations of most parse trees are syntactic incorrect so they cannot match the root/source node S. All parse trees in Fig. 4.8 showed that except the last one (also the correct one), others ended up with broken trees and failed to match the root/source node S again wasted time to parse tree generation.

Let's look at lexicalized and probabilistic parsing as alternatives.

4.6 Lexical and Probabilistic Parsing

4.6.1 Why Using Probabilities in Parsing?

There are two reasons for using probabilities parsing (Eisenstein 2019; Jurafsky et al. 1999): (1) resolve ambiguity and (2) word prediction in voice recognition. For instance:

[4.68] *I saw Jane with the telescope.* (Jane with telescope or I use telescope to see Jane?)
[4.69] *I saw the Great Pyramid flying over Giza plateau* vs.
[4.70] *I saw UFO flying over Giza plateau*

Although both situations have pragmatic problems in which [4.69] is incorrect because the Great Pyramid is an unmovable architecture. It can be solved using probabilities in parsing from a large corpus and knowledgebase (KB) to identify the frequencies of a particular term or constituent is used correctly without pragmatical analysis.

For example, in voice recognition:

[4.71] *Jack has to go* vs.
[4.72] *Jack half to go* vs.
[4.73] *If way thought Jack wood go*

Note: when analyzing N-gram probabilities in Chap. 2 on the N-gram Language Model, *I have, I should, I would* usage and bigram probabilities from *The Adventures of Sherlock Holmes*, they provided directions for one that is more probable and used frequently instead of understanding their exact semantic or pragmatic meanings. So, such a probabilistic method can also be applied to parsing.

4.6.2 Semantics with Parsing

The following examples show how semantic meanings (Bunt et al. 2013; Goddard 1998) affect/determine the validness of sentence/utterance in parsing:

[4.74] *Jack drew one card from a desk* [?] vs.
[4.75] *Jack drew one card from a deck.*
Note: *drew → deck* is clearly a semantic concern.
[4.76] *I saw the Great Pyramid flying over Giza plateau.* [?] vs.
[4.77] *I saw a UFO flying over Giza plateau.*
Note: movable vs. unmovable objects.
[4.78] *The workers dumped sacks into a pin.* [?] vs.
[4.79] *The workers dumped sacks into a bin.*
Note: *dump* looks for a locative complement.

[4.80] *Tom hit the ball with the pen.* [?] vs.
[4.81] *Tom hit the ball with the bat.*
Note: which object can use to hit the ball?
[4.82] *Visiting relatives can be boring.* [?] vs.
[4.83] *Visiting museums can be boring.*

Note: Visiting relatives is genuinely ambiguous. Visiting museums is evident as only animate bodies can visit. There is no need for abstraction with enough data, in other words, sufficient large corpus, databank or dialogue databank can sort out ambiguity problems to work out correct syntax with semantic meaning in many cases.

There are two classical approaches to add semantics into parsing: (1) cascade systems to construct all parses and use semantics for rating tedious and complex; (2) do semantics incrementally.

A modern approach is to forget the meaning and only based on KB and corpus. If a corpus contains sufficient sentences and knowledge, facts about meaning emerge in the probability of observed sentences themselves. It is modern because constructing world models is harder than early researchers realized but there are huge text corpora to construct useful statistics. Here comes the lexical and probabilistic approach of parsing.

4.6.3   What Is PCFG?

A *probabilistic context-free grammar (PCFG)* is a context-free grammar that associates each of its production rules with a probability. It creates the same set of parses for a test that traditional CFG performs but assigns a probability value to each parse. In other words, the probability of a parse generated by a PCFG is the product of probability's rules.

The general format of PCFG production rule is given by:

$$A \rightarrow \beta \left[p \right] \tag{4.5}$$

Another way to interpret it is.

$$P \left(A \rightarrow \beta | A \right) \tag{4.6}$$

Note: the sum of all probabilities of rules with LHS A must be 1.

PCFG extends CFG like how Markov models extend regular grammars. Each production rule is assigned with probability. The probability of a parse is the product of probabilities of productions used in that derivation. These probabilities can be regarded as parameters of the model, and for large NLP problems, it is convenient to learn these parameters via machine learning methods. A probabilistic grammar's validity is constrained by the context of its training dataset.

An efficient PCFG design must weigh scalability, generality factors and issues such as grammar ambiguity must be resolved to improve system performance.

4.6.4 A Simple Example of PCFG

This section used sentences/utterances [4.84] *buy coffee from Starbucks* as example to illustrate how PCFG works. It has simple CFG rules and probabilities in a segment of an AI chatbot dialogue for food ordering at campus as shown in Table 4.5.

The probability of each production rule type must sum up to 1 is one of the most important basic criteria of PCFG as shown. For instance, three production rules of S: S → NP VP (0.82), S → Aux NP VP (0.12) and S → VP (0.06) must sum-up to 1. It is the same as other production rules for NP, Nom, VP, Det, N, V, Aux, Proper-N, and Pronoun. Of course, if the corpus is very large, some of these probability values will be very small, just like N-gram probability evaluation discussed in Chap. 2.

It can apply either a top-down parser or bottom-up parser approach to generate a parse tree with the following PCFG probability evaluation scheme:

$$P(T) = \prod_{n \in T} p(r(n))$$

(4.7)

where *p(r(n))* is the probability that rule *r* will be applied to expand the non-terminal *n*.

Table 4.5 Sample CFG rules and their probabilities in AI chatbot dialogues (food ordering at campus)	CFG rules [Prob]	CFG rules [Prob]
	S → NP VP [0.82]	Det → a[0.12] | that[0.03] | the[0.75] | this[0.10]
	S → Aux NP VP [0.12]	N → coffee [0.75]
	S → VP [0.06]	N → tea [0.13]
	NP → Det Nom [0.21]	N → food [0.12]
	NP → Proper-N [0.37]	V → buy [0.41]
	NP → Nom [0.06]	V → pay [0.27]
	NP → Pronoun [0.36]	V → order [0.32]
	Nom → Noun [0.72]	Aux → do [0.31]
	Nom → N Nom [0.23]	Aux → does [0.26]
	Nom → Proper-N Nom [0.05]	Aux → can [0.43]
	VP → V [0.58]	Proper-N → Starbucks [0.63]
	VP → V NP [0.36]	Proper-N → KFC [0.37]
	VP → V NP NP [0.06]	Pronoun → I[0.42]|you[0.36]| he[0.12]|she[0.10]

So, what required to achieve is

$$\hat{T}(S) = \frac{\arg\max P(T)}{T \in \tau(S)}$$

(4.8)

where $\tau(S)$ denotes the set of all possible parses for S.

Figure 4.9 depicts different meanings and two parse trees for utterances [4.85] *can you buy Starbucks coffee?* The first interpretation regards *Starbucks* and *coffee* are two standalone NPs with equal significance. The second interpretation is to combine *Starbucks* and *coffee* into a single NP constituent which is a brand name, in this case, *can you buy Starbucks coffee* can interpret to buy coffee or non-coffee items. Hence, parse tree probability calculation is also different.

Table 4.6 shows all CFG rules and associated probabilities of these two parse trees. So, PCFG probabilities for parse trees 1 and 2 are

$$P(PT_1) = 0.12^*0.36^*0.06^*0.06^*0.37^*0.72^*0.43^*0.36^*0.41^*0.63^*0.75$$
$$= 1.242 \times 10^{-6}$$

$$P(PT_2) = 0.12^*0.36^*0.36^*0.06^*0.05^*0.72^*0.43^*0.36^*0.41^*0.63^*0.75$$
$$= 1.007 \times 10^{-6}$$

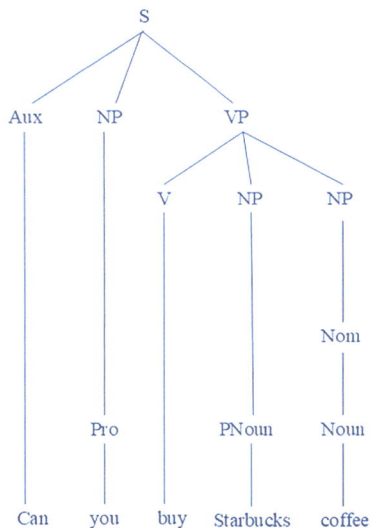
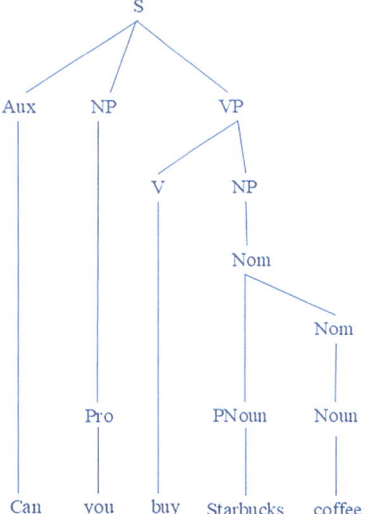

Fig. 4.9 Two possible parse trees for the utterance "Can you buy Starbucks coffee"?

Table 4.6 CFG rules and associated probabilities for two possible parse trees PT1 vs. PT2

CFG rules for TP1 [Prob]	CFG rules for TP2 [Prob]
S → Aux NP VP [0.12]	S → Aux NP VP [0.12]
NP → Pronoun [0.36]	NP → Pronoun [0.36]
VP → V NP NP [0.06]	VP → V NP [0.36]
NP → Nom [0.06]	NP → Nom [0.06]
NP → Proper-N [0.37]	Nom → Proper-N Nom [0.05]
Nom → Noun [0.72]	Nom → Noun [0.72]
Aux → can [0.43]	Aux → can [0.43]
Pronoun → you [0.36]	Pronoun → you [0.36]
V → buy [0.41]	V → buy [0.41]
Proper-N → Starbucks [0.63]	Proper-N → Starbucks [0.63]
N → coffee [0.75]	N → coffee [0.75]

CFG probability algorithm parse tree 1 has a high probability. In other words, it is more possible the meaning is to buy coffee rather than buy other things from Starbucks. It also shows an efficient solution to differentiate which parse tree is more probable, when there are ambiguities in two or more parse trees provided with sufficient lexical probabilities and a corpus to calculate probabilities.

4.6.5 Using Probabilities for Language Modelling

Probability parsing can be considered as the integration of the N-gram probability concept with parse tree formation. Since there are fewer grammar rules than word sequences for N-gram generation, applying this calculation method one can calculate results efficiently instead of N-gram frequencies regardless of syntactic meaning and rules.

Based on this method, the probability of S is the sum of probabilities of all possible parses given by:

$$P(S) = \sum_{T \in \tau(S)} P(T)$$

(4.9)

against N-gram probability calculation with the Markov model.

$$P(S) = P(w_1) * P(w_2|w_1) * P(w_3|w_1w_2) * P(w_4|w_1w_2w_3)\ldots$$

(4.10)

4.6.6 Limitations for PCFG

In many situations, it is adequate to know that one rule is used more frequently than another e.g.

[4.86] *Can you buy Starbucks coffee?* versus [4.87] *Can you buy KFC coffee?*
But often it matters what the context is
For example:

$$S \to NP\ VP$$
$$NP \to Pronoun \quad [0.80]$$
$$NP \to LexNP \quad [0.20]$$

(4.11)

For example, when NP is the subject, the probability of a pronoun may be higher at 0.91. When NP is the direct object, the probability of a pronoun may be lower at 0.34 which means it depends on the NP position in a sentence/utterance. In other words, the probabilities also often depend on lexical options as shown in the following examples:

[4.88] *I saw the Great Pyramid flying over Giza Plateau.* vs.
[4.89] *I saw a UFO flying over Giza Plateau.*
[4.90] *Farmer dumped sacks in the bin.* vs.
[4.91] *Farmer dumped sacks of apples.*
[4.92] *Jack hit the ball with the bag.* vs.
[4.93] *Jack hit the ball with the bat.*
[4.94] *Visiting relatives can be boring.* vs.
[4.95] *Visiting museums can be boring.*
[4.96] *There were boys in park and girls* vs.
[4.97] *There were boys in park and shops.*

For instance, there are two interpretations of utterances [4.98] *boys in park and girls* as shown in Fig. 4.10 showing their syntax ambiguities.

Figure 4.10 shows two possible parse trees for utterance [4.98]. The first is *boys in park* is a noun clause with conjunction NP *girls*. The second is *park and girls* belong to a single NP with boys as single NP. Although structures are different but the mathematization result for two parse trees are identical which means CFG probability calculation can't differentiate which is better or popular. How to fix this problem?

4.6.7 The Fix–Lexicalized Parsing

The lexicon can be considered as an estimation of a knowledgebase (KB) a possible solution to the above ambiguous problem.

Figure 4.11 shows [4.90] applying lexical parsing as an example. Each constituent is a *head word* i.e. *S* using *dumped* as *Head word*. NP and VP are signified by *farmer* and the other signified by head word *dumped* at second tier. From *farmer* it comes up with NNS *farmer*. VP from *dumped* because will come up with VBD, NP, and PP, and VDB is signified by *dumped* as head word and NP is *sack* and PP is *into*.

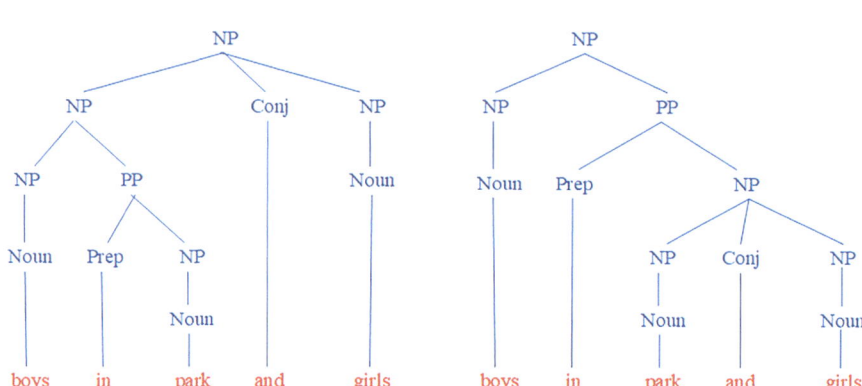

Fig. 4.10 Two interpretations of the utterance "boys in park and girls"

So, *sack* further decomposes in NNS which points to *sacks*, for PP to further decomposes into P and NP into *bin*. This will provide information to further decomposition by combining keywords. So, for the NP *bin* it will further decompose into *the* and *the bin* as head words for DT and NN respectively.

By adding lexical items with production rules:

$$\text{VP}\big(\text{dumped}\big) \rightarrow \text{VBD}\big(\text{dumped}\big)\text{NP}\big(\text{sacks}\big)\text{PP}\big(\text{into}\big) = 8 \times 10^{-10} \quad (4.12)$$

$$\text{VP}\big(\text{dumped}\big) \rightarrow \text{VBD}\big(\text{dumped}\big)\text{NP}\big(\text{cats}\big)\text{PP}\big(\text{into}\big) = 1 \times 10^{-10} \quad (4.13)$$

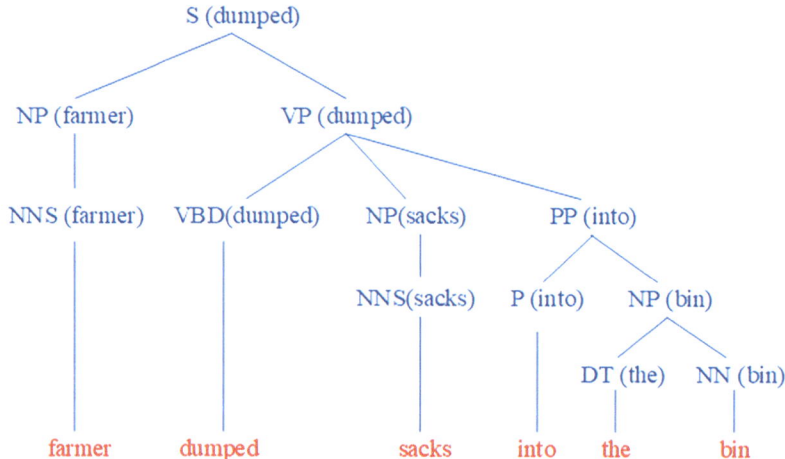

Fig. 4.11 Lexical tree for the utterance "workers dumped sacks into a bin"

$$\text{VP}(\text{dumped}) \rightarrow \text{VBD}(\text{dumped})\text{NP}(\text{stones})\text{PP}(\text{into}) = 2 \times 10^{-10}$$
(4.14)

$$\text{VP}(\text{dumped}) \rightarrow \text{VBD}(\text{dumped})\text{NP}(\text{sacks})\text{PP}(\text{above}) = 1 \times 10^{-12}$$
(4.15)

with lower probabilities means infrequency appeared in the corpus.

This determination method is more efficient as compared with N-gram probability calculation, sample sentences/utterances such as:

[4.99] *The farmer dumped sacks of apples into a bin.* vs.
[4.100] *The farmer dumped sacks of peaches into a bin.* vs.
[4.101] *The farmer dumped all the sacks of apples into a bin.*

But there will be situations that many lexical probabilities come-up with 0 values like N-gram probability evaluation.

A short-cut by considering the following lexical rule as replacement instead of considering the whole lexical rule such as (4.12) can sort out this problem:

$$\text{VP}(\text{dumped}) \rightarrow \text{VBD NP PP }\, p\big(r(n)|n,h(n)\big)$$
(4.16)

By doing so, the lexical probability of certain nodes n with heads h is considered based on two conditions: (1) syntactic type of node n and (2) head of node's mother $h(m(n))$, so lexical rule of (4.16) is split into the following:

$$\text{Given } P\big(h(n) = \text{word}_i |,n|,h\big(m(n)\big)\big)$$
$$\text{VP}(\text{dumped}) \rightarrow \text{PP}(\text{into}), \; p = p1$$
$$\text{VP}(\text{dumped}) \rightarrow \text{PP}(\text{of}), \; p = p2$$
$$\text{NP}(\text{sacks}) \rightarrow \text{PP}(\text{of}), \; p = p3$$
(4.17)

Now the original lexical probability (4.7) becomes:

$$P(T) = \prod_{n \in T} p\big(r(n)|,n|,h(n)\big) * p\big(h(n)|,n|,h\big(m(n)\big)\big)$$
(4.18)

Using Brown corpus as an example, the probability of:

$$p\big(\text{VP} \rightarrow \text{VBD NP PP}|,\text{VP}|,\text{dumped}\big) = 0.67$$
$$p\big(\text{VP}\big) \rightarrow \text{VBD NP}\# \text{ VP},\text{dumped}\big) = 0.0$$
$$p\big(\text{into}|,\text{PP}|,\text{dumped}\big) = 0.22$$
$$p\big(\text{into}|,\text{PP}|,\text{sacks}\big) = 0$$
(4.19)

parse contribution of this part to the total scores for two candidates will be:

$$[\text{dumped into}]0.67 \times 0.22 = 0.147$$
$$[\text{sacks into}]0 \times 0 = 0$$

(4.20)

So, we should consider *dumped into* instead of *sacks into* in this case.

Exercises

4.1. What are the syntax and parsing in linguistics? Discuss why they are important in NLP.

4.2. What is the syntactic rule? State and explain SEVEN commonly used syntactic patterns in English language, with an example each to illustrate.

4.3. Answer (4.2) by applying to other languages such as Chinese, French or Spanish. What is (are) the difference(s) of the syntactic rules between these two languages with examples to illustrate.

4.4. What are constituents in the English language? State and explain three commonly used English constituents, with an example each to illustrate how it works.

4.5. What is Context-Free Grammar (CFG)? State and explain the importance of CFG in NLP.

4.6. State and explain FOUR major CFG components in NLP. Use an example sentence/utterance to illustrate.

4.7. What are TWO major types of CFG parsing scheme? Use an example sentence/utterance [4.102] *Jack just brought an iPhone from Apple store* to illustrate how these parsers work.

4.8. What is PCFG in NLP parsing? Use same example [4.102] *Jack just brought an iPhone from Apple store* to illustrate how it works. Compare with parsers used in (4.7), which one is better?

4.9. What are the advantages and limitations of PCFG in NLP parsing? Use some sample sentences/utterances to support your answers.

4.10. What is lexical parsing in NLP parsing? Discuss and explain how it works by using sample sentence [4.102] *Jack just brought an iPhone from Apple store* for illustration.

References

Allen, J. (1994) Natural Language Understanding (2nd edition). Pearson

Bender, E. M. (2013) Linguistic Fundamentals for Natural Language Processing: 100 Essentials from Morphology and Syntax (Synthesis Lectures on Human Language Technologies). Morgan & Claypool Publishers

Brown, K. and Miller, J. (2020) Syntax: A Linguistic Introduction to Sentence Structure. Routledge.

Bunt, H. et al. (2013) Computing Meaning: Volume 4 (Text, Speech and Language Technology Book 47). Springer.

Eisenstein, J. (2019) Introduction to Natural Language Processing (Adaptive Computation and Machine Learning series). The MIT Press.

Goddard, C. (1998) Semantic Analysis: A Practical Introduction (Oxford Textbooks in Linguistics). Oxford University Press.

Gorrell, P. (2006) Syntax and Parsing (Cambridge Studies in Linguistics, Series Number 76). Cambridge University Press.

Grune, D. and Jacob, C. (2007) Parsing Techniques: A Practical Guide (Monographs in Computer Science). Springer.

Khanam, H. M. (2022) Natural Language Processing Applications: Part of Speech Tagging. Scholars' Press.

Jurafsky, D., Marin, J., Kehler, A., Linden, K., Ward, N. (1999). Speech and Language Processing: An Introduction to Natural Language Processing, Computational Linguistics and Speech Recognition. Prentice Hall.

Tuchong (2022 The milky cat with long tail meowing. https://stock.tuchong.com/image/detail?imageId=896801601105166424. Accessed 17 Dec 2024.

Wagner, R. K., & Torgesen, J. K. (1987). The nature of phonological processing and its causal role in the acquisition of reading skills. Psychological Bulletin, 101(2), 192–212. https://doi.org/10.1037/0033-2909.101.2.192.

Chapter 5
Meaning Representation

5.1 Introduction

While the understanding of sentences and utterances in terms of structure, grammar, and the relationships between words by using *N-gram models* or simple *syntactic rules* has been extensively studied, the actual meaning of sentences or the meanings of individual words within a sentence have not been thoroughly explored. This chapter will focus on how to interpret *meaning*, introducing scientific and logical methods for processing meaning, known as *meaning representation*. Without this foundational understanding, it would be challenging to interpret advanced NLP analyses involving *semantic meaning*, *pragmatic meaning*, and *discourse* in the following chapters. We will begin by exploring *meaning representations*, which integrate linguistic knowledge of the real world with the world of linguistics.

5.2 What Is Meaning?

Language is prodigious in recognizing humans that encode or decode world description from experiences to ideas and interpret others' opinions. It is natural but difficult to utter word strings that match the world into expressions. A way to enrich this activity is to transform essences that wish to convey into meaningful words, clauses, phrases, or sentences/utterances in verbal or written forms for others to listen, understand, inference, and even respond.

In linguistics, *meaning* refers to the message conveyed by words, phrases, and sentences or utterances within a given context. It is often referred to as lexical or semantic meaning. Professor W. Tecumseh Fitch described semantic meaning in *The Evolution of Language* (Fitch 2010) as a branch of language study that is closely linked to philosophy. This connection exists because the study of semantic meaning

© The Author(s), under exclusive license to Springer Nature Singapore Pte Ltd. 2025
R. Lee, *Natural Language Processing*,
https://doi.org/10.1007/978-981-96-3208-4_5

raises numerous fundamental philosophical questions that demand resolution and explanation from philosophers.

A good dictionary provides meaning explanation of a single word in detail and many dictionaries on concept/language translation. Nevertheless, the meanings of sentences or utterances are not simply the combination of individual word's meaning, but usually appeared as phrasal words with specific meanings at the pragmatic level, e.g., [5.1] *off the wagon*.

Semantic meaning is the study of meaning assignment to minimal meaning-bearing elements to form complex and meaningful ideas. Some basic word groups may be aggregated in content relationship called *thematic groups*, and lexical or semantic fields related to *common sense* or *world knowledge*, e.g., the concept of *doctor* in English constitutes the lexical-semantic field in two senses: a medical doctor, or a person with PhD title. Once the meaning of a word (word group) is decrypted or analyzed, a reaction is formed as a response to the event it represents. Words and their meanings are significant informational cues to understand languages. Further, a person's life experience and cultural difference are relevant to linguistic meaning development in the communication process.

5.3   Meaning Representations

This chapter will adopt a similar approach to *syntax* and *morphology analysis* (Bender 2013) to build linguistic input representations and capture their meaning. These linguistic representations are meaning representations of sentences and states of affairs in the real world.

Unlike parse trees, these representations are not primarily a description of the input structure, but a representation of how humans understand, represent anything (such as actions, events, objects, etc.), and try to understand it in our environment— *the meaning of everything*.

There are five types of meaning representation: (1) *categories*, (2) *events*, (3) *time*, (4) *aspect*, and (5) *beliefs, desires*, and *intentions*.

1. *Categories* refer to specific objects and entities, e.g., *company names, locations, objects*.
2. *Events* refer to actions or phenomena experienced, e.g., *eating lunch, watching a movie*. They are relevant to verbs or verb phrases (VPs) expressed in POS.
3. *Time* refers to the exact or reference moment, e.g., *9:30 am, next week, 2025*.
4. *Aspects* refer to.

   (a) *Stative*—to state facts.
       For example, [5.2] Jane knows how to run.
   (b) *Activity*—to describe action.
       For example, [5.3] *Jane is running.*
   (c) *Accomplishment*—to describe completed action without ending terms.
       For example, [5.4] *Jane booked the room.*

(d) *Achievement*—to describe terminated action.
For example, [5.5] *Jane found the book.*
5. *Beliefs, desire, and intention* refer to principles such as:
[5.6] *I think what you are saying is totally correct.*
[5.7] *Jane wants to know why she failed in the test.*
[5.8] *I believe everything happens for a reason.*

These principles are complex because they involve a variety of ideas in philosophy. However, it is important to design appropriate, logical, and computational representations in NLP to facilitate *semantic processing* and express ideas in sentences/utterances.

5.4 Semantic Processing

Semantic processing (Bender 2013; Best et al. 2000; Goddard 1998) performs meaning representation to encode and interpret meaning. These representations allow the following:

1. *Reason* relations with the environment.
For example, [5.9] *Is Jack inside the classroom?*
2. *Answer questions* based on contents.
For example, [5.10] *Who got the highest grade in the test?*
3. *Perform inference* based on knowledge and determine the verity of unknown fact(s), thing(s), or event(s).
For example, [5.11] *If Jack is in the classroom, and Mary is sitting next to him, then Mary is also in the classroom.*

Semantic processing is applied in typical applications, including question-and-answer chatbot systems, where it is necessary to understand meaning, i.e., the ability to answer questions about the context or utterance with knowledge, literal meaning, or even embedded meaning. Examples from our AI Tutor chatbot (Cui et al. 2020) are shown below, involving varying degrees of *semantic processing*:

[5.12] *What is the meaning of NLP?*
– Basic level of semantic processing for the meaning of certain concept.
[5.13] *How does N-gram model work?*
– Requires understanding of facts and meanings to respond.
[5.14] *Is the Turing test still exist?*
– Involves high-level query and inference from previous knowledge.
[5.15] *Why do we need to study self-awareness in AI?*
– Involves high-level information such as world knowledge or common sense aside from AI terminology knowledge base (KB) to respond.
[5.16] *Should I study AI?*
– Involves the highest level information about user's common sense and world knowledge aside from AI concepts learnt by the book.

5.5   Common Meaning Representation

There are four common methods of meaning representation scheme: (1) *first-order predicate calculus (FOPC)*; (2) *semantic networks (semantic net)*, (3) *conceptual dependency diagram (CDD)*, and (4) *frame-based representation*. A sample sentence [5.17] *Jack drives a Mercedes* is used to illustrate how they perform.

5.5.1   First-Order Predicate Calculus

First-Order Predicate Logic (FOPL) (Dijkstra and Scholten 1989; Goldrei 2005) is also known as *predicate logic* or *FOPC*. It is a robust language representation scheme to express the relationship between information objects as *predicates*. For example, FOPC meaning for [5.17] is given by:

$$\exists x, y \, \text{Driving}(x) \wedge \text{Driver}(\text{Jack}, x) \wedge \text{DriveThing}(y, x)$$
$$\wedge \text{CarBrand}(\text{Mercedes}, y) \tag{5.1}$$

This FOPC formulation consists of four predicate calculus segments (*predicates*) in logical terms.

5.5.2   Semantic Networks

Semantic networks (semantic nets) (Jackson 2019; Sowa 1991) are knowledge representation techniques used for *propositional information*. They convey knowledge meanings in a two-dimensional representation. A *semantic net* can be represented as a labeled directed graph. The logic behind this is that a *concept meaning* is connected to other *concepts* and can be represented as a graph. The information in *semantic net* is characterized as a set of *concept nodes* to link up with each other by a set of labeled arcs which characterized the relationship as illustrated in Fig. 5.1 for example sentence [5.17].

Driving is the core concept connected to two *nodes (concepts)*: *Driver* and *DriveThing*, which links to *Jack* as the *Driver* and *Mercedes* as *DriveThing*, respectively.

5.5.3   Conceptual Dependency Diagram

CDD is a theory that describes how sentence/utterance meaning is represented for reasoning. It has been argued that CDD represents an independent representation of the language in which the sentence was originally stated.

Fig. 5.1 Semantic net for example sentence [5.17]

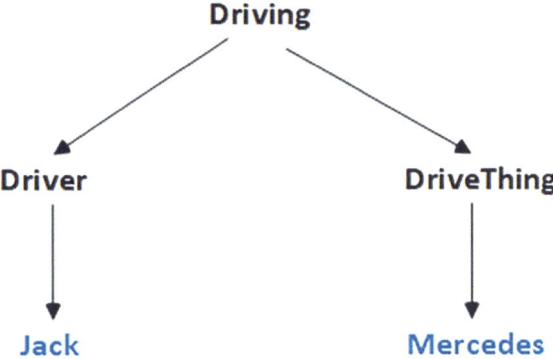

Fig. 5.2 Conceptual dependency diagram for example sentence [5.17]

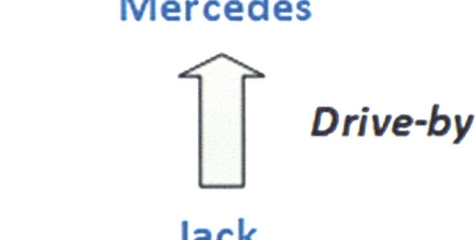

Schank (1972) proposed conceptual dependency (CD) theory as a part of a natural language comprehension project. Sentences/utterances applying CD can translate and express basic concepts as a small set of *semantic primitives*, which can be integrated to represent complex meanings—*conceptualizations*. Figure 5.2 shows a CD diagram for example sentence [5.17].

Mercedes and *Jack* are two concepts linked up by the main concept *Drive-by* using CD representation.

5.5.4 Frame-Based Representation

Frame-based systems use *frames* and *notions* as basic components to characterize domain knowledge introduced by Prof. Marvin Minsky in his remarkable work *A framework for representing knowledge* published in 1975 (Minsky 1975).

A *frame* is a knowledge configuration to characterize a *concept* such as *a car* or *driving a car* attached to certain definitional and descriptive information. There are several constructed knowledge representation systems based on the original model. The vital successor of frame-based representation schemes is *description logs* that encode the declarative part of *frames* using *semantic logics*. Most of these semantic logics are components in the first-order logic that are related to feature logics. A frame-based representation for [5.17] is shown in Fig. 5.3.

The frame-based representation is also invariance to language(s) being used like other meaning representation models.

In summary, these meaning representations indicated that the linguistic meaning for [5.17] describes certain *state-of-affairs* happened in a real world. Different meaning representation models are just different ways to represent the same scenario. For example, FOPC is a kind of mathematical and logical representation of meanings, while semantic nets are graphical representation of such meaning in the form of *directed graphs*.

5.6   Requirements for Meaning Representation

There are three factors to fulfill a meaning representation (Bunt 2013; Butler 2015; Potts 1994): (1) verifiability, (2) ambiguity, and (3) vagueness considerations.

5.6.1   Verifiability

Verifiability refers to determining whether a sentence/utterance has a literal meaning (it expresses a proposition) and whether it is analyzable or empirically verifiable, which means it must provide a link between the meaning representation and facts with the KB, world knowledge, or common-sense comparison methods:

For example, [5.18] *Does Jack drive a Mercedes?*

A verifiable meaning representation asserts to *prove* the correctness of this statement with comparison, matching or inferencing operations.

The answer is *yes* according to statement [5.17].

Fig. 5.3 Frame-based diagram for example sentence [5.17]

Driving

  Driver: Jack

  DriveThing: Mercedes

5.6.2 Ambiguity

Ambiguity is a word, statement, or phrase that consists of more than one meaning. Ambiguous words or phrases can cause confusion, misunderstanding, or even humor situations.

For example, [5.19] *Jack rode a horse in brown outfit.*

This clause may drive readers to wonder whether the horse wore a brown outfit instead of the rider. Likewise, the same words with different meanings induce ambiguity, e.g., *Jack took off his gun at the bank.* It is diverting to confuse the meaning of *bank* that refers to a building or the land alongside of a river or lake. *Context meaning* is important to resolve ambiguity.

5.6.3 Vagueness

Vagueness is to describe borderline cases, e.g., *tall* is a vague term in the sense that a person who is 1.6 m in height is neither *tall* nor *short* since there is no amount of conceptual analysis or empirical investigation can settle whether a 1.6-m person is tall or not without any frame of reference. Here is another example:

[5.20] *He lives somewhere in the south of US,*
which is also vague as to the meaning of location.

Ambiguity and *vagueness* are two varieties of uncertainty which are often discussed together but are distinct in essential features and significances in semantic theory. *Ambiguity* involves uncertainty about mapping between representation levels which have more than a single meaning with different structural characteristics, while *vagueness* involves uncertainty about the actual meaning of terms. Hence, a good meaning representation system should resolve *vagueness* and avoid *ambiguity*.

5.6.4 Canonical Forms

5.6.4.1 What Is Canonical Form?

A *canonical form* refers to entities of resources which can be determined in more than one way, and one of them can be considered as a favorable *canonical (standard) form*.

The *canonical form* of a mathematical entity is a standard way of determining that quantity in mathematical expression. For example, the canonical form of a positive integer in decimal form is a *number sequence* that does not start from zero. It is a class of entity in which an equivalence relation is defined. For example, a *Row*

Echelon Form (REF) and *Jordan Normal Form* are typical canonical forms for matrix interpretation in *Linear Algebra.*

There are many methods to represent the canonical form of the same entity in computer science, for instance, (1) *computer algebra*, which represents mathematical objects and (2) *path* concept in a *hierarchical file system*, where a single file can be referenced in several ways.

5.6.4.2 Canonical Form in Meaning Representation

The canonical form of meaning representation in NLP refers to the phenomena of a single sentence/utterance that can be assigned with multiple *meanings* leading to the same meaning representation. For example:

[5.21] *Jack eats KitKat.*
[5.22] *KitKat, Jack likes to eat.*
[5.23] *What Jack eats is KitKat?*
[5.24] *It's KitKat at that Jack eats.*

All these sentences/utterances have similar meanings with minor variations in *tones* and *thematic* issues.

FOPC, semantic net, CDD, and frame-based representation are good elaborations of how canonical form performs and stores such representations in a KB.

5.6.4.3 Canonical Forms: Pros and Cons

Advantages

1. Simplify reasoning and storage operations.
2. Needless to generate inference rules for all different variations with the same meaning.

Disadvantages

Nevertheless, it may complicate semantic analysis for sentences/utterances with similar meanings, but each has variance in phonemes or high-level semantic meanings like examples [5.21–5.24].

5.7 Inference

5.7.1 What Is Inference?

Inference (Blackburn and Bos 2005) is divided into *deduction* and *induction* with origin dated back to Ancient Greece from Aristotle, 300s BCE. *Deduction* refers to using available information to guess or draw conclusions about facts such as legendary *Sherlock Holmes*' deductive reasoning methods (Doyle 2019). Examples of inference by deduction reasoning are as follows:

[5.25] *Jack is a pilot; he travels a lot.*
[5.26] *Jane's hair is totally soaked; it might be raining outside.*
[5.27] *Mary has been very busy at work and may not be able to come for a gathering this evening.*

Induction is inference from evidence to a universal conclusion. An important fact is that the conclusions may be correct or incorrect.

Examples of inference by inductive reasoning are as follows:

[5.28] *The sun rose in the morning every day for the past 30 years. The sun rises every day (in human history).*
[5.29] *The first two kids I met at my new school were kind to me. The students at this school are kind.*
[5.30] *Our teacher allows us to pick a piece of object out of a box. The first four students got candies. The box must be full of candies.*

An inference is valid in general if it is conformed to sound evidence(s), and the conclusion follows logically from related premises.

5.7.2   Example of Inferencing with FOPC

Inferencing with FOPC is to come up with valid conclusions which leaned on inputs meaning representation and KB. For example:

[5.31] *Does Jack eat KitKat?*

It consists of two FOPC statements:

$$\text{Thing}(\text{KitKat}) \tag{5.2}$$

$$\text{Eat}(\text{Jack},x) \wedge \text{Thing}(x) \tag{5.3}$$

Given the above two FOPC statements are true, it can infer the saying [5.31] as *yes* by using inductive or deductive reasoning.

5.8   Fillmore's Theory of Universal Cases

Case grammar (Fillmore 2020; Mazarweh 2010) is a linguistic system that focuses on the connection between the quantity such as the *subject, object, or valence* of a verb and the grammatical context used in language analysis. This theory was proposed by American linguistic professor Charles J. Fillmore (1929–2014) in his famous book *The Case for Case in Semantic Analysis* published in 1968, also known as *Fillmore's Theory of Universal Cases* (Fillmore 1968). He believed that only a limited number of *semantic roles*, called *case roles*, occur in every sentence/utterance constructed with verbs.

5.8.1   What Is Fillmore's Theory of Universal Cases?

The Fillmore's Theory of Universal Cases (Fillmore 2020; Mazarweh 2010) analyzes the fundamental syntactic structure of sentences/utterances by exploring the association of *semantic roles* such as *agent, benefactor, location, object, or instrument*, which are required by the verb in sentence/utterance. For instance, the verb *pay* consists of semantic roles such as *agent (A), beneficiary (B), and object (O)* for sentence construction. For example:

[5.32] *Jane (A) pays cash (O) to Jack (B).*

According to *Fillmore's Case Theory*, each verb needs a certain number of *case roles* to form a *case-frame*. Thus, *case-frame* determines the vital aspects of semantic valency of verbs, adjectives, and nouns. *Case-frames* are conformed to certain limitations, i.e., a particular *case role* can appear only once per sentence. There are *mandatory* and *optional cases*. *Mandatory cases* cannot be deleted; otherwise, it will produce ungrammatical sentences. For example:

[5.33] *This form is used to provide you.*

This sentence/utterance makes no sense without an additional role that explains *provide you to* or *with* what matter or notion. One possible solution is as follows:

[5.34] *This form is used to provide you with the necessary information.*

The association between nouns and their structures contain both syntactic and semantic importance. The syntactic positional relationship between forms in a sentence varies from language to language, so grammarians can observe, examine semantic values in these nouns, and provide information to consider case role in a specific language.

One of the major tasks of semantic analysis in *Fillmore's Theory* is to offer a possible mapping between syntactic constituents of a parsed clause and their semantic roles associated with the verb. The term case role is widely used for purely semantic relations, including theta and thematic roles. The *theta role (θ-role)* refers to a formal device for representing syntactic argument structure required syntactically by a particular verb. For instance:

[5.35] *Jack gives the toy to Ben.*

Statement [5.35] shows the verb *give* has three arguments, whereas *Jack* is determined as the external *theta role* of *agent*, *toy* is determined as the *theme role*, and *to Ben* is determined as the *goal role*.

Thematic role, also called *semantic role*, refers to case role that a noun phrase (NP) may deploy with respect to *action* or *state* used by the *main verb*. For example:

[5.36] *Jack gets a prize.*

Statement [5.36] shows *Jack* is the *agent* as he is *doer* to *get*, the *prize* is the *object* being received, so it is a *patient*.

5.8.2   Major Case Roles in Fillmore's Theory

There are six major roles in *Fillmore's Theory*:

1. *Agent*—doer of action, attribute intention.
2. *Experiencer*—doer of action without intention.
3. Theme—thing that undergoes change or being acted upon with
4. *Instrument*—tool being used to perform the action.
5. *Beneficiary*—person or thing for which the action being acted on or performed to.
6. *To/At/From Loc/Poss/Time*—to possess thing(s), place, location, or time.

For example:

[5.37] *Jack cut the meat with a knife.*
[5.38] *The meat was cut by Jack.*
[5.39] *The meat was cut with a knife.*
[5.40] *A knife cut the meat.*
[5.41] *The meat is cut.*
[5.42] *Jack lent Jane the CD.*
[5.43] *Jack lent the CD to Jane.*

These examples can conclude that:

1. *Agent*—*Jack* is the *doer* revealed in [5.37, 5.38, 5.42 and 5.43] that performs the *action*.
2. Theme—*meat* and *CD* are *things (objects)* being acted upon or undergoing change as revealed in [5.38–5.43] accordingly
3. *Instrument*—*knife* is the *tool* to complete an action as revealed in [5.37, 5.39 and 5.40].
4. *To-Poss*—*Jane* is the one that *possesses the CD* as revealed in [5.42 and 5.43] driven by Jack, the giver.

Syntactic choices intuition is largely a reflection of underlying semantic relationships, which means that identical meanings can descend to articles, e.g., [5.37] can also be presented in [5.38 and 5.39], or simplified versions can also be presented in [5.40 and 5.41]. *Syntax* can have several syntactic options that are related to same meanings in semantic meanings. *Semantic analysis* is a major task to offer a suitable linkage between constituent of a *parsed clause* and associated semantic roles related to the *main verb*.

5.8.3   *Complications in Case Roles*

There are four types of complications in case role analysis:

1. Syntactic constituents' ability to indicate semantic roles in several cases, e.g., subject position: *agent* vs. *instrument* vs. *theme*:
   [5.44] *Jack cut the fish.*
   [5.45] *The knife cut the fish.*
   [5.46] *The fish is cut.*
2. Syntactic expression option availability, e.g., *agent* and *theme* in different configurations:
   [5.47] *Jack cut the fish.*
   [5.48] *It was the fish that Jack cut.*
   [5.49] *The fish was cut by Jack.*
3. Prepositional ambiguity not always introduces the same role, e.g., proposition *by* may indicate either *agent* or *instrument*:
   [5.50] *The meat was cut by Jack.*
   [5.51] *The meat was cut by a knife.*
4. Role options in a sentence:
   [5.52] *Jack cut the fish with a knife.*
   [5.53] *The fish was cut by Jack.*
   [5.54] *The fish was cut with a knife.*
   [5.55] *A knife cut the fish.*
   [5.56] *The fish was cut.*

It seems that *semantic roles* act like a *musical conductor* in an orchestra with old syntactic constituents and leave them out at times, but it isn't as bad as it seems. There are regularities to consider sets of rules which are the beauty of human languages to describe the same idea in different styles and configurations.

There are possible rules in case role, such as:

$$
\boxed{
\begin{array}{l}
\text{If } \exists \text{ Agent, it becomes Subject} \\
\quad \text{Elseif } \exists \text{ Instrument it becomes Subject} \\
\quad\quad \text{Elseif } \exists \text{ Theme it becomes Subject} \\
\text{Agent preposition is BY} \\
\text{Instrument preposition is BY if no agent, else WITH}
\end{array}
}
\qquad (5.4)
$$

Note that:

1. They are general rules; some verbs may have exceptions.
2. Every syntactic constituent can only fill-in one case at a time.
3. No case role can appear twice in the same rule.
4. Only NPs of the same case role can be cojoined in the rule.

5.8.3.1   Selectional Restrictions

Selectional restrictions are methods to restrict types of certain roles to be used for semantic consideration. For instance:

1. *Agents* must be *animate objects*, i.e., a living thing such as a person, *Jack*.
2. *Instruments* must be *inanimate objects*, i.e., nonliving things such as *rock*.
3. Themes are types that may be *dependent on verbs* e.g. *window* relates to the verb *break*.

Such constraints can be applied to the following examples to check whether they make sense or not:

[5.57] *Someone <u>assassinated</u> the President* vs
[5.58] *The spider <u>assassinated</u> the fly.* ⊠

Nevertheless, additional rules can be deployed to state that *assassinate* has *intentional* or *political killing* such that [5.58] may be incorrect. In fact, such a method is usually applied for semantic analysis to be discussed in Chap. 6.

5.9   First-Order Predicate Calculus

5.9.1   FOPC Representation Scheme

FOPC (Dijkstra and Scholten 1989; Goldrei 2005) can be used as a framework to derive semantic representation of a sentence/utterance. Although it is imperfect, it is still the most straightforward mechanism to interpret meanings as other alternatives are finite and complex for implementation. In most cases, they become notational variants in which the quintessential parts are the same regardless of the variant to select.

FOPC supports:

1. *Reasoning* in *truth* condition analysis to respond *yes* or *no* questions.
2. *Variables* in general cases through variable binding at responses and storage.
3. *Inference* to respond beyond KB storage on new knowledge.

This choice is neither arbitrary nor determined by practical application. FOPC reflects natural language semantics as it was designed by humans.

5.9.2   Major Elements of FOPC

FOPC consists of four major elements: (1) terms, (2) predicates, (3) connectives, and (4) quantifiers.

1. **Terms**

   Terms are object names with three representations: (a) constants, (b) functions, and (c) variables.

   Constants refer to the specific object described in sentence/utterance, e.g., Jack, IBM.

   Functions refer to concepts expressed as genitives such as brand name, location, e.g., *Brandname(Mercedes), LocationOf(KFC)* can be regarded as single-argument predicate.

   Variables refer to objects without reference which object is referred to, like variables *x, y, and z* used in a mathematical equation $x + y = z$ (e.g., *a, b, c, x, y, z*) They are frequently used in FOPC for query and inferencing operations.

2. **Predicates**

   Predicates (Epstein 2012) refer to a predicate notion in traditional grammar that traces back to *Aristotelian logic* (Parry and Hacker 1991). A *predicate* is regarded as the property of a subject that has or is characterized by. It can be considered as the expression of fact to the relations that link up some fixed number of objects in a specific domain, e.g., *he talks, she cries, Jack plays football*, etc. *Predicates* are often represented with capital letters like *Buy* or *Play* in FOPC and combine with object names to form a proposition, e.g., *Drive(Mercedes), Drive(Mercedes, Jack), Drive(Mercedes, x), Drive(Mercedes, Jack, UIC, Starbucks), Drive(car, x, org, dest)*.

3. **Connectives**

   Connectives refer to proposition combinations. *Conjunctions* (*and* as in English, written as & or ∧), *disjunctions* (*or* as in English, written as ∨), and *implications* (*if-then* as in English, written as → or ⊃). *Negation* (*not* as in English, written as ¬ or ~) is also regarded as a *connective*, even though it operates on a single proposition.

4. **Quantifiers**

   Quantifiers refer to generalizations. There are two major kinds of quantifiers: *universal* (*all* as in English, written as ∀) and existential (*some* as in English, written as ∃). The term *first-order* in FOPC means that this logic only uses quantifiers to generalize *objects*, but never onto *predicates*.

   A FOPC *context-free grammar* (CFG) specification is shown in Fig. 5.4.

5.9.3   Predicate-Argument Structure of FOPC

The *semantics* of human languages usually exhibit certain *predicate-argument* structure by *variables*, e.g., indefinites in generic cases and inferencing. It also uses *quantifiers*, e.g., *every, some* to create FOPC flexibility for sentence structures and partial compositional semantics, e.g., *sort of*.

Predicate-argument structure refers to *actions, events,* and *relations* that can be determined and represented by *predicates* and *arguments*. Languages exhibit a certain *division-of-labor* in which words/constituents are served as *predicates* and

Formula → AtomicFormula

      | Formula Connective Formula
      | Quantifier Variable; : : : Formula
      | ¬ Formula
      | (Formula)

AtomicFormula → Predicate(Term, ...)

Term → Function(Term, ...)
     | Constant
     | Variable

Connective → ∧ | ∨ →
Quantifier → ∀ | ∃
Constant   → IBM | Tesla | USA | Jack | A | ...
Variable   → x | y | z | ...
Predicate → Drive | Buy | Find | ...
Function   → LocationOf | Brandname | ...

Fig. 5.4 Context-free grammar (CFG) specification of FOPC

arguments, e.g., *predicates* to manifest *verb*, and *arguments* to manifest *different cases of the verb*.

    Predicates are primarily *verbs (V), VPs, prepositions, adjectives,* and *sentences/ utterances*, and sometimes can be *nouns* and even *NPs*. For instance:

[5.59] *Helen cries.*
[5.60] *Helen speaks to Mary.*
[5.61] *Helen speaks loudly.*
[5.62] *Helen speaks loudly in the classroom.*

    Arguments are primarily nouns, nominals, and NPs, but can be other constituents which rely upon the actual context of sentence/utterance. For instance:

[5.63] *Jack goes to the bank* vs.
[5.64] *He goes to the bank.*

The following shows an FOPC formulation example:

[5.65] *Jack gave a pen to Jane.*

$$\text{Giving}(\text{Jack}, \text{Jane}, \text{Pen}) \qquad (5.5)$$

Note that the corresponding FOPC formulation (5.5) is precisely in Fillmore's case role theory that *give* conveys a three-argument predicate: (1) *Agent* which is *Jack* as the *giver*; (2) *Possess* which is *Jane* as the *recipient*; and (3) *Theme* which is the *pen* as the *direct object*.

It can have other configurations to describe the same predicate logic; for example:

$$\text{Giving}(\text{Jack},\text{Pen},\text{Jane}) \tag{5.6}$$

$$\text{Gave}(\text{Jack},\text{Pen},\text{Jane}) \tag{5.7}$$

Here are some complex cases with additional constituents:

[5.66] *Jack gave Jane a pen for Susan.*

$$\text{Giving}(\text{Jack},\text{Jane},\text{Pen},\text{Susan}) \tag{5.8}$$

[5.67] *Jack gave Jane a pen for Susan on Monday.*

$$\text{Giving}(\text{Jack},\text{Jane},\text{Pen},\text{Susan},\text{Monday}) \tag{5.9}$$

[5.68] *Jack gave Jane a pen for Susan in class on Monday.*

$$\text{Giving}(\text{Jack},\text{Jane},\text{Pen},\text{Susan},\text{in class},\text{Monday}) \tag{5.10}$$

Note that all these predicates should be treated individually as their arguments have different overall meanings.

5.9.4   Meaning Representation Problems in FOPC

A *predicate* that represents a *verb* meaning, e.g., *give*, has the same argument numbers present as its syntactic categorization frame. It is still difficult to (1) determine the correct *role* numbers for an event, (2) manifest facts about case role(s) associated with the event, and (3) ensure correct inference(s) is/are derived from meaning representation.

According to the above considerations, the FOPC formulation stated in (5.5) is not as useful as it seems it would be preferable if *roles* or *cases* are separated and flexible when deciding the whole FOPC statement like this:

$$\exists x, y \, \text{Borrowing}(x) \wedge \text{Borrower}(\text{Jack}, x) \wedge \text{Borrowed}(y, x)$$
$$\wedge \text{Borrow}\_\text{to}(\text{Jane}, x) \wedge \text{Isa}(y, \text{Pen}) \tag{5.11}$$

Note: Corresponding to Fillmore's case role theory, *Borrower* = *Agent*, *Borrowed* = *Theme*, *Borrow_to* = *To-Poss*.

Although the notion of predicate relation becomes complicated, it allows more flexibility for sentence/utterance construction.

It may further generalize (5.11) into the following formulation:

$$\exists x, y, z \, \text{Borrowing}(x) \land \text{Borrower}(w, x)$$
$$\land \text{Borrowed}(y, x) \land \text{Borrow\_to}(z, x) \tag{5.12}$$

By doing so, it can generate other complicated clauses by applying different predicate combinations. The semantics of NPs and PPS in a sentence plug into slots provided by the template which can allow flexibility to variable argument number associated with an event (predicate).

Event has many roles to cement input with specific category (e.g., pen) on categories and instances declaration. For example:

$$\text{Isa}\big(\text{MobyDick}, \text{Novel}, \text{AKO}(\text{Novel,Literature})\big) \tag{5.13}$$

Note: Just like *Isa()* to serve as predicate *Is a*, *AKO()* is a useful predicate to serve as the meaning *a kind of*. In fact, FOPC materializes events so that they can be quantified, and related to other events and objects through a defined set of relationships, and logical connections between closely related instances without meaning assumptions.

5.9.5   Inferencing Using FOPC

Inference is an important process in FOPC which has the capability to *validate* or *prove* whether a proposition is *true* or *false* from a KB. *Modus Ponens* (MP) is a fundamental inferencing method used in FOPC.

MP is a mode of reasoning from a hypothetical proposition. If the *antecedent* is *true*, then the *consequent* should be also *true*. In other words, MP is a kind of *deductive reasoning* in the form of: *P implies Q*, i.e., *If P is true, then Q must also be true*. Its rule may be written in sequent notation as:

$$P \rightarrow Q \quad \text{or} \quad P \vdash Q \tag{5.14}$$

where P, Q, and $P \rightarrow Q$ are statements or propositions in a formal language, and $\vdash$ is a metalogical symbol, meaning that Q is a syntactic consequence of P and $P \rightarrow Q$ in a logical system. MP rule justification in a classical two-valued logic is given by a *truth table* as shown in Table 5.1

The following example uses a *Tesla car* to demonstrate how FOPC works in *logic inference*. It has three statements to process:

Table 5.1 Truth table of MP
in two-valued logic

P	Q	$P \rightarrow Q$
T	T	T
T	F	F
F	T	T
F	F	T

$$
\begin{array}{l}
\text{ElectricCar}\left(\text{Tesla}\right) \\
\forall x \;\; \text{ElectricCar}\left(x\right) \rightarrow \text{Fuel}\left(x,\text{Electricity}\right) \\
\hline
\qquad \text{Fuel}\left(\text{Tesla},\text{Electricity}\right)
\end{array}
\qquad (5.15)
$$

Note: The first statement says *Tesla* is an *electric car*, the second statement says for all electric cars *x*, if a car is an electric car, the fuel being used must be electricity.

The above predicate *ElectricCar (Tesla)* matches the *antecedent* of the rule, so based on simple MP deduction to conclude that *Fuel(Tesla, Electricity)* is a *True* statement.

In fact, MP can be applied in *Forward* and *Backward Reasoning* modes.

Forward Reasoning (FR), also called *normal mode,* is used in normal situations by adding all facts into a KB to invoke all applicable implication rules to examine clause correctness or new knowledge addition.

Backward Reasoning (BR) is MP that operates in reverse mode to prove specific proposition or called *query* in computer science, i.e., to examine whether the query formula is true by its presence in KB, or without negative implication or facts on return query results.

Exercises

5.1. What is *meaning representation*? Explain why *meaning representation* is important in NLP. Give one or two examples to support your answer.

5.2. State and explain five major categories of *meaning representation*. Give an example to support your answer.

5.3. State and explain four common types of meaning representation in NLP. For each type, use the following sample sentence/utterance [5.70] [5.69] *Jack buys a new flat in London* to illustrate how they work for meaning representation.

5.4. What are the three basic requirements for meaning representation? Give two examples for each requirement to support your answer.

5.5. What is canonical form? How canonical form is applied to meaning representation. For sample sentence/utterance [5.70] [5.69] *Jack buys a new flat in London*, give five variations of this sentence and work out the canonical form in the forms of FOPC and semantic net.

5.6. What is inference? Explain why inference is vital to NLP and the implementation of NLP applications such as Q&A chatbot.

5.7. What is Fillmore's Theory of universal cases? State and explain six major case roles of Fillmore's Theory in meaning representation. Use an example to illustrate.

5.8. What is the complication of Fillmore's Theory in meaning representation by using several examples? Explain how it can be solved.

5.9. What are the four basic components of FOPC? State and explain their roles and function in FOPC formulation.

5.10. What is MP in inferencing? In addition to MP, state and explain other possible inferencing methods that can be applied to FOPC in meaning representation.

References

Bender, E. M. (2013) Linguistic Fundamentals for Natural Language Processing: 100 Essentials from Morphology and Syntax (Synthesis Lectures on Human Language Technologies). Morgan & Claypool Publishers

Best, W., Bryan, K. and Maxim, J. (2000) Semantic Processing: Theory and Practice. Wiley.

Blackburn, P and Bos, J. (2005) Representation and Inference for Natural Language: A First Course in Computational Semantics (Studies in Computational Linguistics). Center for the Study of Language and Information.

Bunt, H. (2013) Computing Meaning: Volume 4 (Text, Speech and Language Technology Book 47). Springer.

Butler, A. (2015) Linguistic Expressions and Semantic Processing: A Practical Approach. Springer.

Cui, Y., Huang, C., Lee, Raymond (2020). AI Tutor: A Computer Science Domain Knowledge Graph-Based QA System on JADE platform. World Academy of Science, Engineering and Technology, Open Science Index 168, International Journal of Industrial and Manufacturing Engineering, 14(12), 543 - 553.

Dijkstra, E. W. and Scholten, C. S. (1989) Predicate Calculus and Program Semantics (Monographs in Computer Science). Springer. Advanced Reasoning Forum.

Doyle, A. C. (2019) The Adventures of Sherlock Holmes (AmazonClassics Edition). AmazonClassics.

Epstein, R. (2012) Predicate Logic. Advanced Reasoning Forum.

Fillmore, C. J. (1968) The Case for Case. In Bach and Harms (Ed.): Universals in Linguistic Theory. New York: Holt, Rinehart, and Winston, 1-88.

Fillmore, C. J. (2020) Form and Meaning in Language, Volume III: Papers on Linguistic Theory and Constructions (Volume 3). Center for the Study of Language and Information.

Fitch, W. T. (2010). The evolution of language. Cambridge University Press.

Goddard, C. (1998) Semantic Analysis: A Practical Introduction (Oxford Textbooks in Linguistics). Oxford University Press.

Goldrei, D. (2005) Propositional and Predicate Calculus: A Model of Argument. Springer.

Jackson, P. C. (2019) Toward Human-Level Artificial Intelligence: Representation and Computation of Meaning in Natural Language (Dover Books on Mathematics). Dover Publications.

Mazarweh, S. (2010) Fillmore Case Grammar: Introduction to the Theory. GRIN Verlag.

Minsky, M. (1975). A framework for representing knowledge. In P. Winston, Ed., The Psychology of Computer Vision. New York: McGraw-Hill, pp. 211-277.

Parry, W. T. and Hacker, E. A. (1991) Aristotelian logic. Suny Press.

Potts, T. C. (1994) Structures and Categories for the Representation of Meaning. Cambridge University Press.

Schank, R. C. (1972). Conceptual dependency: A theory of natural language processing. Cognitive Psychology, 3, 552–631.

Sowa, J. (1991) Principles of Semantic Networks: Explorations in the Representation of Knowledge (Morgan Kaufmann Series in Representation and Reasoning). Morgan Kaufmann Publication.

Chapter 6
Semantic Analysis

6.1 Introduction

6.1.1 What Is Semantic Analysis?

Semantic analysis (Cruse 2011; Goddard 1998; Kroeger 2019) can be considered as the process of identifying meanings from texts and utterances by analyzing grammatical structure relationships between words, tokens of written texts, or verbal communications in NLP.

Semantic analysis tools can assist organizations to extract meaningful information from unstructured data automatically such as emails, conversations, and customers' feedback. There are many ways ranging from complete ad-hoc domain-oriented techniques to some theoretical but impractical methods. It is a sophisticated task for a machine to perform interpretation due to complexity and subjectivity involved in human languages. Semantic analysis of natural language captures text meaning with contexts, sentences, and grammar logical structures (Bender and Lascarides 2019; Butler 2015).

Semantic analysis is a process to transform linguistic inputs to meaning representation and stamina for machine-learning tools like text analysis, search engines, and chatbots. From the computer science perspective, semantics can be considered as a group of words, phrases, or clauses that provide concern-specific context to language, or clues to word meanings and relationships. For instance, a successful semantic analysis will base on quantity methods such as word frequency and context on location to generate cognitive connection between the clause *giant panda is a portly folivore found in China* and its semantic meaning instead of just the name *panda* it stands for.

© The Author(s), under exclusive license to Springer Nature Singapore Pte Ltd. 2025
R. Lee, *Natural Language Processing*,
https://doi.org/10.1007/978-981-96-3208-4_6

6.1.2   The Importance of Semantic Analysis in NLP

Semantic analysis (Goddard 1998; Sowa 1991) is important to conscious of knowledge relevance and information about *meaning*, e.g., *giant panda* characteristics, comparisons with other panda species, evolution history, related news and information.

It ensures that the contents are relevant to the understanding of (1) *user*, (2) *content*, and (3) *context* presence in NLP. The problem with establishing relationships between contents and context is that most data-driven technology cannot comprehend contextual message of the sentence (phrase or clause) it conveys. If the understanding of context and user's behavior has a deep semantic level, it can produce content relevance and resonant experience.

There are many automatic classification systems today with a purely *bag-of-words* approach to identify relevant features and determine document meanings. Few use *correlation* and *collocation* to account for words that have several meanings based on the context. Nevertheless, none uses full semantic analysis for words' meanings. But this is markedly required to interpret a document correctness because language, especially English language, is ambiguous. English nouns have an average of five to eight synonyms; e.g., *run* has more than 100 common meanings like *running toward the finish line, run to a meeting, run a company, the machine is running, tears ran down her face, ran for president, run him a couple thousand dollars,* etc. If a bag of words is used as features, the software will never be able to distinguish between important facts and irrelevant information leading to imprecise classification results and ambiguities.

6.1.3   How Human Is Good in Semantic Analysis?

Humans extract abstract ideas and notions like breathing without awareness. Take the meaning of *apple* as an example, when discussing the concept of *apple*, they referred to *fruit* consumed regularly. But now, a majority refer to the brand name *Apple* that dominates mobile phone and computer industry. In other words, humans are competent to extract context surrounding *words, phrases, objects, scenarios* and compare information with *prior experience, common sense*, and *world knowledge* to construct overall *meanings* in a text or conversation. These analyses outputs will be used to predict outcome with incredible accuracy, but algorithm and computer capacity upgrades had modified habitual practices to fit in with machine learning and NLP allowing machine-driven semantic analysis becomes reality. Such machine-learning-based semantic analysis schemes can assist in revealing the meanings in online messages and conversations and determine answers to questions without manually extracting relevant information from large volumes of unstructured data. The truth is semantic analysis aims to *make sense of everything* from words to languages in daily life.

6.2 Lexical vs. Compositional Semantic Analysis

6.2.1 What Is Lexical Semantic Analysis?

Lexical semantic analysis (Cruse 1986) is a subfield of *linguistic semantics* to study word's compositionality, grammar, structure mechanisms, and the relationships between *word senses* and their usages.

The analytical unit in *lexical semantics* is called *lexical unit*, which includes not only words, but also partial words, affixes (subunits), compound words, and phrases, collectively referred to as *lexical terms. Lexical units* are catalog of words called *lexicon* of a language. *Lexical semantics* can be interpreted as the relationship between *lexical terms, sentence/utterance syntax*, and *its meaning*.

Lexical semantics analyzes the meaning of lexical items in relation to language and syntactic structure. This field of study involves the following:

1. Classifying and decomposing lexical terms and tokens.
2. Examining the similarities and differences in lexical semantic structures across languages.
3. Reviewing the correlation between a sentence's lexical and syntactic meaning with its semantic meaning.

Lexical relation in lexical semantic involves the analysis of meaning or word relevance in lexical level and includes homonymy, polysemy, metonymy, synonyms, antonyms, and hyponymy and hypernymy to be studied in word sense and relation section.

6.2.2 What Is Compositional Semantic Analysis?

Compositionality is a concept in the philosophy of language that posits the meaning of a complex expression in a sentence or utterance depends not only on the meanings of individual words but also on their syntactic structure and arrangement. From a linguistic perspective, a sentence or utterance can be considered compositional if its meaning arises from both the meaning of its constituent words and how those words are syntactically linked together.

In a compositional language, the meaning of a sentence or utterance is derived solely from the meanings of the words that comprise it and their syntactic relationships. Thus, compositional semantics focuses on investigating the meaning of a sentence or utterance as a whole, rather than analyzing the individual words in isolation. The underlying logic is that words collectively create the overall meaning of the sentence or utterance, rather than merely combining their individual meanings. For example:

[6.1] *Andrew likes Jane = > likes (Andrew, Jane)* vs.
[6.2] *Jane likes Andrew = > likes (Jane, Andrew)*

Although individual meaning of every single word in these sentences/utterances is the same due to different words' arrangement, their meanings and predicate logics can be different.

Compositional semantics is to study the meaning of complex language units such as sentences, paragraphs, or documents. It is vital to transform the information represented by language units into a formal representation which consists of (1) symbolic and (2) vectorial representations.

Symbolic representations are meanings expressed as a logical formula by inferential mechanisms, or graph-based representations expressed by graphical transformation.

Vectorial representations are methods based on *distributional semantics* such as word embeddings to represent meaning as word vectors in multidimensional space.

Currently, only *vectorial representations* are widely used, as it is challenging to ensure the consistency of large sets of logical propositions based on textual input due to problematic inferential mechanisms. Moreover, there is no consensus on suitable graph-based representations, such as semantic networks, for expressing the meanings of linguistic entities, nor are there appropriate operations for applying these representations.

6.3   Word Senses and Relations

6.3.1   *What Is Word Sense?*

Word sense is a crucial concept for interpreting the meanings of words in linguistics. For example, the word "bank" can have over 20 different word senses in a dictionary, each with a distinct meaning depending on the context and syntactic structure in which it is used. Some of these senses include the following:

1. Financial organizations that accept deposits and use funds for lending operations (noun).
   [6.3] *Jack goes to the bank and withdraws some money.*
2. Inventory or stock that keeps for emergencies (noun).
   [6.4] *Jack goes to the food bank to acquire some food.*
3. A container with an opening on top to store money (noun).
   [6.5] *His coin bank was empty.*
4. A sloping land besides a slope or body of water (noun).
   [6.6] *Jack stands beside the bank of a river (noun).*
5. A long plie or ridge (noun).
   [6.7] *Jack digs a bank of earth.*
6. Enclose with a bank (verb).
   [6.8] *bank roads.*
7. Cover with ashes to control the flames (verb).
   [6.9] *Bank a fire.*

8. Tip laterally (verb).
   [6.10] *The pilot had to bank the aircraft.*
9. A fighter maneuvers the aircraft to tip laterally (noun).
   [6.11] *F19 fighter went into a steep bank.*
10. Similar objects are arranged in a row (noun).
   [6.12] *He operated a bank of switches.*

6.3.2 Types of Lexical Semantics

There are six types of commonly used lexical semantics: (1) homonymy, (2) polysemy, (3) metonymy, (4) synonyms, (5) antonyms, and (6) hyponymy and hypernymy.

6.3.2.1 Homonymy

Homophones are words that are spelled and pronounced the same but have different meanings. The word *homonym* comes from prefix *homo-* that stands for *same* and suffix *-nym* that stands for *name*.

Example 1: bank$_1$: financial institution vs. bank$_2$: slopping land:
[6.13] *He went to the bank and withdrew some cash.*
[6.14] *He was standing at the bank of the lake in the forest.*
Example 2: bat$_1$: a sporting club for ball hitting vs. bat$_2$: a kind of flying mammal:
[6.15] *He handles his bat skillfully during the game.*
[6.16] *Bats live the longest as compared with other species of similar size.*
Example 3: play$_1$: light-hearted recreational activity for amusement vs. play$_2$: the activity of doing something in an agreed succession:
[6.17] *This Shakespeare play is excellent.*
[6.18] *It is still my play.*

There are two related concepts with *homonymy*: (1) *homographs* are usually defined as words that have the same spelling with different pronunciations and (2) *homophones* are words that share same pronunciation regardless of spellings as examples above. Further, *homographs* are words with the same spellings, and *heterographs* are words that share the same pronunciation but different spellings, e.g., *chart vs. chat,* peace vs. piece, right vs. write.

Homonymy often causes problems in the following NLP applications:

1. Information retrieval confusion, e.g., *cat scan.*
2. Machine translation confuses foreign languages' meanings:
e.g., bank$_1$—financial institution; bank (English) → la banque (French).
[6.19] *He goes to the bank and withdraws some cash.* (English)
[6.20] *Il va à la banque et retire de l'argent.* (French)

e.g., bank₂—sloping land, bank (English) → la rive (French)

[6.21] *He lived by the bank of the lake*. (English)

[6.22] *Il habitait au bord du lac*. (French)

3. text-to-speech confusion:

e.g., *bass* (string instrument) vs. *bass* (fish).

6.3.2.2 Polysemy

Polysemy are words with the same spellings but different in meanings and context. The difference between *homonymy* and *polysemy* is delicate and subjective.

For example, bank

[6.23] *The bank was built in 1866*. (financial building)

[6.24] *He withdrew some money from the bank early this morning*. (financial organization)

In fact, many commonly used words are polysemy with multiple contexts and meanings in different sentence situations.

For example, *get* is a commonly used word that has at least three distinct meanings.

[6.25] *I get an apple from the basket*. (have something)

[6.26] *I get it*. (understand)

[6.27] *She gets thinner*. (reach or cause to a specified state or condition)

6.3.2.3 Metonymy

Metonymy is a kind of *figure-of-speech* in which one word or phrase is replaced by another association.

It is also a rhetorical strategy to describe the periphery of nucleus indirectly, as in describing someone's outfit to individual's characteristics. It is regarded as a systematic relationship between senses, or systematic polysemy, e.g., *college, hospital,* and *museum* can all stand for building with semantic relationship between that building and an institution.

Metonymy and *metaphor* have fundamental differences in functions. *Metonymy* is about referring a method of designation or component identification or symbolic linkage with association, e.g., *crown* for monarchy or *royalty*. *Metaphor* is about understanding and interpretation in contract. It is a means to understand or explain a phenomenon by another description. For instance:

[6.28] *Her business rises like phoenix*.

6.3.2.4 Zeugma Test

Zeugma is the usage of a word(s) that make(s) sense in one way but not the other. Examples of *zeugma* that caused conflicts in semantics:

[6.29] *Wage neither war nor peace.*
– There is a term to *wage war* but is literally incorrect to say to *wage peace*.
[6.30] *He watched the brightness of lightning and the pounding of thunderstorm.*
– He can only *watch lightning* but not *thunder*.

The *zeugma test* in semantic analysis consists of using a putatively ambiguous expression in a sentence in which several of its putative meanings are crowded together, whether it makes sense or not. Let's use the word *serve* as example:

[6.31] *Which United Airlines flights serve dinner?*
[6.32] *Does Jack serve the Army?*
[6.33] *Do United Airlines flights serve dinner and the Army?*

It showed that there are two different senses of *serve* though [6.33] may sound odd.

6.3.2.5 Synonyms

Synonyms are words with the same meaning in some or all contexts. They usually appear in language in different contexts, such as formal and informal language, daily conversations, and business correspondence. Synonyms have modest meaning when used, although they have the same meaning, e.g., *create/make*, *start/begin*, *big/huge*, *attempt/try*, *house/mansion*, *pretty/beauty*. Synonyms have two *lexemes* if they are interchangeable in all cases and retain the same meaning.

However, there are very few *truths* synonymy in the real-world situation as to whether two words are truly *synonyms*. The logic behind this is if they are different words, then they must mean something else or have some context differences in usage and cannot be the same in all situations. In many cases, two words are not exactly interchangeable when they appear, even though many aspects of the meaning are the same. These words are used and mean differently due to concepts of politeness, slang, register, genre, etc.

For example, *large vs. big* (are they *exactly* the same?)
[6.34] *This building is very big* vs.
[6.35] *This building is very large.*
[6.36] *Janet is her big sister* vs.
[6.37] *Janet is her large sister.*

Although both words have same meanings in the description of *size*, the word *big* has an additional notion of older in terms of *seniority* description.

6.3.2.6 Antonyms

Antonyms are the word sense between words with opposite context meanings. It is a place which other sense relations do not occupy synonym regardless of human tendency to categorize experience in dichotomous contrast that is not easily judged.

However, the notion of antonyms is immeasurable. Humans understand the concept of *opposite* from childhood, encounter them in daily life, and even use *antonyms* as a kind of a cognitive method to organize notions, concepts, and experiences, e.g., *big vs. small, dark vs. bright, hot vs. cold, in vs. out.* Antonyms can also use to interpret binary, scale, or position *opposition* such as *long vs. short, fast vs. slow, and up vs. down.*

6.3.2.7 Hyponymy and Hypernymy

Hyponym is a word sense of another word if the first word sense is specific, denoting a subclass of the other sense in linguistics, e.g., *truck* is a hyponym of *vehicle,* *mango* is a hyponym of *fruit,* and *chair* is a hyponym of *furniture*; or conversely *hypernym/superordinate* (hyper is super), e.g., *vehicle* is a hypernym of *truck, fruit* is a hypernym of *mango*, and *furniture* is hypernym of *chair.*

It is interesting to know that hyponymy is not only limited to nouns, but it can also be found in verbs, e.g., *gaze, glimpse,* and *stare* are all regarded to specific moment of *seeing.*

Hyponymy and *hypernymy* relationship between word sense and relation is regarded as the relationship between class and subclass concepts in *object-oriented programming* (OOP) from the computer science perspective, e.g., the class of *vehicle* has three subclasses: *car, lorry,* and *bus*, while the class of *fruit* can have numerous subclasses such as *apple, orange,* and *mango*; or in reverse manner, the concept *vehicle* is the superclass of *car,* and the concept *fruit* is the superclass of *mango.*

Further, words that have *hyponyms* of the same broader term are hypernym known as *co-hyponyms.* The semantic relationship between each of the more specific words, e.g., *daisy* and *rose,* and the broader term, e.g., *flower,* is called *hyponymy* or *inclusion,* which has the same situation for word sense relation of *co-hypernymy.*

Hyponymy has (1) *extensional,* (2) *entailment,* (3) *transitive,* and (4) *IS-A hierarchy* characteristics:

1. *Extensional* is the class represented by the parent extension, including the class represented by hyponym, e.g., the relations between vehicle and truck.
2. *Entailment* is a hyponym sense A of sense B if A entails B.
3. *Transitive means if* A entails B and B entails C, then A entails C, e.g., *truck, vehicle, transport* where *truck* is a hyponymy of *vehicle* and *vehicle* is a hyponymy of *transport*, so *truck* is a hyponymy of *transport.*
4. *IS-A hierarchy* where *A IS-A B (or A IsA B)*, and *B subsumes A* in OOP.

6.3.2.8 Hyponyms and Instances

Hyponyms have notions of *instance* and *class.* In linguistics, an *instance* can be considered as a proper noun with a unique entity. For example, *New York* is an instance of *city*; *USA* is an instance of *country.* It is regarded as the relationship between *class vs. object* in object programming.

In short, *class* is the notion of things and objects, whereas *object* is the instance of class, e.g., *person* is a *class* concept to describe an individual person, while *Jack* is an *object*, which is an *instance* of that *class* concept.

A simple test: the relationship between *car* and *Tesla*, are they *class-object* relationship or *class-subclass* relationship?

6.4 Word Sense Disambiguation

6.4.1 What Is Word Sense Disambiguation (WSD)?

Word sense disambiguation (WSD) (Agirre and Edmonds 2007) is a well-known challenge in computational linguistics that involves the identification for *correct semantic meaning* of words used in sentences/utterances. WSD is the ability to determine which meaning of a word is activated when a word is used in a specific context of NLP.

Lexical ambiguity is one of the initial problems that any NLP system may encounter. In summary, POS tagging is applied to resolve *syntactic ambiguity*, while WSD is applied to resolve *semantic ambiguity*. However, it is always difficult to resolve semantic rather than syntactic ambiguity. Consider distinct sense for the word *bass* examples:

[6.38] *Jane hates to hear the bass sound.*
[6.39] *Jack is eating fried bass.*

It has completely different word sense in which [6.38] represents a musical instrument and [6.39] represents a type of fish. So, by using WSD the two sentences can be interpreted as follows:

[6.40] *Jane hates to hear the bass/instrument sound.*
[6.41] *Jack is eating fried bass/fish.*

6.4.2 Difficulties in WSD

There are five major concerns in WSD: (1) difference meaning across dictionaries, (2) POS tagging, (3) inter-judge variance, (4) pragmatics (discourse), and (5) sense discreteness.

1. Difference **meaning across dictionaries.**
A problem with WSD is sense decision as dictionaries and thesauri offer several word divisions into senses. Many WSD research papers have commonly used WordNet (WordNet 2022a) as the reference word sense corpus for English. It can be considered as a comprehensive lexicon that is composed of word concepts and their semantic relations with other concepts (e.g., synonyms). For example, the concept of *car* is interpreted as *{car, auto, automobile, machine,*

motorcar}. BabelNet (2022) is a recent multilingual encyclopedic dictionary with multilingual WSD.

2. **POS tagging**.

WSD and POS tagging involve disambiguation or tagging with words. However, algorithms used for one tend not to work well for the other, mainly because a word's POS is primarily determined by adjacent one to three words versus word sense determined by more distant words in many cases. For example, the success rate of POS tagging algorithms is around 96% versus 75% in WSD with supervised learning (SL) current research and findings (Agirre and Edmonds 2007).

3. **Inter-judge variance**.

WSD system test results on a task are usually compared to ones by humans. While it is easy to attribute POS to texts, it is difficult in training to mark word senses. Since human performance serves as the standard, it is an upper limit for computer performance. However, humans fared much better at coarse-grained discrimination than at fine-grained discrimination and it is the reason for the research of the former to put the test in recent WSD evaluation exercises.

4. **Pragmatics (discourse).**

Pragmatics and *discourse* are complex problems in NLP. Many AI researchers believe that one cannot analyze meanings of words without some form of sensible *ontology analysis* and *world knowledge* at a pragmatic level. Also, *common sense* is sometimes required to distinguish words such as pronouns in anaphors or cataphors of the text.

5. **Sense discreteness**.

The *notion* of word sense is sometimes unpredictable and controversial. Most can agree on semantic interpretation at the level of *coarse-grained homographs*, but going down to *fine-grained polysemy* can lead to disagreement. For example, Senseval-2 (Preiss 2006) uses fine-grained sensory distinctions, with only 85% of the annotated words that can agree with. Word meanings are infinitely variable, in principle are dependent on context, and cannot be easily broken down into distinct or separate submeanings.

6.4.3   Method for WSD

WSD commonly used methods include (1) knowledge base (KB), (2) SL, (3) semi-supervised learning, and (4) unsupervised learning (UL) (Agirre and Edmonds 2007; Preiss 2006).

1. *KB* is a method mainly based on dictionaries, thesauri, and lexical knowledge databases. They don't need corpus evidence for disambiguation. The Lesk method (Lesk 1986) is a pioneering dictionary-based method introduced by Prof. Michael Lesk in 1986. The Lesk definition and its algorithm aim to measure the overlap between the meaning definitions of all words in a context. Kilgarriff and Rosenzweig (2000) simplified the Lesk definition to measure the

overlap between meaning definition of a word and current context, meaning the correctness of identifying one word at a time, the current context being the set of words in surrounding sentence/utterance or paragraph (Ayetiran and Agbele 2016).

2. *SL* methods are standard machine-learning techniques applying semantically annotated corpora to train disambiguation. These methods assume that context alone can provide sufficient evidence to clarify meaning, so verbal knowledge and reasoning are considered unnecessary. A context is interpreted as a set of word features that contains information about surrounding words. Support Vector Machines (SVMs) and memory-based learning are commonly used SL methods for WSD. However, they are usually computationally intensive and require large manually labeled corpora to produce satisfactory results.

3. Since many WSD problems lack training corpora, *semi-supervised methods* are applied on both labeled and unlabeled data, which require only amount of annotated text and a large amount of plain unannotated text, as well as bootstrapping from starting data. The bootstrapping method starts with a small amount of starting data for each word, either with manually labeled training examples or with a small set of triggering decision rules. The seed value is intended to train an initial classifier with some supervised method. This classifier is then applied on the unlabeled portion of corpus to extract a larger training set with the safest classification. This process is repeated to train each new classifier until the entire corpus is exhausted or the maximum number of iterations is reached. Other semi-supervised techniques apply large unlabeled corpora to provide co-occurrence information to complement labeled corpus perspectives to help supervised models adapt to different domains.

4. *UL* methods assume that similar meanings appear in similar contexts, that is why *perceptions* can be induced from texts by clustering word occurrences using a similarity measure of context. This task is called *word sense induction* or *discrimination*. UL methods can overcome knowledge acquisition bottlenecks due to their independence from manual work. Although the performance is lower than other methods mentioned above, fair comparison is hard as the induced senses should link up to a known word sense dictionary.

6.5 WordNet and Online Thesauri

6.5.1 What Is WordNet?

WordNet (WordNet 2022a) is a lexical corpus of words with over 200 languages with adjectives, adverbs, nouns, and verbs grouped into a set of synonyms where each word in WordNet has a distinct concept. It is organized by concepts and meanings against a dictionary in alphabets. Since traditional dictionaries were created by humans, a lexical resource is required for computers effecting WordNet that is

Table 6.1 WordNet basic statistical information

Category	Unique strings
Noun	117,798
Verb	11,529
Adjective	22,479
Adverb	4481

applicable in NLP. It is available for public access and free download with statistical information as shown in Table 6.1.

WordNet's structure is an integral tool for computational linguistics and NLP implementations. It resembles a thesaurus and group words by meanings. However, they have basic differences: (a) WordNet indicates *word senses* in addition to *word forms*. As a result, words that are found near one another in the network are semantically related or even synonym with each other. (b) WordNet encodes semantic relations among words, whereas words in a thesaurus do not follow a distinct pattern other than the similarity in surface meaning.

6.5.2   What Are Synsets?

WordNet can be considered as a network of words connected by lexical and semantic relations. Nouns, verbs, adjectives, and adverbs are combined into a group of cognitive synonyms called *synsets* with each expressing a specific concept. *Synsets* are associated with conceptual semantics and lexical relationships such as hyponyms and antonyms. WordNet contains over 117,000 *synsets*. Each of these *synsets* is associated with other in a small number of conceptual relationships.

A *synset* contains a short definition called a *gloss*, and one or more short sentences describing how members of synset are used in most contexts. Word forms with many different meanings are represented in different synsets. This is the form of each form-meaning pair in WordNet. Each synonym group is a synset within a WordNet term, and synonyms that are part of a synset are lexical variants of that concept. Figure 6.1 shows a synset tree for the synset concept *book* and all concept relationships with all other related synsets. Meaningful related words and concepts in the generated network can be browsed from the WordNet browser (WordNet 2022b).

6.5.3   Knowledge Structure of WordNet

A WordNet structure is concepts of word relationship in a WordNet network to arrange same concepts in similar interchange contexts in Fig. 6.2. These words are unordered sets grouped into synsets and linked with small conceptual relations. An example of synset structure *benefit* arrayed synonyms *profit* with definitions and

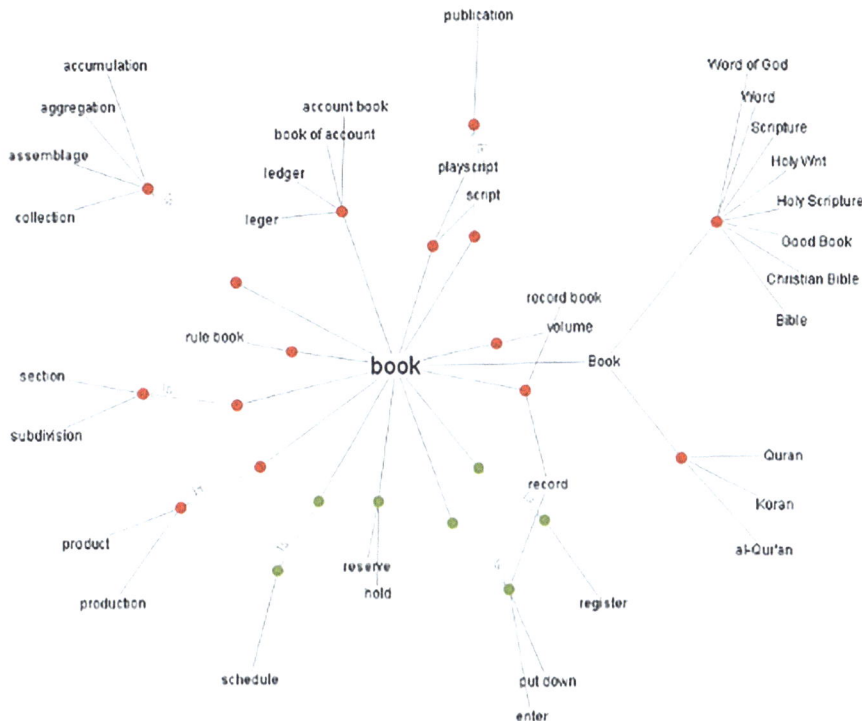

Fig. 6.1 Synset concept of *book* in WordNet

examples as shown in Fig. 6.3. Benefit(profit) is defined as an advantage or profit gain from something. For example, *He receives benefits of computers trade.*

6.5.4 What Are Major Lexical Relations Captured in WordNet?

Super-subordinate relation, also called *hypernymy, hyponymy*, or *IS-A relation* is a frequently used relation among synsets. It links generic synsets such as {*furniture, piece_of_furniture*} to subconcepts like {*chair*} and {*armchair*}. Thus, WordNet indicates that synset *furniture* consists of synset *chair*, which in turn includes synset *armchair*; conversely, synsets like *chair* or *armchair* make up the synset *furniture*. In fact, the synset tree goes up to *root-node* {*entity*}.

As said, such hyponym relation is transitive in nature, e.g., if an *armchair* is a kind of *chair* and if a *chair* is a kind of *furniture*, then an *armchair* is a kind of *furniture*. WordNet distinguishes between types (general nouns) and instances (specific people, countries, and geographic entities), e.g., *book* is a type of *publication*,

Fig. 6.2 Basic knowledge structure of WordNet

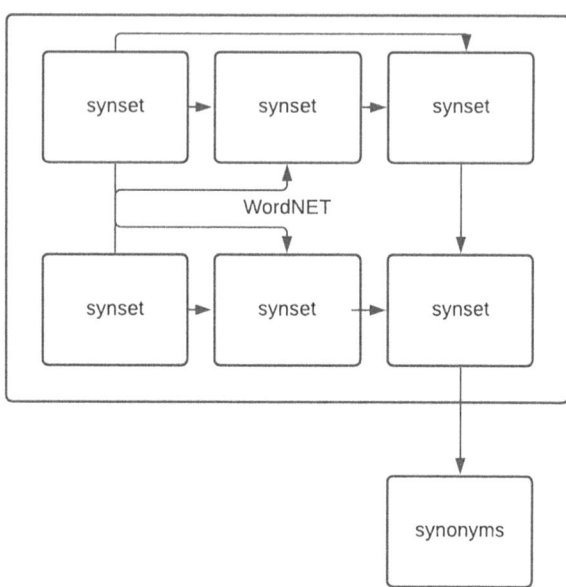

and *Abraham Lincoln* is an instance of *President*. Instances are always denoted as *leaves (terminal nodes)* in synset tree hierarchies.

Major lexical relations include the following:

* *Synonymy*: words with similar meaning.
* *Polysemy*: words with more than single sense.
* *Hyponymy/Hypernymy*: *IS-A* relation between words.
* *Meronymy/Holonymy*: *part-whole* relation between words.
* *Antonymy*: opposite meanings between words.
* *Troponymy*: applicable for verbs, e.g., whisper is troponym of speak.

Table 6.2 shows the major lexical relation capture in WordNet with examples.

6.5.5   Applications of WordNet and Thesauri?

WordNet and *Thesauri* applications include information extraction, information retrieval, question answering, medical informatics, and machine translation. WordNet has another common usage to determine word similarity with algorithms proposed, including to measure the distance(s) among words in WordNet synset graphs (trees), e.g., counting the number of edges among synsets. Intuitive words or synonyms are close to meaning. Many WordNet-based world similarity algorithms are implemented in a Perl package called WordNet::Similarity and a Python package using NLTK and SpaCy will be explored in the second part of NLP workshops.

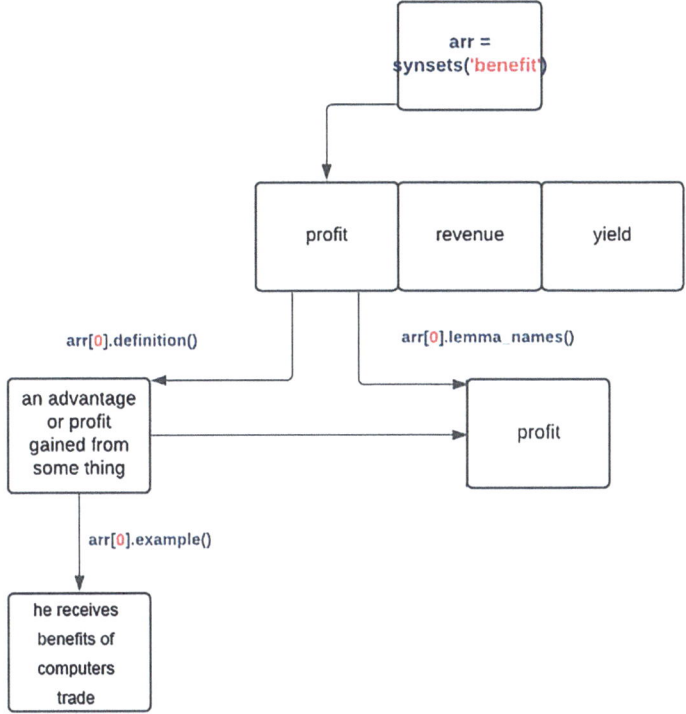

Fig. 6.3 Example of knowledge structure of synset *benefit*

Table 6.2 Major lexical relations captured in WordNet with examples

Semantic relation	Syntactic category	Examples
Synonymy (similar)	N, V, Aj, Av	pipe, tube rise, ascend sad, unhappy rapidly, speedily
Antonymy (opposite)	Aj, Av, (N, V)	wet, dry powerful, powerless friendly, unfriendly rapidly, slowly
Hyponymy (subordinate)	N	sugar maple, maple maple, tree tree, plant
Meronymy (part)	N	brim, hat gin, martini ship, fleet
Troponomy (manner)	V	march, walk whisper, speak
Entailment	V	drive, ride divorce, marry

Note: N = Nouns, Adj = Adjectives, V = Verbs, Av = Adverbs

6.6 Other Online Thesauri: MeSH

6.6.1 What Is MeSH?

Medical Subject Thesaurus, aka. MeSH (MeSH 2022) is a hierarchically organized vocabulary for indexing, cataloging, and searching biomedical- and health-related information created by the US National Library of Medicine (NLM). *MeSH*

contains subject headings that appear in MEDLINE/PubMed, NLM catalog, and other NLM databases. It consists of 177,000 entries and 26,142 biomedical titles, and continues to soar as the literature expands. The 2020 edition contains more than 25,000 subject headings, 4400 approximately more since its launch in 1960. These headings are organized into an 11-level hierarchy with 83 subheadings. *MeSH* can be freely used via US NLM's online *MeSH browser* (MeSH 2022). *MeSH* headings are organized in a knowledge tree with 16 major branches:

A. Anatomy, B. Organisms, C. Diseases, D. Chemicals and Drugs, E. Analytical Diagnostics and Therapeutic Techniques and Equipment, F. Psychiatry and Psychology, G. Phenomena and Processes, H. Disciplines and Occupations, I. Anthropology, Education, Sociology and Social Phenomena, J. Technology, Industry, Agriculture, K. Humanities, L. Information Science, M. Named Groups, N. Health Care, V. Publication Characteristics, and Z. Geographicals.

MeSH glossary contains several entry terms intended to be synonyms for canonical title terms in addition to a hierarchical set of canonical terms.

6.6.2   Uses of the MeSH Ontology

MeSH ontology usage includes the following:

1. Synonyms as entry terms, e.g., *sucrose and saccharose*.
2. Hypernyms from hierarchy, e.g., *sucrose is a glycosyl glycoside*.
3. Index in *MEDLINE/PubMed* databases such as bibliographic database NLM contains 20 million journal articles with 10–20 *MeSH terms* manually assigned to each article.

6.7   Word Similarity and Thesaurus Methods

6.7.1   Introduction

A *synonym* can be considered as a binary relationship between two *synonyms* or *non-synonyms*. *Similarity* or distance is a looser measure when two words share more semantic features with each other. *Similarity* is a relationship between sensations, e.g., *bank* is usually not like *slope*, but in some cases, they may have the same meaning, e.g., $bank_1$ is similar to $fund_3$, and $bank_2$ is similar to $slope_5$, in which the similarity can be calculated by word sense relationship in a sentence.

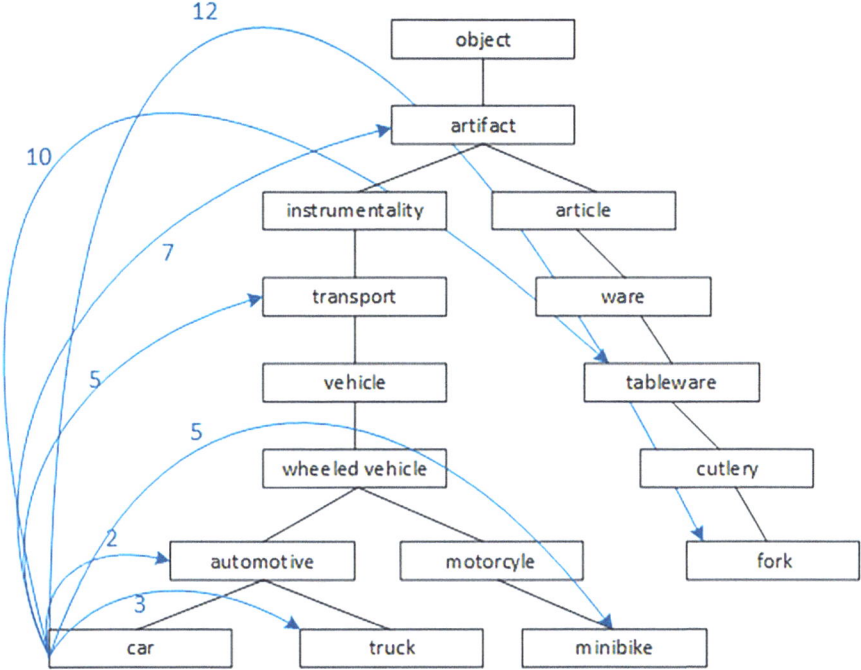

Fig. 6.4 Path-based similarity for a concept related to *car*

Word similarity is important because a good measure can be used in information retrieval, question answering, machine translation, natural language generation, language modeling, automatic paper scoring, and even plagiarism detection.

The difference between *word similarity* and *word relation* is that similar words are almost synonyms, e.g., *car, bicycle* are similar in concept but not a kind of *Is-A* relation, whereas related words can be related in any way, e.g., *car, gasoline* are highly related but not similar in semantic meaning.

There are two types of *similarity algorithms*: (1) *thesaurus-based algorithms* and (2) *distributional algorithms*. *Thesaurus-based algorithms* are designed to examine adjacent words in a *hypernym hierarchy* with similar annotations or definitions. *Distribution algorithms* are designed to examine words with similar *distributional contexts*.

6.7.2   Path-Based Similarity

Path-based similarity aims to examine two concepts in general. The two concepts are similar if they are in the vicinity of thesaurus hierarchy. *Synset tree* (graph), the distance (path) between two synset nodes, can provide a good indication of semantic similarity between two concepts. This evaluation method is known as *path-based similarity* measurement. Figure 6.4 shows an example of path-based similarity for the concept *car*. Note that all concepts have a path value of 1 point to themselves.

For example:
pathlen(car, car) = 1
pathlen(car, automotive) = 2
pathlen(car, truck) = 3
pathlen(car, minibike) = 5
pathlen(car, transport) = 5
pathlen(car, artifact) = 7
pathlen(care, tableware) = 10
pathlen(car, fork) = 12

In general:

$$\text{pathlen}(c_1, c_2) = 1 + \text{nos.of edges in the shortest}$$
$$\text{path at hypernym graph between sense nodes } c1 \text{ and } c2 \qquad (6.1)$$

where pathlen(c_1, c_2) ranges from 0 to 1.
   The path-based similarity simpath(c_1, c_2) of two nodes (concepts) is given by:

$$\text{simpath}(c_1, c_2) = \frac{1}{\text{pathlen}(c_1, c_2)} \qquad (6.2)$$

$$\text{wordsim}(w_1, w_2) = \max\left(\text{simpath}(c_1, c_2)\right)$$

$$\forall c_1 \in \text{senses}(w_1), \quad c_2 \in \text{senses}(w_2) \qquad (6.3)$$

Using *car* concept as example:
simpath(car, car) = 1/1 = 1.0.
simpath(car, automotive) = 1/2 = 0.50.
simpath(car, truck) = 1/3 = 0.33.
simpath(car, minibike) = 1/5 = 0.20.
simpath(car, transport) = 1/5 = 0.20.
simpath(car, artifact) = 1/7 = 0.14.
simpath(car, tableware) = 1/10 = 0.10.
pathlen(car, fork) = 1/12 = 0.08.

6.7.3   Problems with Path-Based Similarity

Let's assume every link denotes a uniform distance. It seems that *car* to *minibike* is closer than *car* to *transport* because higher synsets are more *abstract* in synset tree, e.g., *object* is abstract than *artifact*, *transport* is abstract than *vehicle*.

Although *simpath(car, minibike)* and *simpath(car, transport)* have identical values, their *semantic relationship* between each other is different; naturally, synsets in other branch of the synset tree are less related in concept, e.g., *car* vs. *tableware* or even *fork*.

Hence, it is suggested to have a metric that can represent the cost of each edge independently so that words associated with abstract nodes should have less similarity scores.

6.7.4 Information Content Similarity

The *information content similarity* metric uses information content (IC) to assess semantic similarity in taxonomy that was first proposed by Prof. Philip Resnik, whose distinguished work *Using information content to evaluate sematic similarity in taxonomy* was published in 1995 (Resnik 1995).

Let's define $P(c)$ as the probability of a random word in corpus for an instance of concept c. There is a unique random variable ranging from words formally associated with each concept in the hierarchy. For a given concept, each observed noun is either a member of the concept with probability $P(c)$ or is not a member of the concept with probability $1 - P(c)$. All words are members of the root node entity, i.e., $P(root) = 1$; lower nodes in the hierarchy have lower probability.

Information content similarity is generally determined by counting against the corpus. When applying to *car* concept example, each instance of *car* counts toward frequency of *automotive, wheeled vehicle, vehicle*, etc. So given *word(c)* is the collection of all words that are children of node c, the probability of information content similarity $P(c)$ in a corpus is given by Eq. 6.4:

$$P(c) = \frac{\sum_{w \in \text{words}(c)} \text{count}(w)}{N} \tag{6.4}$$

Thus, (1) *words(transport) = {transport, wheeled vehicle, automotive, car, truck, motorcycle, minibike}* and (2) *words(automotive) = {car, truck}*.

A synset tree of *car* associated with $P(c)$ up to transport level in each corpus is shown in Fig. 6.5.

IC is given by:

$$IC(c) = -\log P(c) \tag{6.5}$$

where the lowest common subsume (LCS) is given by:

$$LCS(c_1, c_2) = \text{the lowest common subsumer} \tag{6.6}$$

Fig. 6.5 Synset tree of
"*car*" with associated *P(c)*
(up to *transport* level in the
corpus)

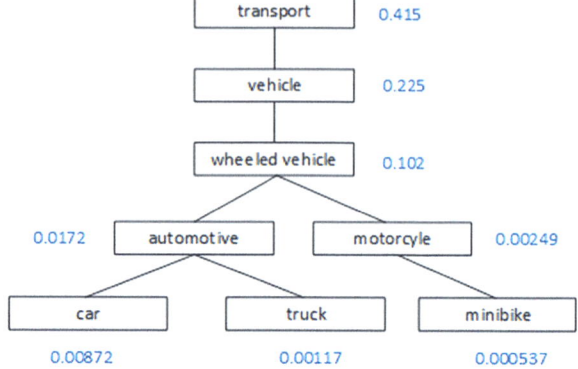

That is, the lower node in hierarchy that *subsumes* (is a hypernym of) both c_1 and c_2 is ready to apply IC as a similarity metric.

6.7.5 The Resnik Method

The Resnik method (Resnik 1995, 1999) refers to the similarity between two words that are in the vicinity of their common information. It is defined to measure the most informative common ICs, i.e., (lowest) *subsumer* (MIS/LCS) of two nodes, given by:

$$\mathrm{sim}_{\text{Resnik}}\left(c_1,c_2\right) = -\log P\left(\mathrm{LCS}\left(C_1,C_2\right)\right) \tag{6.7}$$

6.7.6 The Dekang Lin Method

The Dekang Lin method was proposed by Prof. Dekang Lin with his work *Information-Theoretic Definition of Similarity* at ICML in 1998 (Lin 1998). It determines not only the similarity between concepts A and B what they have in common but also the differences between them. It concerns (1) *commonality* and (2) *difference*. *Commonality*, denoted by *IC(common(A,B))*, means A and B are more in common that has more *similarity*. *Difference*, denoted by *IC(description(A,B) – IC (common(A,B))*, means more differences between A and B that has less *similarity*.

Similarity theorem is similarity between A and B measured by the ratio between amount of information required to state commonality of A and B, and the information required to describe what A and B are:

$$\mathrm{simLin}\left(A,B\right) - \log P\left(\text{common}\left(\text{A,B}\right)\right) / \log P\left(\text{description}\left(\text{A,B}\right)\right) \tag{6.8}$$

He further modified the *Resnik method* demonstrating that information in common is twice the LCS IC given by:

$$\text{SimLin}(c_1, c_2) = \frac{2x \log P\left(\text{LCS}(c_1, c_2)\right)}{\log P(c_1) + \log P(c_2)} \qquad (6.9)$$

Using *car* concept as example:

$$\text{SimLin}(\text{car}, \text{minibike}) = \frac{2x \log P(\text{wheeled vehicle})}{\log P(\text{car}) + \log P(\text{minibike})}$$

$$= \frac{2x \log P(0.102)}{\log P(0.00872) + \log P(0.000537)} = 0.372$$

$$\text{SimLin}(\text{car}, \text{truck}) = \frac{2x \log P(\text{automotive})}{\log P(\text{car}) + \log P(\text{truck})}$$

$$= \frac{2x \log P(0.0172)}{\log P(0.00872) + \log P(0.00117)} = 0.707$$

This calculation showed that *car* is related to *truck* than *minibike* at *hierarchy tree* in Table 6.5

6.7.7   The (Extended) Lesk Algorithm

The (extended) Lesk algorithm uses a thesaurus-based algorithm to measure glosses which contain similar words for concept similarity. For instance, *drawing paper* is a type of paper for *drafting, including the art of transferring designs from specially prepared paper to a glass, wood, or even metal surface.*
    For all *n*-word phrases which appear in two glosses:

1. Add a score of n^2.
2. *Paper* and specially prepare for $1 + 2^2 = 5$.
3. Evaluate the overlaps for other relations which define glosses of hypernyms and hyponyms.

The extended Lesk for similarity ($\text{sim}_{\text{eLesk}}$) is given by:

$$\text{sim}_{\text{eLesk}}(c_1, c_2) = \sum_{r, q \in \text{RELS}} \text{overlap}\left(\text{gloss}(r(c_1)), \text{gloss}(q(c_2))\right) \qquad (6.10)$$

6.8   Distributed Similarity

6.8.1   *Distributional Models of Meaning*

Distributional models of meaning can be considered as a kind of *vector-space models* of meaning. Prof. Zellig Harris (1909–1992) claimed that *oculist and eye-doctor... occur in almost the same environments...* (Harris 1954), which means A & B is *synonym* if *A and B have almost identical environments*. Sir John R Firth (1890–1960) also stated that *you shall know a word by the company it keeps!* (Firth 1957):

[6.41] *A bottle of Baileys is on the table.*
[6.42] *Many coffee drinkers like Baileys.*
[6.43] *Baileys will make you drunk.*
[6.44] *We make Baileys out of Irish whiskey and cream.*

Humans can guess *Baileys* from context words is an *alcoholic coffee beverage flavored with cream and Irish whiskey*. This means that two words are *semantically similar* if they are similar in the context of the word being used for algorithm interpretation.

6.8.2   *Word Vectors*

Word vector is a vector of *weights*. In a simple *1-of-N* encoding, every element in the vector is associated with a word in vocabulary. Word encoding is vector where the corresponding element is set to one, and other elements are zero.

Given a target word w, assume there is a binary feature f_i for each N word in lexicon v_i, the word vector is given by:

$$W = \left(f_1, f_2, f_3, \ldots f_N \right) \tag{6.11}$$

Apply to above *Baileys'* example, if $w = Baileys$, $f_1 = coffee$, $f_2 = whiskey$, $f_3 = beer$, $f_4 = cream$, ...

$$w = \left(1,1,0,1,\ldots \right) \tag{6.12}$$

6.8.3   *Term-Document Matrix*

Text data is denoted as a matrix in this method. The rows represent sentences from the data to be analyzed, and columns represent words of the matrix. Each cell is the counting of term t in a document d:$tf_{t,d}$, and each document is a counter vector in $\mathbf{N}^v$.

Table 6.3 shows a *term-document matrix* to investigate the relationships of four important words: *battle*, *soldier*, *fool*, and *trick* from six famous literatures: *As You Like It, Henry V, Julius Caesar,* and *Twelfth Night* extracted from *The Complete Works of Shakespeare* by William Shakespeare (1564–1616) (Shakespeare 2021), *The Adventures of Sherlock Holmes* (Doyle 2019) and *Moby Dick* by Herman Melville (1819–1891) (Melville 2012).

It showed that:

1. Two documents *Julius Caesar* and *Henry V* are similar if their term-document vectors are similar as in Table 6.4.
2. Each word is a count vector in $\mathbf{N}^D$ as a row. Table 6.5 shows row vector for the word *fool* across these six documents.
3. Two words are semantically similar if their word vectors are similar, e.g., *fool* and *trick*. It makes sense because they are related to each other *semantically* as compared to *battle* and *soldier* as shown in Table 6.6.

Table 6.3   Term-document matrix of six famous English literature

	As You Like It	Twelfth Night	Julius Caesar	Henry V	Adv of Sherlock Holmes	Moby Dick
battle	1	1	8	15	1	20
soldier	2	2	12	36	0	4
fool	37	58	1	5	3	7
trick	1	3	1	1	3	3

Table 6.4   Term-document matrix comparison by document vectors

	As You Like It	Twelfth Night	Julius Caesar	Henry V	Adv of Sherlock Holmes	Moby Dick
battle	1	1	8	15	1	20
soldier	2	2	12	36	0	4
fool	37	58	1	5	3	7
trick	1	3	1	1	3	3

Table 6.5   Illustration of count vector for six document domain

	As You Like It	Twelfth Night	Julius Caesar	Henry V	Adv of Sherlock Holmes	Moby Dick
battle	1	1	8	15	1	20
soldier	2	2	12	36	0	4
fool	37	58	1	5	3	7
trick	1	3	1	1	3	3

Table 6.6 Sample of two similar words by vector comparison across six documents

	As You Like It	Twelfth Night	Julius Caesar	Henry V	Adv of Sherlock Holmes	Moby Dick
battle	1	1	8	15	1	20
soldier	2	2	12	36	0	4
fool	37	58	1	5	3	7
trick	1	3	1	1	3	3

A term-context matrix can be formed using smaller context, e.g., *a set of ten successive words from a paragraph or search engine*. A word is now defined by a vector over the number of context words, which can be an entire document, literature, or a list of words in a search engine, etc.

There is an argument as to whether raw counts can be used. *tf-idf (term-frequency and inverse document-frequency)* are commonly used in place of raw term counts for *term-document matrix*, whereas *Positive Pointwise Mutual Information (PPMI)* method is used in place of raw term counts for *term-context matrix*.

6.8.4   Pointwise Mutual Information

Pointwise Mutual Information (PMI) is to evaluate whether events x and y co-occur more if they are independent, which is given by:

$$\text{PMI}(X,Y) = \log_2 \frac{P(x,y)}{P(x)P(y)} \tag{6.13}$$

For word similarity measurement application, Church and Hanks (1990) proposed PMI between two words which is given by:

$$\text{PMI}(\text{word}_1, \text{word}_2) = \log_2 \frac{P(\text{word}_1, \text{word}_2)}{P(\text{word}_1)P(\text{word}_2)} \tag{6.14}$$

Niwa and Nitta (1994) proposed *Positive PMI (or PPMI)* by replacing all PMI values less than zero into zero values, which is now commonly used in PMI calculations for document similarity comparison.

6.8.5   Example of Computing PPMI on a Term-Context Matrix

Given matrix F with C columns (contexts) and W rows (words) and f_{ij} is the number of times w_i occurs in context c_j, *Positive PMI(PPMI)* between $word_1$ and $word_2$ is given by:

$$PPMI(word_1, word_2) = \max\left(\log_2 \frac{P(word_1, word_2)}{P(word_1)P(word_2)}, 0\right) \quad (6.15)$$

where

$$\begin{cases} PMI(W,C), & \text{if } PMI(W,C) > 0 \\ \qquad 0, & \text{if } PMI(W,C) < 0 \end{cases} \quad (6.16)$$

$$p(W,C) = \frac{f_{ij}}{\sum_{i=1}^{W}\sum_{j=1}^{C}f_{ij}}, \quad p(W_i) = \frac{\sum_{j=1}^{C}f_{ij}}{N}, \quad p(C_j) = \frac{\sum_{i=1}^{W}f_{ij}}{N} \quad (6.17)$$

in which

- $p(W, C)$ is the probability of considering *target word* W and *context word* C together.
- $p(W)$ and $p(C)$ are the probability of occurring *target word* W and *context word* C, if they are independent, f_{ij} is the number of times W_i occurs in context C_j.

Let's use the previous document term matrix of six English literatures as example to calculate word and context of total counts and probabilities as shown in Tables 6.7 and 6.8.

$$P(W = fool, C = As You Like It) = 37 / 225 = 0.164$$

Table 6.7 Term-context matrix of six contexts with word and context total counts

	As You Like It	Twelfth Night	Julius Caesar	Henry V	Adv of Sherlock Holmes	Moby Dick	Word
battle	1	1	8	15	1	20	46
soldier	2	2	12	36	0	4	56
fool	37	58	1	5	3	7	111
trick	1	3	1	1	3	3	12
Context	41	64	22	57	7	34	225

Table 6.8 Term-context matrix of six contexts with word and context total probabilities

Context	As You Like It	Twelfth Night	Julius Caesar	Henry V	Adv of Sherlock Holmes	Moby Dick	Word
battle	1	1	8	15	1	20	0.204
soldier	2	2	12	36	0	4	0.249
fool	37	58	1	5	3	7	0.493
trick	1	3	1	1	3	3	0.053
Context	0.182	0.284	0.098	0.253	0.031	0.151	1

Table 6.9 Term-context matrix of six contexts with PPMI values

	As You Like It	Twelfth Night	Julius Caesar	Henry V	Adv of Sherlock Holmes	Moby Dick
battle	0.000	0.000	0.576	0.252	0.000	1.057
soldier	0.000	0.000	0.785	0.931	–	0.000
fool	0.604	0.608	0.000	0.000	0.000	0.000
trick	0.000	0.000	0.000	0.000	2.084	0.503

$$P(W = \text{fool}) = 111 / 225 = 0.493$$

$$P(C = \text{As You Like It}) = 41 / 225 = 0.182$$

Let's calculate PMI score for the word *fool* co-occurred with context from $C1 = $ As You Like It based on the above information from Table 6.8.

Using $\text{PMI}(W,C) = \log \dfrac{p(W,C)}{p(W)p(C)}$

$$\text{PMI}(\text{fool}, C1) = \log \frac{0.164}{0.493 * 0.182} = 0.604 \tag{6.18}$$

Similarly, the rest of PMI values for this term-context matrix are calculated as follows in Table 6.9:

Note that: $\text{PPMI}(W,C) = \left\{ \begin{matrix} \text{PMI}(W,C), \text{if PMI}(W,C) > 0 \\ 0, \text{if PMI}(W,C) < 0 \end{matrix} \right\}$ from (6.16).

6.8.6 Weighing PMI Techniques

It is noted that PMI is biased toward infrequent events from above matrix, e.g., rare words have high PMI values. There are two possible methods to improve PMI values: (1) apply *add-k smoothing*, e.g., *add-1 smoothing* and (2) assign rare words with higher probabilities.

6.8.7 Add-K Smoothing in PMI Computation

Since PMI is usually biased with infrequent events, add-K smoothing method can be solution. For example, apply *add-2 smoothing* (i.e., set $k = 2$) in every cell of co-occurrence matrix as in Table 6.10 and see how it works.

The corresponding probability matrix after *add-2 smoothing* is shown in Table 6.11.

Table 6.10 Term-context matrix of six contexts with word and context total count with add-2 smoothing

	As You Like It	Twelfth Night	Julius Caesar	Henry V	Adv of Sherlock Holmes	Moby Dick	Word
battle	3	3	10	17	3	22	58
soldier	4	4	14	38	2	6	68
fool	39	60	3	7	5	9	123
trick	3	5	3	3	5	5	24
Context	49	72	30	65	15	42	273

Table 6.11 Term-context matrix of six contexts with word and context total prob. with add-2 smoothing

	As You Like It	Twelfth Night	Julius Caesar	Henry V	Adv of Sherlock Holmes	Moby Dick	Word
battle	0.011	0.011	0.037	0.062	0.011	0.081	0.212
soldier	0.015	0.015	0.051	0.139	0.007	0.022	0.249
fool	0.143	0.220	0.011	0.026	0.018	0.033	0.451
trick	0.011	0.018	0.011	0.011	0.018	0.018	0.088
Context	0.179	0.264	0.110	0.238	0.055	0.154	1.000

Table 6.12 Term-context matrix of six contexts with PPMI values with add-2 smoothing

	As You Like It	Twelfth Night	Julius Caesar	Henry V	Adv of Sherlock Holmes	Moby Dick
battle	0.000	0.000	0.450	0.208	0.000	0.902
soldier	0.000	0.000	0.628	0.853	0.000	0.000
fool	0.569	0.615	0.000	0.000	0.000	0.000
trick	0.000	0.000	0.129	0.000	1.333	0.303

The *term-context matrix* with PPMI values after applying *add-2 smoothing* is shown in Table 6.12.

It may have certain improvement in PPMI values giving the rate context words theoretically.

However, there were not many improvements in this case.

Another method to achieve this is by raising context probabilities to a certain factor α, say 0.8.

$$\text{PPMI}_\alpha\left(w,c\right) = \max\left(\log\frac{P\left(w,c\right)}{P\left(w\right)P_\alpha\left(c\right)}, 0\right) \tag{6.19}$$

where $P_\alpha\left(c\right) = \dfrac{\text{count}\left(c\right)^\alpha}{\sum_c \text{count}\left(c\right)^\alpha}$

Table 6.13 Term-context matrix of six contexts with PPMI values with $\alpha = 0.80$

	As You Like It	Twelfth Night	Julius Caesar	Henry V	Adv of Sherlock Holmes	Moby Dick
battle	0.000	0.000	0.144	0.000	0.000	0.625
soldier	0.000	0.000	0.435	0.581	–	0.000
fool	0.369	0.373	0.000	0.000	0.000	0.000
trick	0.000	0.000	0.000	0.000	1.315	0.000

Table 6.14 Term-context matrix of six contexts with PPMI values with $\alpha = 0.90$

	As You Like It	Twelfth Night	Julius Caesar	Henry V	Adv of Sherlock Holmes	Moby Dick
battle	0.000	0.000	0.460	0.137	0.000	0.941
soldier	0.000	0.000	0.711	0.858	–	0.000
fool	0.587	0.592	0.000	0.000	0.000	0.000
trick	0.000	0.000	0.000	0.000	1.798	0.217

For example: say $P(a) = 0.95$ and $P(b) = 0.05$:

$$P_\alpha\left(a\right) = \frac{0.95^{0.8}}{0.95^{0.8} + 0.05^{0.8}} = 0.913, \quad P_\alpha\left(b\right) = \frac{0.05^{0.8}}{0.95^{0.8} + 0.05^{0.8}} = 0.083. \quad (6.20)$$

Results using $\alpha = 0.8$ and 0.9 are shown in Tables 6.13 and 6.14 respectively.

6.8.8 Context and Word Similarity Measurement

When applying context and world similarity measurement against context and *word vector*, remember that cosine for computing similarity is given by:

$$\cos\left(\vec{v}, \vec{w}\right) = \frac{\vec{v} \cdot \vec{w}}{|\vec{v}||\vec{w}|} = \frac{\sum_{i=1}^{N} v_i w_i}{\sqrt{\sum_{i=1}^{N} v_i^2} \sqrt{\sum_{i=1}^{N} w_i^2}} \quad (6.21)$$

where v_i is PPMI value for *word v* in *context i*; w_i is PPMI value for *word w* in *context I*; and $\cos(v,w)$ is the cosine similarity of v and w.

Context and *word similarity* measurement of six literatures is shown in Table 6.15.

For context comparison, cosine similarity measurement is performed *between C1 As You Like It* and other five literatures, in which *cosine (C1, C2)* have the highest *0.453* as compared to others ranging from *0.044 (C3:Julius Caesar)* to *0.157 (C6:Moby Dick)*. It showed that it makes sense as the context of *As You Like It* has theme similarity with *Twelfth Night* than other literatures.

Table 6.15 Context and word similarity from six sample literatures

	C1:As You Like It	C2:Twelfth Night	C3:Julius Caesar	C:4Henry V	C5:Adv of Sherlock Holmes	C6:Moby Dick	Wx * Wx	W4 * Wx	Sim(W4. Wx)
W1:battle	1	1	8	15	1	20	692	90	0.077
W2:soldier	2	2	12	36	0	4	1462	68	0.035
W3:fool	37	58	1	5	3	7	1511	247	0.124
W4:trick	1	3	1	1	3	3	24		
Cx * Cx	1375	3378	210	1547	19	474			
C1 * Cx		2154	70	273	115	290			
Sim(C1, Cx)		0.453	0.044	0.093	0.082	0.157			

For word comparison, comparison is performed at *W4: trick* with three other words across six literatures, in which cosine *W4:trick, W3:fool* have the highest similarities among other two words *W1:battle* and *W2:Solder* which in fact they are related in meanings and English usage.

It also showed other possible similarity measurements including *Jaccard, Dice,* and *JSs* methods given by:

$$\text{sim}_{\text{Jaccard}}\left(\vec{v},\vec{w}\right) = \frac{\sum_{i=1}^{N}\min\left(v_i,w_i\right)}{\sum_{i=1}^{N}\max\left(v_i,w_i\right)} \tag{6.22}$$

$$\text{sim}_{\text{Dice}}\left(\vec{v},\vec{w}\right) = \frac{2x\sum_{i=1}^{N}\min\left(v_i,\ w_i\right)}{\sum_{i=1}^{N}\left(v_i+w_i\right)} \tag{6.23}$$

$$\text{sim}_{JS}\left(\vec{v}\ \vec{w}\right) = D\left(\vec{v}\mid\frac{\vec{v}+\vec{w}}{2}\right) + D\left(\vec{w}\mid\frac{\vec{v}+\vec{w}}{2}\right) \tag{6.24}$$

6.8.9   Evaluating Similarity

Like N-grams, similarity methods have (1) *intrinsic* and (2) *extrinsic evaluation schemes*. *Intrinsic evaluation* refers to the correlation between similarity scores of algorithms and human words. *Extrinsic evaluation*, also called *task-based* or *end-to-end evaluation*, refers to detect misspellings, *WSD*, and use in grading essays or TOEFL multiple-choice vocabulary tests.

Exercises

6.1. What is semantic analysis? State and explain the importance of semantic analysis in NLP. Give two examples to illustrate.

6.2. State and explain how humans are good in semantic analysis. Give two examples to support your answers.

6.3. What is the difference between lexical vs. compositional semantic analysis? Give two examples for each to support your answers.

6.4. What is word sense in linguistic? State and explain any five basic types of lexical semantics and their word senses. Give two examples for each to illustrate.

6.5. What is zeugma is linguistic and why is important in NLP? Give two examples to illustrate how zeugma test is used for testing semantic correctness of sentences/utterances.

6.6. What are the major concerns and difficulties encountered in WSD. Give an example for each concern to support your answers.

6.7. State and explain four major methods to tackle WSD. Which one(s) is(are) commonly used in NLP application nowadays to tackle WSD? Why?

6.8. What are synsets in WordNet framework? Give two examples on how it works to support your answers.

6.9. What is path-based similarity in semantic analysis? Use *book* as the basic synset to construct a synset tree like Table 6.4 and calculate all the related path-based similarity between different concepts related to *book*.

6.10. Based on the synset tree created in question 6.9, calculate the similarity values by using: (1) the Resnik method and (2) the Dekang Lin method, and compare them with the ones calculated in 6.9.

6.11. What is distributed similarity? State and explain methods used for distributed similarity measurement.

6.12. Use four famous literatures: (1) Moby Dick (Melville 2012), (2) Little Women by Louisa Mary Alcott (1832–1888) (Alcott 2017), (3) The Adventures of Sherlock Holmes (Doyle 2019), and (4) War and Peace by Leo Tolstoy (1828–1910) (Tolstoy 2019) as context documents, and select any four words (wisely) to illustrate how term-context matrix, PMI, and PPMI are used for document and word similarity measurement in semantic analysis.

6.13. Repeat question 6.12 by using the add-K smoothing method for PMI/PPMI calculations (with $k = 1$ and 2) and different values of α and compare them with results found in 6.12. Explain why it can/cannot be improved.

References

Agirre, E. and Edmonds, P. (Eds) (2007) Word Sense Disambiguation: Algorithms and Applications (Text, Speech and Language Technology Book 33). Springer.

Alcott, L. M. (2017) Little Women. AmazonClassics.

Ayetiran, E. F., & Agbele, K. (2016). An optimized Lesk-based algorithm for word sense disambiguation. Open Computer Science, 8(1), 165-172.

BabelNet (2022) BabelNet official site. https://babelnet.org/. Accessed 25 July 2022.

Bender, E. M. and Lascarides, A. (2019) Linguistic Fundamentals for Natural Language Processing II: 100 Essentials from Semantics and Pragmatics (Synthesis Lectures on Human Language Technologies). Springer.

Butler, A. (2015) Linguistic Expressions and Semantic Processing: A Practical Approach. Springer.

Church, K. W., & Hanks, P. (1990). Word association norms, mutual information, and lexicography. Computational Linguistics - Association for Computational Linguistics, 16(1), 22-29.

Cruse, A. (2011) Meaning in Language: An Introduction to Semantics and Pragmatics (Oxford Textbooks in Linguistics). Oxford University Press

Cruse, A. (1986) Lexical Semantics (Cambridge Textbooks in Linguistics). Cambridge University Press.

Doyle, A. C. (2019) The Adventures of Sherlock Holmes (AmazonClassics Edition). AmazonClassics.

Firth, J. R. (1957). Papers in Linguistics 1934-1951. London: Oxford.

Goddard, C. (1998) Semantic Analysis: A Practical Introduction (Oxford Textbooks in Linguistics). Oxford University Press.

Harris, Z. S. (1954). Distributional structure. Word (Worcester), 10(2-3), 146-162. https://doi.org/1 0.1080/00437956.1954.11659520.

Kilgarriff, A. and Rosenzweig, J. (2000). Framework and results for english SENSEVAL. Computers and the Humanities, 34(1/2), 15-48.

Kroeger, P. (2019) Analyzing meaning: An introduction to semantics and pragmatics (Textbooks in Language Sciences). Freie Universität Berlin.

Lesk, M. (1986). Automatic sense disambiguation using machine readable dictionaries: How to tell a pine cone from an ice cream cone. ACM Special Interest Group for Design of Communication: Proceedings of the 5th annual international conference on Systems documentation. ACM; 24-26. https://doi.org/10.1145/318723.318728.

Lin, D. K. (1998) An Information-Theoretic Definition of Similarity. In Proceedings of the Fifteenth International Conference on Machine Learning (ICML'98). Morgan Kaufmann Publishers Inc., San Francisco, CA, USA, 296–304.

Melville, H. (2012). Moby-dick. Penguin English Library.

MESH (2022) MeSH browser official site. https://www.nim.nih.gov/mesh/meshome.html. Accessed 25 July 2022.

Niwa, Y. and Nitta. Y. (1994). Co-Occurrence Vectors From Corpora vs. Distance Vectors From Dictionaries. In COLING 1994 Volume 1: The 15th International Conference on Computational Linguistics, Kyoto, Japan. https://aclanthology.org/C94-1049.pdf.

Preiss, J. (2006). A detailed comparison of WSD systems: An analysis of the system answers for the SENSEVAL-2 english all words task. Natural Language Engineering, 12(3), 209-228.

Resnik, P. (1995). Using information content to evaluate semantic similarity in a taxonomy. Cornell University Library. https://arxiv.org/abs/cmp-lg/9511007.

Resnik, P. (1999). Semantic Similarity in a Taxonomy: An Information-Based Measure and its Application to Problems of Ambiguity in Natural Language. JAIR 11, 95-130.

Shakespeare, W. (2021) The Complete Works of Shakespeare (AmazonClassics Edition). AmazonClassics.

Sowa, J. (1991) Principles of Semantic Networks: Explorations in the Representation of Knowledge (Morgan Kaufmann Series in Representation and Reasoning). Morgan Kaufmann Publication.

Tolstoy, L. (2019) War and Peace. AmazonClassics.

WordNet (2022a) WordNet official site. https://wordnet.princeton.edu/. Accessed 25 July 2022.

WordNet (2022b) WordNet browser official site: http://wordnetweb.princeton.edu/parl/webwn. Accessed 25 July 2022.

Chapter 7
Pragmatic Analysis and Discourse

7.1 Introduction

Pragmatics and discourse analysis (Bender and Lascarides 2019; Cruse 2011; Goddard 1998; Kroeger 2019) focus on the study of language in its *contextual meaning*, distinguishing it from earlier discussions on word-level *semantics, syntax, grammatical relations, meaning representation*, and *semantic analysis*.

Pragmatics analysis focuses on *context meaning*. Discourse analysis studies social context in written and spoken language. They consist of structured, coherent, and cohesive sets of sentences or utterances to reflect what constitutes an utterance versus a set of unrelated sentences and how the text is related.

There are two types of *discourse* in daily life: (1) *monologu*e and (2) *dialogue*. A *monologue* is a one-way communication between a *speaker (writer)* and an *audience (reader)*, e.g., read or write a book, watch a TV show or a play, listen to a speech, attend a presentation or a lecture that depends on the deposition of dialogue. *Dialogue* refers to participation in turn to speaker and hearer. It has a two-way or multiple ways of communications.

There are also two types of *dialogue* (1) *human-to-human*, e.g., daily conversations, group discussions, and (2) (a) *human-to-computer interaction* (HCI), e.g., conversational agent, chatbot in NLP, and (b) *computer-to-computer interaction* (CCI), e.g., cross-machine verbal communication in smart city and intelligent transportation system, multi-agent-based bargain and negotiation systems.

7.2 Discourse Phenomena

There are many discourse phenomena that can be solved naturally by humans, but some like *coreference resolutions (CR)* require a lot of effort by machines to solve.

R. Lee, *Natural Language Processing*,
https://doi.org/10.1007/978-981-96-3208-4_7

7.2.1   Coreference Resolution

CR (Bender and Lascarides 2019; Goddard 1998) is the task of identifying all linguistic expressions, also known as *mentions*, that correspond to real-world entities described in a text. These mentions are assembled and replaced with the correct pronouns and noun phrases (NPs). It's simple for humans, but machines make mistakes all the time. For example:

[7.1] *Jack saw Andrew in the examination hall. He looked nervous.*
[7.2] *Jack saw the student in the examination hall. He looked nervous.*

Humans and machine will likely consider the first subject mentioned in foregoing sentence or utterance as reference to pronoun of the following sentence. For instance, *He* in [7.1] will refer to *Jack*. However, coreference resolution from human perspective in [7.2] will consider *He* may not refer to *Jack* but *the student* as it is natural and logical to relate *student* with examination.

Example below is more obvious:

[7.3] *Jane talked to Amy about her examination result. She looked worried.*
[7.4] *Jane talked to Amy about her examination result. She felt sorry about it.*

She in [7.3] should refer probably to *Amy* who is worried as she participated in the examination instead of *Jane*.

She in [7.4] should probably refer to *Jane* instead of *Amy* participating in the examination but *Jane* is more likely to *feel sorry* as empathy to *Amy*.

Humans can discern the above naturally by context, common sense, or world knowledge but confound computers to develop judgment.

7.2.2   Why Is It Important?

Let's look at some standard situations prior to complex coreference resolution cases:

[7.5] *Jack gives Ian 1000 dollars. He is generous.* (original sentence).
[7.6] *Jack gives Ian 1000 dollars. Jack is generous.* (with coreference resolution)
or compact cases handled by computer satisfactorily:
[7.7] *I voted for Jack as he is more aligned to my values, Ian said.* (original sentence).
[7.8] *Ian voted for Jack as Jack is more aligned to Ian's values, Ian said.* (with coreference resolution)

From The Adventures of Sherlock Holmes (Doyle 2019):

[7.9] *I was seized with a keen desire to see Holmes again, and to know how was employing his extraordinary powers.* (original sentence).
[7.10] *Watson was seized with a keen desire to see Holmes again, and to know how Holmes was employing Holmes' extraordinary powers.* (with coreference resolution)

or more challenging sentences of famous discourse from *A Scandal in Bohemia*:

[7.11] *To Sherlock Holmes, she is always "the woman." I have seldom heard him mention her under any other name.* (original sentence).

[7.12] *To Sherlock Holmes, Irene Adler is always "the woman." Watson has seldom heard Holmes mention Irene Adler under any other name.* (with coreference resolution)

[7.11] is more challenging as the reference name *Irene Adler* for *she* did not occur, but after two sentences, there is no emotion akin to affection for *Irene Adler*.

This phenomenon is called *cataphor* to acquire meaning from a subsequent word or phrase in linguistics.

The subsequent phrase (or word group) is called *antecedent* or a referent against anaphora, a rhetorical term for a phrase (or word group) repetition at the start of consecutive sentences/utterances used in many English sentences' construction; i.e., [7.5], [7.7], [7.9] are reference terms mentioned repetitively prior to pronoun replacement.

CR is a versatile tool suitable for many NLP applications, including *text understanding and analysis*, *information retrieval and extraction*, *text summarization*, *machine translation*, and even *sentiment analysis*. This is a great way to get unambiguous sentences that computers can understand.

7.2.3   Coherence and Coreference

7.2.3.1   What Is Coherence?

In linguistics, *coherence* (Bender and Lascarides 2019; Goddard 1998) refers to *meaning relationships* between individual units, which can be *sentences (discourses)* or *textual statements*. Texts appear to have logical and semantical consistency for reader or hearer due to these relations.

Coherence-oriented text analysis is primarily concerned with the construction and configuration of meaning in a text; that is, how various components are connected to make the text meaningful to recipient as a random sequence of disjointed phrases and clauses.

In other words, if a text has coherence, its parts are well connected and head for the same direction. Without coherence, a discussion or utterance may neither make sense nor be followed by the audience. It has both verbal and written language significance.

Here are some coherence examples:

[7.13] *History reveals that humans have come a long way from birth. They have invented many new technologies that improve the standard of living. However, technologies that are supposed to provide us a better world sometimes end-up to* *disaster, such as the invention of <u>nuclear weapons</u>, environmental pollution, and the extinction of some animal species.*

In [7.13], coherence terms *History* → *humans* → *They* → *technologies* → *nuclear weapons* with repetitive terms and concepts provide a stream of idea flow and knowledge for hearer or reader to understand the message conveyed in this passage.

7.2.3.2   What Is Coreference?

Coreference (*coreference*) appears when two (or group of) terms refer to the same person or thing with a unified reference to achieve linguistic coherence. For example:

[7.14] *Jack said Helen would arrive soon, and she did.*
  – *Helen* and *she* refer to the same person.

Conference is not always trivial to determine, e.g.:

[7.15] *Jack said he would join the term* vs.
[7.16] *Jack told Ian to come, he smiled.*

When comparing [7.15] vs. [7.16], [7.15] is trivial as there is only one subject (noun) that *he* can refer to (i.e., *Jack*), while *he* in [7.16] can refer to either *Jack* or *Ian*.

Determining *coreferential expressions* is important in many NLP applications, such as *information retrieval and extraction, text summarization,* and *conversation understanding* in *question-and-answer chatbot systems.*

7.2.4   *Importance of Coreference Relations*

To understand the meaning of a coreference relationship, let's look at how to extract key information or summarize the following text:

[7.17] *XYZ bank is continuing to struggle with severe financial problems. According to the finance news report, their CEO Charles Smith will announce to step-down at the press conference tomorrow morning.*

The texts in [7.17] are coherent with well-structured coreference in a typical news article. Coherence concept terms are also used to extract information:

[XYZ bank] → [financial problem] → [CEO] → [Charles Smith] → [step down] → [press conference] → [tomorrow morning].

A reasonable text summary may be:

[7.18] *The CEO of XYZ bank Charles Smith will announce his step-down at tomorrow morning's press conference.*

This example shows the coherent relationships between text segments, where the first sentence provides context weights of the second sentence.

Remarks: A well-structured text summarization/information extraction case will and should match with *Fillmore's case role theory* with well-defined agent, patient, location, time, purpose, beneficiary, possessor, instrument, etc.; in other words, a well coherence text message and utterance regard the first sentence as the *opening* of a speech followed by *elaboration* of an open statement in coreference relation with a *thematic relation* like watching a movie or a TV show.

Further to elaboration and thematic relation, *coreference relation* has another type called *inference type*. It regards the first sentence/utterance as *claims* followed by *explanation* of claims sentence. For inference argument, the first sentence is the *effect* followed by *cause(s)* of the following sentences:

[7.19] *Jack keeps Ian's car key. He was drunk last night.* (coherence) vs.
[7.20] *Jack keeps Ian's car key. He wants to see a movie tonight.* (without coherence)

Coherence occurred in [7.19] as the first statement has relevance to the second statement with *pragmatic meaning*, whereas the second statement is *probably* an explanation, or a *cause* of the event where *Jack* keeps *Ian's* car key because *Ian* was drunk by common sense/world knowledge. Thus, *He* should be *Ian* instead of *Jack* by inference.

While two statements in [7.20] have neither *coherence* nor logic *cause-effect* relationship between them, it is difficult to judge whether *He* in the second statement should refer to *Jack* or *Ian*. Thus, *Jack* regards as the subject and the referent *He* in usage of English although it may be incorrect.

7.2.5   Entity-Based Coherence

Let's look at the following examples:

[7.21] *Helen went to the superstore to buy a cello.*
[7.22] *She had frequented the store for a long time.*
[7.23] *She was delighted to buy the cello finally.*
[7.24] *She just discovered that the store is closed.*
[7.25] *It was the store Helen had frequented for a long time.*
[7.26] *She was delighted to buy that cello.*
[7.27] *The music generated by it is beautiful.*
[7.28] *It was closed when Helen arrived.*

Entity-based coherence models measure coherence to track salient central entities across utterances. *Centralization theory* (Grosz et al. 1995) is a remarkable entity-based coherence theory for tracking whether entities (so-called *Central Entity*, CE) are prominent at each point in a discourse model. For cases from [7.21] to [7.23] is *Helen* who will be the reference for *she* in these statements naturally. While CE in [7.25] is shifted from *the superstore* to *cello* in [7.26] and [7.27], CE is shifted back to *the store* in [7.28] to make it more complex.

7.3   Discourse Segmentation

7.3.1   What Is Discourse Segmentation?

Discourse segmentation is the task of determining the smallest nonoverlapping discourse units, known as *elementary discourse units (EDUs),* which can be further categorized into (1) *sentence segmentation* and (2) *sentence-level discourse segmentation.* The main purpose of discourse segmentation is to divide a text document (set of utterances) into a list of subtopics. This is often a higher level simplification structure of a discourse. For example, an academic article is usually segmented into *abstract, introduction, methodology, implementation, results, discussion, conclusion,* etc., to comprehend.

There are (1) *unsupervised* and (2) *supervised discourse segmentation* methods. The applications of automatic discourse segmentation include (1) *information extraction or retrieval* and (2) *text summarization* on each segment separately.

7.3.2   Unsupervised Discourse Segmentation

Unsupervised discourse segmentation is a class usually presented as a linear segmentation of raw data and segmentation into multiple paragraph subtopics. *Unsupervised* means that the task is not given training data as examples to understand linear segmentation task. These examples involve splitting the text into multiparagraph units to represent paragraphs of the original text. These algorithms rely on cohesion, which can be defined as the linguistic means of linking units of text together.

Cohesion-based approach involves dividing text into subtopics, where sentences or paragraphs cohere to each other, and reveal the relationship between two or more words in two units like synonyms.

Cohesion is the linking of text units based on linguistic means. Lexical cohesion is the use of similar words to link units of text with the same word, synonym, or hypernym. For instance:

[7.29] *Yesterday was Jane's birthday. Betty and Mary went to buy a present from the gift shop. Mary intended to buy a purse. "Don't do that," mentioned Betty. "Jane already got one. She will ask you to return it."*

The non-lexical cohesion approach is the use of anaphora.

[7.30] *Peel, core and slice <u>peaches</u> and <u>pineapples</u>, then place <u>these fruits</u> in the skillet.*

Unsupervised discourse segmentation was proposed by Prof. Marti Hearst in her classical works on *TextTiling* in early 1990.

7.3.3   Hearst's TextTiling Method

Hearst's TextTiling (Hearst 1997) is a typical *discourse segmentation algorithm* to subdivide explanatory text into multiple paragraphs or automatically grouped sub-topic segments representing in the original text.

Hearst's TextTiling method is a typical unsupervised method that no training dataset and prior knowledge base are required. Hearst's original work used articles from *Stargazers*, a science magazine with a *TextTiling method* to characterize article text messages into subtopics.

For example, consider a 21-paragraph science news article extracted from the magazine with a topic focused on reports of life on Earth and other plants; its contents are characterized into the following subtopic discussions (Hearst 1997):

[Para 1–3] *Introduction—the search of life in space*
[Para 4–5] *The moon's chemical composition*
[Para 6–8] *How early earth-moon proximity shaped the moon*
[Para 9–12] *How the moon helped life evolve on earth*
[Para 13] *Improbability of the earth-moon system*
[Para 14–16] *Binary/trinary star systems make life unlikely*
[Para 17–18] *The low probability of nonbinary/trinary systems*
[Para 19–20] *Properties of earth's sun that facilitate life*
[Para 21] *Summary*

TextTiling is a technique to divide a full-length text document into coherent multi-paragraph units that correspond to a series of subtopic paragraphs as shown in the example above. The algorithm assumes that during a subtopic discussion, a set of words is used, and subtopics change significant parts of vocabulary accordingly.

The distribution of terms extracted from the *Stargazers* text is assigned with a single-digit frequency for each sentence number, with spaces for zero frequencies (Hearst 1997) as shown in Fig. 7.1. It revealed that terms:

1. Occurred frequently throughout the text; e.g., *moon* and *planet* are often indicative of main topic(s) of the text.
2. Occurred less common but evenly distributed; e.g., *scientists* and *form* are both generic to create a subtopic title.
3. Like *space* and *star* occurred more frequent from sentences 5 to 20 and 60 to 90, while term *life* to *planet* occurred more frequently from sentences 58 to 78 which may create two distinct clusters of subtopic discussion.
4. Like *life* to *species* have similar phenomena occurred to create a natural cluster between sentences 35–55 and conform with human judgment as subtopic discussion of *How the moon helped life evolve on earth*.

These results suggested that the logic behind sentences or paragraphs in subtopics are consistent with each other but not with paragraphs in adjacent topics.

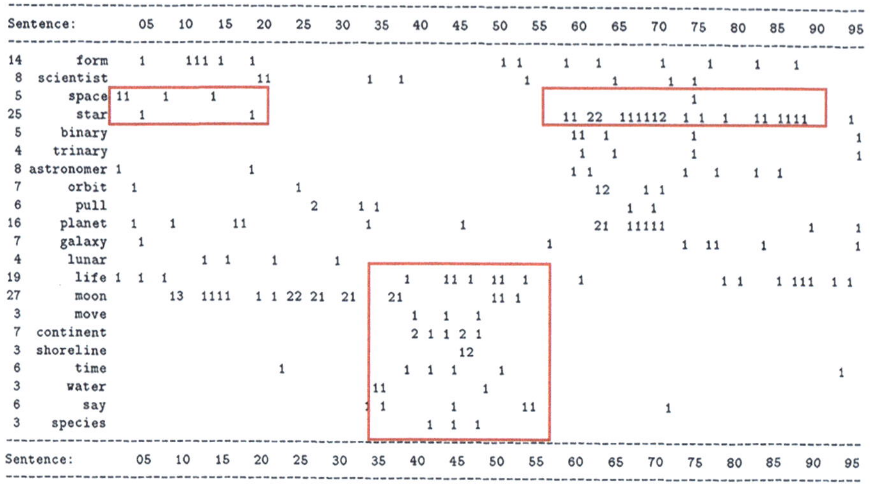

```
Sentence:    05   10   15   20   25   30   35   40   45   50   55   60   65   70   75   80   85   90   95
-----------------------------------------------------------------------------------------------------------
14    form    1       111 1                                    1 1    1 1         1       1       1    1
 8  scientist          11                    1    1               1         1    1 1
 5     space 11   1    1                                                           1
25      star  1             1                               11 22  111112  1 1  1    11 1111              1
 5    binary                                                11  1                1                        1
 4    trinary                                                1    1              1                        1
 8 astronomer 1                  1                          1 1              1   1    1 1
 7     orbit  1                        1                                12    1 1
 6      pull                       2    1 1                             1 1
16    planet  1    1        11                     1              1      21  11111
 7    galaxy  1                                              1                   1   11      1            1
 4     lunar           1  1    1         1                                                                1
19      life 1  1  1                        1    111 1  11 1    1                 1 1       1 111  1 1
27      moon      13  1111    1 1 22 21  21   21              11 1
 3      move                                 1    1    1
 7  continent                            2 1 1 2 1
 3  shoreline                                12
 6      time                   1              1 1  1    1                                                 1
 3     water                           11          1
 6       say                    1            1         11                  1
 3    species                            1    1  1
-----------------------------------------------------------------------------------------------------------
Sentence:    05   10   15   20   25   30   35   40   45   50   55   60   65   70   75   80   85   90   95
```

Fig. 7.1 Distribution of selected terms in the *Stargazers* text (blanks mean zero frequency)

7.3.4 TextTiling Algorithm

TextTiling algorithm (Hearst 1997) for discourse segmentation and subtopic structure characterization using term repetition consists of three processes: (1) *tokenization*, (2) *lexical score determination*, and (3) *boundary identification*.

Tokenization includes converting words to lowercase, removing stop words and root words, and converting words into pseudo-sentences with the same length such as 15 words.

Lexical score determination includes calculating lexical cohesion scores for each gap between pseudo-sentences. This lexical cohesion score represents word similarity. For instance, take ten pseudo-sentences each before and after *gap*, followed by the computation of cosine similarity between word vectors which is given by:

$$\text{sim}_{\text{cosine}}\left(\vec{b},\vec{a}\right)=\frac{\vec{b}\cdot\vec{a}}{\left|\vec{b}\right|\left|\vec{a}\right|}=\frac{\sum_{i=1}^{N}b_i\times a_i}{\sqrt{\sum_{i=1}^{N}b_i^2}\sqrt{\sum_{i=1}^{N}a_i^2}} \tag{7.1}$$

Boundary identification involves assigning a boundary distance to identify a new segment. Similarity is first created, and the depth value of *similarity valley* $(a-b)+(c-b)$ is calculated as shown in Fig. 7.2; then, *segmentation* is performed if the depth score value is greater than the threshold as shown in Fig. 7.3.

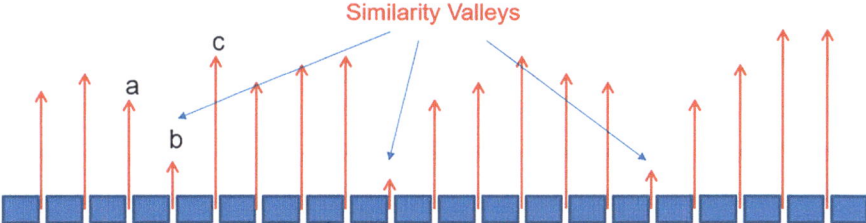

Fig. 7.2 Lexical score determination with similarity valleys

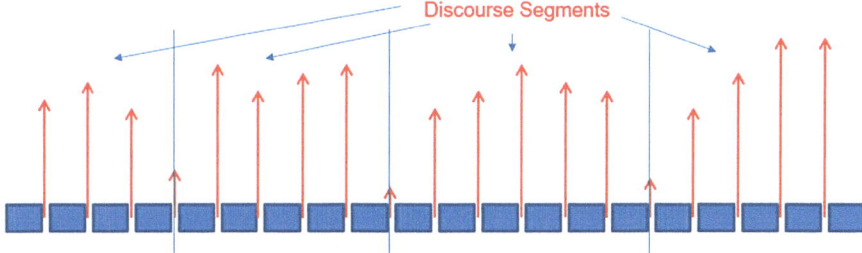

Fig. 7.3 Boundary identification with discourse segments

7.3.5   *Supervised Discourse Segmentation*

It is relatively easy to collect bounded training data using *supervised discourse segmentation* such as news reports from TV shows, paragraph segmentation in text, or dialogue to find paragraphs in speech recognition output.

Several classifiers can be used to achieve *supervised segmentation*, one is called *feature set* which is a superset for unsupervised segmentation with often domain-specific utterance tokens and keywords.

Supervised discourse segmentation is also a model. It is (1) a classification task that uses one of the supervised classifier methods, such as *SVM, naïve Bayer,* and *maximum entropy* to distinguish whether sentence boundaries have paragraph boundaries, or (2) a sequence labeling task to label sentences with or without paragraph borders. It uses cohesive features including word overlap, word cosine similarity, anaphora, and additional features such as discourse markers or keywords.

Discourse tokens or keywords/phrases indicate discourse structure, e.g., *good evening, join our broadcast news now, or join the company at the beginning/end of the segment.* They can be manual codes or automatically determined by feature selection.

However, measuring *precision, recall,* and *F-measure* is not always good evaluation ideas as they are insensitive to near misses. Pevzner and Hearst (2002) proposed a good and effective evaluation metric for text segmentation called the *WindowDiff* method.

7.4 Discourse Coherence

7.4.1 What Makes a Text Coherent?

A *text coherent* refers to the application of:

1. A coherent relationship between a subfield of discourse called *rhetorical structure* and a whole theory called *rhetorical structure theory* (RST). It is a text organization theory that describes the relationships that exist between parts of a text. It was proposed by Mann and Thompson (1988) in their remarkable paper *Rhetorical structure theory: toward a functional theory of text organization*, published in 1988. The theory was developed as part of research on computer-aided text generation in text summarization and applications used by NLP researchers.
2. The ordering of subsections of discourse called *discourse topic structure*. It is the key to discourse cohesion and embodies the essence of discourse analysis. It has been extensively adopted in the past decades and has become a key component in text analysis. Linearly segmenting text into appropriate topic structures can reveal valuable information such as the overall topic structure of the text, which can be used for text analysis tasks such as text summarization, information retrieval, and discourse analysis.
3. A *Referring Expression (RE)* is any NP or a substitute for an NP whose function in spoken, and signed or written text is to single out a single person, place, object, or group of people, places, objects, etc.

7.4.2 What Is Coherence Relation?

Coherence relation refers to discourse properties that make each discourse meaningful (or have appropriate meaning) in the context. It refers to common denominator to identify possible connections between utterances in a series of statements or discourses about the same topic.

These sense relations in discourse analysis named *Coherence Relations* by Prof. Jerry R. Hobbs in his works *Coherence and Coreference* published by *Cognitive Science* in 1979 (Hobbs 1979) had been further developed by other linguistics including Sanders et al. (1992) and Kehler (2002) into a well-defined theory.

These meaning relationships, called *propositional relations* defined by Mann and Thompson (1986), are encoded in text recognized by the reader trying to understand the text and its components, and to see why the speaker or author added the sentence. Coherent relationships are sometimes referred to as types of thematic development such as the narrative of a movie or TV show involving *cause-and-effect* story type in sense relations development.

7.4.3   Types of Coherence Relations

There are five major types of coherence relations: (1) *parallel*, (2) *elaboration*, (3) *cause-and-effect*, (4) *contrast*, and (5) *occasion*.

1. *Parallel* infers $p(a_1, a_2, ...)$ from the assertion of S_0 and $p(b_1, b_2...)$ from the assertion of S_1, where a_i and b_i are similar for all i.
   [7.31] Rich man wants more power. Poor man wants more food.
   They are frequently used in describing two sense relations with similar situation (meaning) but different in object, reference, and scenario.
2. *Elaboration* infers the same proposition P from the assertions of S_0 and S_1.
   [7.32] *Dorothy was from Kansas. She lived in the great Kansas prairies.*
   [7.33] *Nicola Tesla was a genius. He invented over hundreds of things in his life.*
   They are frequently used in discourse construction; the successive sentences/utterances are further elaboration of the previous one.
3. *Cause-and-effect* are S_0 and S_1 if S_1 infers S_0, i.e., $S_1 \rightarrow S_0$.
   [7.34] *Jack cannot afford to buy the car. He lost his job.*
   [7.35] *Nicola Tesla invented over hundreds of things in his life. He was a genius.*
Cause-and-effect discourse relation that can refer to animate or inanimate subjects in [7.35] is the reverse of elaboration statement [7.33] but does not always occur.
4. *Contrast* infers S_0 and S_1 if P_0 and P_1 infer from S_0 and S_1 with one pair of elements that are contrast with each other, where other elements are similar in context.
   [7.36] *Hope for the best. Prepare for the worst.*
   [7.37] *Jack is meticulous while Bob is sloppy.*
   Contrast coherence relations can exist within a sentence or in successive sentences/utterances. It often refers to two subjects, or events with contrast sense relations.
5. *Occasion* is the alteration of state that can infer from the assertion of S_0, where final state can infer from S_1, or the alteration of state can infer from the assertion of S_1, whose initial state can infer from S_0.
   [7.38] *Jane put the books into a schoolbag, she left the classroom with Helen.*
   [7.39] *Jack failed the exam. He started to work hard.*
   State change invokes new action.

7.4.4   Hierarchical Structure of Discourse Coherence

Discourse coherence can also be revealed by the hierarchy between coherent relations. For example:

[7.40] *Jack went to town to buy a toy.*
[7.41] *He took a bus to the shopping mall.*
[7.42] *He needed to buy a toy for his child.*
[7.43] *It is Jane's birthday.*

[7.44] *He also wanted to buy some books for weekend reading.*

A hierarchical structure of discourse coherence is shown in Fig. 7.4. [7.40]–[7.44] can be organized in a hierarchy tree structure; e.g., *occasion* consists of two expressions, one is expression e_1 (statement [7.40]), and the other is an explanatory clause which in turn consists of expression e_2 (statement [7.41]) and a parallel clause which consists of two entities, one is explanatory expression e_3 and the other is expression e_5 (statement [7.44]); e_3 is further divided into statements [7.42] and [7.43], respectively.

7.4.5   Types of REs

RE is a surrogate for any NP or NP whose function in utterance is to identify some discrete objects. There are five frequently used REs in discourse coherence: (1) *indefinite NPs*, (2) *definite NPs*, (3) *pronouns*, (4) *demonstratives,* and (5) *names.*

1. *Indefinite NPs* introduce entities into context that are new to listener, e.g., *a policeman, some apples, a new iPad.*
   [7.45] *I go to the electronic store to buy a new notebook computer.*
2. *Definite NPs* refer to entities recognizable by listener such as abovementioned combination of beliefs about the world, e.g., *a furry white cat, and the cat.*
   [7.46] *Don't look at the sun directly with bare eyes, it will hurt yourself.*
3. *Pronouns* are another form of definite designation, usually with stronger restrictions than standard designation, e.g., *s/he, it, they.*
   [7.47] *I go to the electronic store to buy a new notebook computer. This computer is rather light and fast.*
4. *Demonstratives* are pronouns that can act alone or as determiners, e.g., *this, that.*
   [7.48] *That book seems to be very interesting and worth buying it.*
5. *Names* are common methods to refer to people, organizations, and locations.
   [7.49] *I bought lunch at KFC today.*

Fig. 7.4 Hierarchical structures in discourse coherence

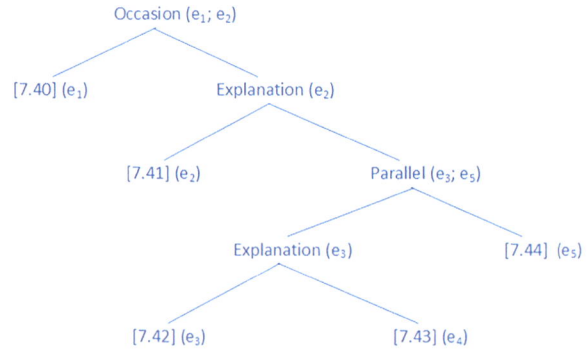

7.4.6 Features for Filtering Potential Referents

There are four common features to filter potential references in discourse coherence: (1) *number agreement*, (2) *person agreement*, (3) *gender agreement*, and (4) *binding theory constraints*.

1. *Number agreement* refers to pronouns, and references must agree in number, e.g., *single or plural*.
   [7.50] *The children are playing in the park. They look happy.*
2. *Person agreement* refers to the first, second, or third person.
   [7.51] *Jane and Helen got up early. They needed to take an exam this morning.*
3. *Gender agreement* refers to male, female, or nonperson, e.g., *he, she or it*.
   [7.52] *Jack looked tired. He didn't sleep last night.*
4. *Binding theory constraints* refer to constraints imposed by syntactic relations between denotative expressions and possible preceding NPs in the same sentence.
   [7.53] *Jane purchased herself an iPad.* (*herself* should be *Jane*)
   [7.54] *Jane purchased her an iPad.* (*her* may not be *Jane*)
   [7.55] *She claimed that she purchased Mary an iPad.* (*She* and *she* may not be *Mary*)

7.4.7 Preferences in Pronoun Interpretation

There are six types of preferences in pronoun interpretation: (1) *recency*, (2) *grammatical role*, (3) *repeated mention*, (4) *parallelism*, (5) *verb semantics*, and (6) *selectional restrictions*.

1. *Recency* refers to entities from recent utterances:
   [7.56] *Tim went to see a doctor at the clinic. He felt sick. It might be influenza.*
2. *Grammatical role* is to emphasize the hierarchy of entities according to grammatical position of the terms that represent them, e.g., subject and object.
   [7.57] *Jane went to Starbucks to meet Jackie. She ordered a hot mocha.* (*She* should be *Jane*)
   [7.58] *Jane discussed with Jackie about her exam results. She felt so nervous about it.* (*She* should be *Jackie* instead of *Jane*)
   [7.59] *Jane discussed with Jackie about her exam results. She felt so sorry about it.* (*She* should be *Jane* instead of *Jackie*)
3. *Repeated mention* refers to mentioning about the same thing.
   [7.60] *Jane went to supermarket to buy some food. It turned out it was closed.*
4. *Parallelism* refers to subject-to-subject or object-to-object kind of expression:
   [7.61] *Mary went with Jane to Starbucks. Ian went with her to the bookstore afterward.* (*her* should probably be *Jane* instead of *Mary*)
5. *Verb semantics* are *verbs seem to emphasize one of their argument positions:*
   [7.62] *Jane warned Mary. She might fail the test.*
   [7.63] *Jane blamed Mary. She lost the watch.*

In [7.62], *She* should be *Mary* as *Mary* is the one being warned about failing the test. For [7.63], *She* should be *Jane* who suffered. It is a pragmatic phenomenon because it involves common sense by word meaning *blamed* to understand correct coreference in the second statement.

6. *Selectional restrictions* refer to another semantic knowledge playing a role:

   [7.64] *Mary lost her iPhone in the shopping mall after carrying it the whole afternoon.*

Note that [7.64] involves high-level semantics or common sense understanding of *it* can mean iPhone or shopping mall but it has been carried for the whole afternoon, so it cannot be an unmovable object except iPhone.

7.5   Algorithms for Coreference Resolution

7.5.1   Introduction

CR is the task of finding all linguistic expressions (called mentions) in any text involving real-world entities. After finding these mentions and grouping them, they can be resolved by replacing pronouns with NPs.

There are three fundamental algorithms for conference resolution: (1) *Hobbs' algorithm*, (2) *centering algorithm*, and (3) *log-linear model*.

7.5.2   Hobbs' Algorithm

7.5.2.1   What Is Hobbs' Algorithm?

Hobbs' algorithm was one of the early approaches to pronoun resolution proposed by Prof. Jerry R. Hobbs in 1978 (Hobbs 1978) and further consolidated as well-known algorithm for coreference resolution in his remarkable work *Coherence and Coreference* published in Cognitive Science, 1979 (Hobbs 1979).

The original work proposed two CR algorithms, a simple algorithm based purely on grammar, and a complex algorithm that incorporated semantics into parsing methods (Hobbs 1978, 1979).

Unlike other algorithms, Hobbs' algorithm does not turn to a discourse model for parsing because its parse tree and grammar rules are the only information used in pronoun parsing. Let's look at how it works.

7.5.2.2   Hobbs' Algorithm

Hobbs' algorithm assumes a parse tree where each NP node has an N-type node below it as the parent of a lexical object. It operates as follows:

1. Start with the node of noun phrase (NP) that directly dominates the pronoun.
2. Go up the tree to the first NP or sentence (S) node visited; denote this node as X and name the path being applied to reach it as p.
3. Visit all branches under node X to the left of path p, breadth first, from left to right, taking any NP node found as an antecedent; there is an NP or S node between it and X.
4. If node X is the highest S node in sentence, visit the surface parse trees of previous sentences in the text with the most recent first; each tree is then visited in a left-to-right and breadth-first manner. When an NP node is encountered, it is recommended as an antecedent. If X is not the first S node in the set, go to step 5.
5. Climb up from node X to the first NP or S node encountered; denote this new node as X and name the path as p.
6. If X is an NP vertex, and if the path p to X does not pass through a nominal vertex immediately dominated by X, then denote X as an antecedent.
7. Visit all branches under node X to the left of path p, breadth-first manner, from left to right, denoting each NP node encountered as an antecedent.
8. If X is an S node, visit all branches of node X to the right of path p from left-to-right and breadth-first manner, but do not visit below any NP or S being encountered as the antecedent.
9. Return to step 4.

7.5.2.3 Example of Using Hobbs' Algorithm

Statement [7.65] is a classic paper (Hobbs 1978) to demonstrate how Hobbs' algorithm works as shown in Fig. 7.5.

[7.65] *The castle in Camelot remained the residence of the king until 536 when he moved it to London.*

Example—What does *it* stand for?

1. Start with node NP_1, step 2 climbs up to node S_1.
2. Step 3 searches the left part of S_1's tree but fails to locate any eligible NP node.
3. Step 4 fails to apply.
4. Step 5 climbs up to NP_2 which step 6 proposes 536 as antecedent of *it*.
5. The algorithm can be further improved by applying simple selectional constraints, such as
   Date can't move.
   Places can't move.
   Large or fixed objects can't move.
6. After NP_2 is rejected, steps 7 and 8 turn up nothing, and control is returned to step 4 which fails to apply.
7. Step 5 climbs up to S_2 which step 6 fails to apply.
8. In step 7, the breadth-first search recommends the NP_3 where *the castle* is rejected by the constraint number 3.

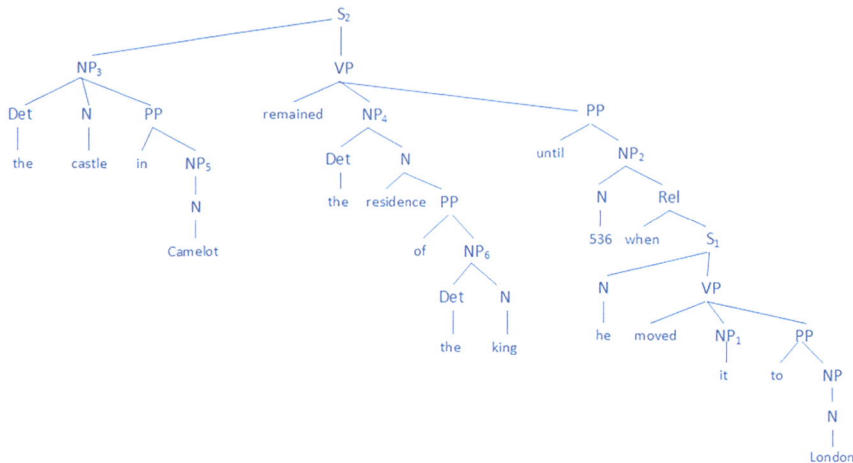

Fig. 7.5 Parse tree for statement [7.65]

9. The algorithm continues to visit NP$_4$ where it correctly recommends *the residence* as antecedent.

Exercise: How to check coreference resolution of *he* as *the king*?

7.5.2.4   Performance of Hobbs' Algorithm

In the original work, Hobbs manually analyzed 100 consecutive examples from 3 different texts, assuming correct parsing was available, and the algorithm was 72.7% correct (Hobbs 1978); which is quite impressive for such simple algorithm. If the algorithm is integrated with syntactic constraints when resolving pronouns as shown in Fig. 7.5, the performance can be even higher.

However, Hobbs' algorithm experiences two major problems.

1. When looking for the antecedent of a pronoun within a sentence, it goes sequentially further up the tree to the left of pronoun such an error is looked for in the previous sentence.
2. This algorithm does not assume a discourse segmentation structure and may revert to arbitrarily far of the text to find an antecedent.

Nevertheless, as he concluded in his original paper, naïve-based approach on coreference resolution did provide a high baseline and works in many usual situations in discourse analysis, and is still being used as a benchmark in related CR research nowadays (Cornish 2009; Kehler et al. 2008; Lata et al. 2022; Wolna et al. 2022).

7.5.3   Centering Algorithm

Centering theory (CT) was proposed by Profs Barbara J. Grosz and Candace L. Sidner in their distinguished work *Attention, Intentions, and the Structure of Discourse*, as part of its main theory of discourse analysis (Grosz and Sidner 1986). It is a theory of discourse structure that models the interrelationships between foci or centers as the choice of reference terms and the perceived coherence of discourse.

The basic idea is as follows:

1. A discourse has a focus, or center.
2. The center typically remains the same for a few sentences, then shifts to a new object.
3. The center of a sentence is typically pronominalized.
4. Once a center is determined, there is a strong inclination for subsequent pronouns to continue referring to it.

In centering algorithm, utterances from a discourse have a *backward-looking center* (C_b) and a set of *forward-looking centers* (C_f). The C_f set of an utterance U_0 is the set of utterance units elicited by that utterance. C_f set is ranked by discourse emphasis, the most accepted ranking is by grammatical role. The highest-ranked element in this list is called the *preferred center* (C_p), which represents the highest-ranked element among previous utterances found in the current utterance and serves as a link between these utterances. Any sudden shifts in the topic of utterances are reflected in changes in C_b between utterances.

7.5.3.1   What Is Centering Algorithm?

Centering algorithm (Grosz and Sidner 1986; Tetreault 2001) consists of three parts: (1) initial settings, (2) constraints, and (3) rules and algorithm.

7.5.3.2   Part I: Initial Setting

- Let U_n, U_{n+1} be 2 successive utterances.
- Backward-looking center of U_n, written as $C_b(U_n)$, denotes focus after U_n is interpreted.
- Forward-looking centers of U_n, written as $C_f(U_n)$, form an ordered list of entities in U_n that can serve as $C_b(U_{n+1})$.
- $C_b(U_{n+1})$ is the highest-ranking element of $C_f(U_n)$ mentioned in U_{n+1}.
- Order of entities in $C_f(U_n)$: in which subject > existential predicate nominal > object > indirect object > demarcated adverbial PP.
- Let $C_p(U_{n+1})$ be the highest-ranked forward-looking center.

7.5.3.3 Part II: Constraints

For each utterance U_i (i = …m) in a discourse segment D:

- There is precisely one C_b.
- Every element of C_f-list for U_i must be realized in U_i.
- The center, C_b (U_i, D), is the highest-ranked element of C_f (U_{i-1}, D) realized by U_i.

7.5.3.4 Part III: Rules and Algorithm

For each utterance U_i (i = …m) in a discourse segment D:

Rule 1: If some elements of C_f (U_{i-1}, D) are realized as a pronoun in U_i, then so is C_b (U_i, D).

Rule 2: Transition states, defined as follows, are ordered such that the sequence of *Continue* is preferred over the sequence of *Retains*, which are preferred over *Smooth-Shift* and then *Rough-Shift*.

The relationship between C_b and C_p of two utterances determines coherence between words. CT ranks the coherence of adjacent utterances with transitions determined by:

1. C_b is the same from U_{n-1} to U_n or not.
2. This entity coincides with C_p of U_n or not.

Table 7.1 shows the criteria for each transition in the centering algorithm. The algorithm based on these rules and conditions is defined as follows:

1. Create all possible C_b-C_f combinations.
2. Filter these combinations by constraints and centering rules.
3. Rank remaining combinations by transitions.

7.5.3.5 Example of Centering Algorithm

U_1: *Jane heard some beautiful music at the CD store.*
U_2: *Jane played it to Mary.*
U_3: *She bought it.*

By applying grammatical role hierarchy to construct C_f. So, for U_1 will have:

Table 7.1 Criteria for each transition in centering algorithm

	$C_b(U_{n+1}) = C_b(U_n)$ or undefined $C_b(U_n)$	$C_b(U_{n+1}) \neq C_b(U_n)$
$C_b(U_{n+1}) = C_p(U_{n+1})$	Continue	Smooth-shift
$C_b(U_{n+1}) \neq C_p(U_{n+1})$	Retain	Rough-shift

$C_f(U_1)$: *{Jane, music, CD store}.*
$C_p(U_1)$: *Jane.*
$C_b(U_1)$: *Undefined.*

U2 has two pronouns: *She* and *it*. *She* is compatible (in syntax) with *Jane*, while *it* is compatible with either *music* or *CD store*.

Since *Jane* is the highest $C_f(U_1)$ ranked member, $C_b(U_2)$ should be referred to *Jane* by comparing result transitions for every possible referent of *it*.

If *it* is assumed to *music*, the result will be:

$C_f(U_2)$: *{Jane, music, Mary}.*
$C_p(U_2)$: *Jane.*
$C_b(U_2)$: *Jane.*

Result: *Continue* (since $C_p(U_2) = C_b(U_2)$ and $C_b(U_1)$ is *undefined*).
On the other hand, if *it* is assumed to *CD store*, the result will be:

$C_f(U_2)$: *{Jane, CD store, Mary}.*
$C_p(U_2)$: *Jane.*
$C_b(U_2)$: *Jane.*

Result: *Continue* (since $C_p(U_2) = C_b(U_2)$ and $C_b(U_1)$ is *undefined*).
As both are *Continue*, it will be set referring to *music* instead of *CD store*.
Next, let's look at U_3.

For U_3, *She* is compatible with either *Jane* or *Mary*, while *it* is compatible with *music*. So, if *she* refers to *Jane*, i.e., $C_b(U_3) = Jane$, the result will be:

$C_f(U_3)$: *{Mary, music}.*
$C_p(U_3)$: *Mary.*
$C_b(U_3)$: *Mary.*

Result: *Smooth-Shift* (since $C_p(U_3) = C_b(U_3)$ but $C_b(U_3) \neq C_b(U_2)$).
Since *Continue* is preferred to *Smooth-shift* using Rule 2, *Jane* should be assigned as the referent, so centering algorithm works in this situation.

7.5.3.6   Performance of Centering Algorithm

Clearly, the *centering algorithm* implicitly accounts for grammatical roles, recency, and repeated-mention preference in pronoun interpretation.

However, the grammatical role hierarchy affects emphasis indirectly because the final conversion type specifically determines the final reference assignment. Confusion can arise if the former leads to a high-level transformation in this case, where a referent in a low-level grammatical role prefers a referent in a high-level role. For instance:

U_1: *Jane opened a new music store in the city.*
U_2: *Mary entered the store and looked at some CDs.*
U_3: *She finally bought some.*

In this example, common sense indicates that *She* in U_3 should refer to *Mary* instead of *Jane*. However, by applying the c*entering algorithm* in this case, it will assign *she* to *Jane* incorrectly because $C_h(U_2) = Jane$ becomes *Continue* while *Mary* becomes a *Smooth-shift*. While applying Hobbs' algorithm, *Mary* will still be assigned as the referent.

Obviously, such situation occurs usually depending on situation and thematic scenario, as Prof. Marilyn A. Walker in her work *"A corpus-based evaluation of centering and pronoun resolution"* (Walker 1989) compared a version of centering to Hobbs on 281 examples from 3 genres of text in 1989 with 77.6% and 81.8% accuracy, respectively.

7.5.4   Machine-Learning Method

7.5.4.1   What Is Machine-Learning Method?

Machine-learning (ML) method is a simple supervised ML by using either stochastic or AI approach. It trains classifier by using manual labeled corpus markers: (1) positive samples are antecedents marked with each pronoun and (2) negative (derived) samples are pairing pronouns with non-antecedent NPs.

In a typical supervised ML scenario, the ML system trains on a set of features and produces a pro-antecedent pair to predict *1* if they corefer and *0* otherwise. A typical example by applying the *log-linear model* for pronominal anaphora resolution is introduced with the following features:

- Strict number [*true or false*].
- Compatible number [*true or false*].
- Strict gender [*true or false*].
- Compatible gender [*true or false*].
- Sentence distance [*0, 1, 2, 3, …*] from pronoun.
- Hobbs' distance [*0, 1, 2, 3, …*] (non-groups).
- Grammatical role [*subject, object, PP*] (taken by potential antecedent).
- Linguistic form [*definite, indefinite and proper pronouns*].

Example for *Pronominal Anaphora Resolution*:

U_1: *Jack saw a beautiful Mercedes GLB300 at a used car dealership.*
U_2: *He showed it to Jim.*
U_3: *He bought it.*

A table of feature vector values for sentence U_2 is shown in Table 7.2.

Table 7.2 Table of feature vector values for sentence U_2: *He showed it to Jim*

Feature	He(U2)	it(U2)	Jim(U2)	Jack(U1)
Strict number	1	1	1	1
Compatible number	1	1	1	1
Strict gender	1	0	1	1
Compatible gender	1	0	1	1
Sentence distance	1	1	1	2
Hobbs distance	2	1	0	3
Grammatical role	Subject	Subject	PP	Subject
Linguistic form	Pronoun	Pronoun	Proper	Proper

7.5.4.2   Performance of the Log-Linear Model

A *log-linear model* trains on vectors and filters out pleonastic *it* as in *it is raining*. It results in weights for each and the combination of features. Most of the time, it is rigid and harder and must decide if any two NPs corefer.

New features can be added to improve model performance such as:

- Anaphor edits distance.
- Antecedent edits distance.
- Alias [*true or false*] (based on the named entity tagger).
- Appositive [*true or false*].
- Linguistic form [*proper, definite, indefinite, pronoun*].

7.5.4.3   Other Advanced ML Models

Big data and *AI* offer advancement for current ML models. CR research focuses on *convolutional neural networks (CNN)* (Auliarachman and Purwarianti 2019), *recurrent neural networks (RNN)* (Afsharizadeh et al. 2021), *long short-term memory networks (LSTM)* (Li et al. 2021), *transformers*, and *BERT models* (Joshi et al. 2019), which will be discussed in Chap. 9.

7.6   Evaluation

From performance perspective, commonly used methods emphasize on coreference chain evaluation as forming a set of facts A, B, and C that are assigned with A, B, and C classes. They consist of two data types: (1) reference/true chain is correct or true coreference chain occurred in an entity and (2) hypothesis chain/class is assigned with the entity by a coreference algorithm.

For instance, *precision* of the system can be evaluated according to:

$$\frac{\text{weighted sum of correct elments in hypothesis chain}}{\text{Number of elements in hypothesis chain}} \qquad (7.2)$$

and *recall* can be evaluated according to:

$$\frac{\text{Number of correct elements in hypothesis chain}}{\text{Number of elements in reference chain}} \qquad (7.3)$$

Like previous chapters on N-gram and semantic analysis, CR model evaluation can be achieved by using: (1) intrinsic (using prototype and model itself) vs. (2) extrinsic (task-based, end-to-end) evaluation schemes.

Exercises

7.1. What is *pragmatic analysis* and *discourse* in linguistics? Discuss their roles and importance in NLP.

7.2. What is the difference between *pragmatic analysis* and *semantic analysis* in terms of their functions and roles in *natural language understanding* (NLU)?

7.3. What is *CR* in linguistics? Why it is important in NLP? Use two examples to illustrate and support your answer.

7.4. State and explain the differences between the concept of *coherence* vs. *coreference* in *pragmatic analysis*. Give two examples to support your answer.

7.5. What is *discourse segmentation*? State and explain why it is vital to *pragmatic analysis* and the implementation of NLP application such *Q&A chatbot*. Give two examples to support your answer.

7.6. State and explain *Hearst's TextTiling* technique on *discourse segmentation*. How can it be further improved by using nowadays' AI and ML technology?

7.7. What is *coherence relation*? State and explain five basic types of *coherence relations*. For each type, give an example to illustrate.

7.8. What is *referencing expression* in *pragmatic analysis*? State and explain five basic types of *referencing expressions*. For each type, give an example to illustrate.

7.9. State and explain *Hobbs' algorithm* for coreference resolution. Use a sample sentence/utterance (other than the one given in the book) to illustrate how it works.

7.10. State and explain the pros and cons of *Hobbs' algorithms* for *CR*. Use example(s) to support your answer.

7.11. State and explain *centering algorithm* for coreference resolution. Use a sample sentence/utterance (other than the one given in the book) to illustrate how it works.

7.12. Compare pros and cons between *Hobbs' algorithm* vs. *centering algorithm*. Use example(s) to support your answer.

7.13. What is ML? State and explain how ML can be used for coreference resolution. Use example(s) to support your answer.

7.14. Name any three types of ML models for *CR*. State and explain how they work.

7.15. Name any two types of evaluation method/metrics for *CR* model in *pragmatic analysis*. State and explain how they work.

References

Afsharizadeh, M., Ebrahimpour-Komleh, H., and Bagheri, A. (2021). Automatic text summarization of COVID-19 research articles using recurrent neural networks and coreference resolution. Frontiers in Biomedical Technologies. https://doi.org/10.18502/fbt.v7i4.5321

Auliarachman, T., & Purwarianti, A. (2019). Coreference resolution system for Indonesian text with mention pair method and singleton exclusion using convolutional neural network. Paper presented at the 1-5. https://doi.org/10.1109/ICAICTA.2019.8904261

Bender, E. M. and Lascarides, A. (2019) Linguistic Fundamentals for Natural Language Processing II: 100 Essentials from Semantics and Pragmatics (Synthesis Lectures on Human Language Technologies). Springer.

Cornish, F. (2009). Inter-sentential anaphora and coherence relations in discourse: A perfect match. Language Sciences (Oxford), 31(5), 572-592.

Cruse, A. (2011) Meaning in Language: An Introduction to Semantics and Pragmatics (Oxford Textbooks in Linguistics). Oxford University Press

Doyle, A. C. (2019) The Adventures of Sherlock Holmes (AmazonClassics Edition). AmazonClassics.

Goddard, C. (1998) Semantic Analysis: A Practical Introduction (Oxford Textbooks in Linguistics). Oxford University Press.

Grosz, B. J., Joshi, A. K., and Weinstein, S. (1995). Centering: A framework for modeling the local coherence of discourse. Computational Linguistics - Association for Computational Linguistics, 21(2), 203-225.

Grosz, B. J., and Sidner, C. L. (1986). Attention, intentions, and the structure of discourse. Computational Linguistics - Association for Computational Linguistics, 12(3), 175-204.

Hearst, M. A. (1997). TextTiling: Segmenting text into multi-paragraph subtopic passages. Computational Linguistics - Association for Computational Linguistics, 23(1), 33-64.

Hobbs, J. R. (1979) Coherence and Coreference. Cognitive Science 3, 67-90.

Hobbs, J. R. (1978) Resolving pronoun references. Lingua, 44:311–338.

Joshi, M., Levy, O., Weld, D.S., and Zettlemoyer, L. (2019) BERT for Coreference Resolution: Baselines and Analysis. In Proc. of Empirical Methods in Natural Language Processing (EMNLP) 2019. https://doi.org/10.48550/arXiv.1908.09091

Kehler, A. (2002) Coherence, Reference, and the Theory of Grammar. Stanford, Calif.: CSLI Publishers.

Kehler, A., Kertz, L., Rohde, H., and Elman, J. L. (2008). Coherence and coreference revisited. Journal of Semantics (Nijmegen), 25(1), 1-44.

Kroeger, P. (2019) Analyzing meaning: An introduction to semantics and pragmatics (Textbooks in Language Sciences). Freie Universität Berlin.

Lata, K., Singh, P., & Dutta, K. (2022). Mention detection in coreference resolution: Survey. Applied Intelligence (Dordrecht, Netherlands), 52(9), 9816-9860.

Li, Y., Ma, X., Zhou, X., Cheng, P., He, K. and Li, C. (2021). Knowledge enhanced LSTM for coreference resolution on biomedical texts. Bioinformatics, 37(17), 2699-2705. https://doi.org/10.1093/bioinformatics/btab153

Mann, W. C. and Thompson, S. A. (1988) Rhetorical Structure Theory: Toward a functional theory of text organization. Text & Talk, 8, 243 - 281.

Mann, W. C. and Thompson S. A. (1986) Relational Propositions in Discourse. Discourse Processes 9: 57-90.

Pevzner, L., and Hearst, M. A. (2002). A critique and improvement of an evaluation metric for text segmentation. Computational Linguistics - Association for Computational Linguistics, 28(1), 19-36.

Sanders, T., Spooren, W. and Noordman, L.G. (1992). Toward a taxonomy of coherence relations. Discourse Processes, 15, 1-35.

Tetreault, J. R. (2001). A corpus-based evaluation of centering and pronoun resolution. Computational Linguistics. Association for Computational Linguistics, 27(4), 507-520.

Walker, Marilyn A. (1989). Evaluating discourse processing algorithms. In Proceedings of the 27th Annual Meeting of the Association for Computational Linguistics, pp. 251-261.

Wolna, A., Durlik, J., and Wodniecka, Z. (2022). Pronominal anaphora resolution in polish: Investigating online sentence interpretation using eye-tracking. PloS One, 17(1), e0262459-e0262459.

Chapter 8
Transfer Learning and Transformer Technology

8.1 What Is Transfer Learning?

Transfer learning (TL) involves solving a problem by leveraging acquired knowledge and applying that knowledge to address another related problem (Pan and Yang 2009; Weiss et al. 2016; Zhuang et al. 2020). It can be likened to two students learning to play the guitar, where one already has musical knowledge while the other does not. Naturally, the student with a background in music can apply that knowledge to the new learning process. In traditional *machine learning (ML)*, each task is associated with its own isolated dataset and trained model. In contrast, TL allows for the learning of a new task by building on the knowledge gained from previously learned tasks, often utilizing larger datasets, as illustrated in Fig. 8.1.

8.2 Motivation of TL

Traditional ML datasets and trained model parameters cannot be reused. They involve enormous, rare, inaccessible, time-consuming, and costly training processes in NLP tasks and computer vision. For example, if a task is text sentiment review predictions on laptops, there are large amounts of labeled data, target data, and training data from these reviews.

Traditional ML can work well on correlated domains, but when there are large amounts of target data like food reviews, the inference results will be unsatisfactory due to domain differences. Nevertheless, these domains are correlated in some sense to bear same domain reviews as language characteristics and terminology expressions, which makes TL possible to apply in a high-level approach to the prediction task. This approach enables source domains to become a target domain and determine its subdomain correlations as shown in Fig. 8.2.

R. Lee, *Natural Language Processing*, https://doi.org/10.1007/978-981-96-3208-4_8

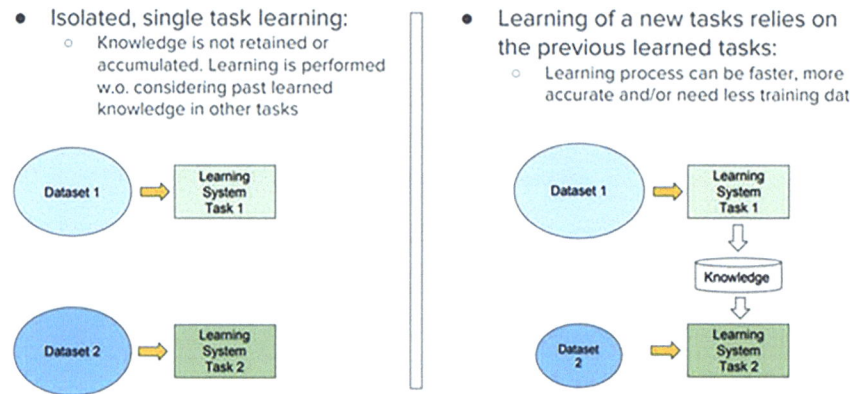

Fig. 8.1 Traditional machine learning vs. transfer learning

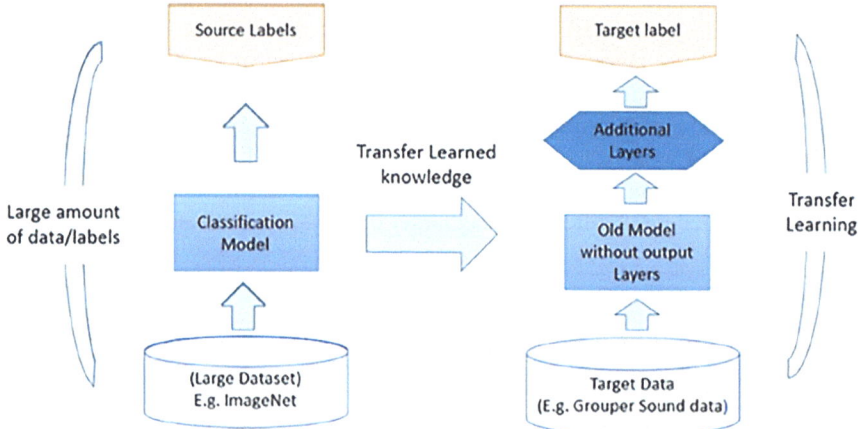

Fig. 8.2 Transfer learning

TL has been implemented to several ML applications such as image and text sentiment classifications.

8.2.1 Categories of TL

The domain is to be assigned with a definition by feature space X and marginal probability distribution $P(X)$ where $X = \{x_1, x_2, x_3, ..., x_n\} \in X$.

If a feature space X and distribution $P(X)$ between two domains are different, they are different domains.

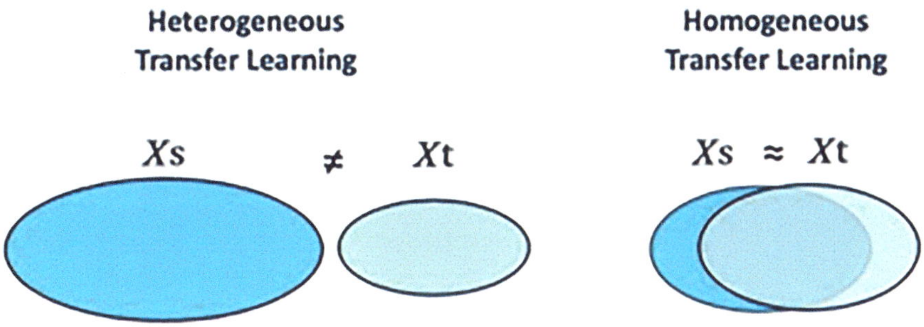

Fig. 8.3 Two categories of transfer learning

If a task is defined by a label space Y with a predictive function $f(\cdot)$, $f(\cdot)$ is represented by a conditional probability distribution given by (8.1):

$$f(x_i) = P(y_i|x_i) \tag{8.1}$$

If a function $f(\cdot)$ and label space Y between two tasks are different, they are different tasks.

Now TL can give a new representation by above definitions that have D_s as source domain and T_s as source learning task. D_t represents target domain, and T_t represents target learning task. Given two domains are unidentical or have two different tasks, TL aim is to improve the results $P(Y_t|X_t)$ of D_t when T_s and D_s knowledge can be obtained. There are two types of TL: (1) *heterogeneous* and (2) *homogeneous* as shown in Fig. 8.3.

Heterogeneous TL: when source feature space and feature space are different which means that $Y_t \neq Y_s$ and/or $X_t \neq X_s$. Under the condition of same domain distributions, the strategy of resolution is to adjust feature space smaller and transform it to homogeneous so that the differences between marginal or conditional of source, and target domains will be reduced.

Homogeneous TL: when there are conditions $X_t = X_s$ and $Y_t = Y_s$, the difference between two domains lies in data distributions. Three strategies are commonly used to tackle homogeneous TL problems: (1) reduction in the differences of $P(X_t) \neq P(X_s)$, (2) reduction in the differences of $P(Y_t|X_t) \neq P(Y_s|X_s)$, and (3) the combination of strategies (1) and (2).

8.3 Solutions of TL

There are four methods to solve problems produced by homogeneous and heterogeneous TL: (1) *instance-based*, (2) *feature-based*, (3) *parameter-based*, and (4) *relational-based methods*.

8.3.1   Instance-Based Method

This method reweights samples from source domains and uses them as target domain data to bridge the gap of marginal distribution differences which works best when conditional distributions of two tasks are equal.

8.3.2   Feature-Based Method

This method works for both heterogeneous and homogeneous TL problems. For homogeneous types, it is to bridge the gap between conditional and marginal distributions of target and source domains. For heterogeneous types, it is to reduce the differences between source and target feature spaces. It has two approaches (a) asymmetric and (b) symmetric.

(a) *Asymmetric feature transformation* aims to modify the source domain and reduce the gap between source and target instances by transforming one of the source and target domains to the other as shown in Fig. 8.4. It can be applied when Y_s and Y_t are identical.
(b) *Symmetric feature transformation* aims to transform source and target domains into their shared feature space, starting from the idea of discovering meaningful structures between domains. The feature space they share is usually low-dimensional. The purpose of this approach is to reduce the marginal distribution distance between destination and source. The difference between symmetric and asymmetric feature transformation is shown in Fig. 8.5.

8.3.3   Parameter-Based Method

This method transfers learned knowledge by sharing parameters common to the models of source and target learners. It applies to the idea that two related tasks have similarity in model structure. The trained model is transferred from source domain

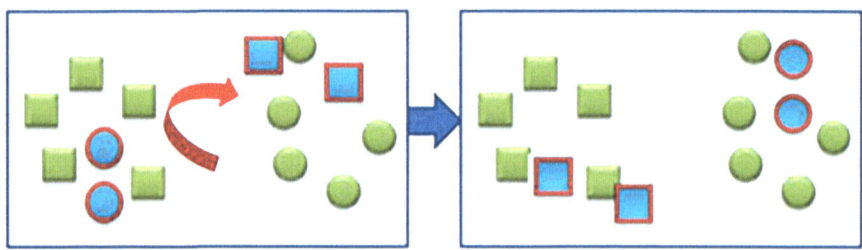

Fig. 8.4 Asymmetric feature transformation

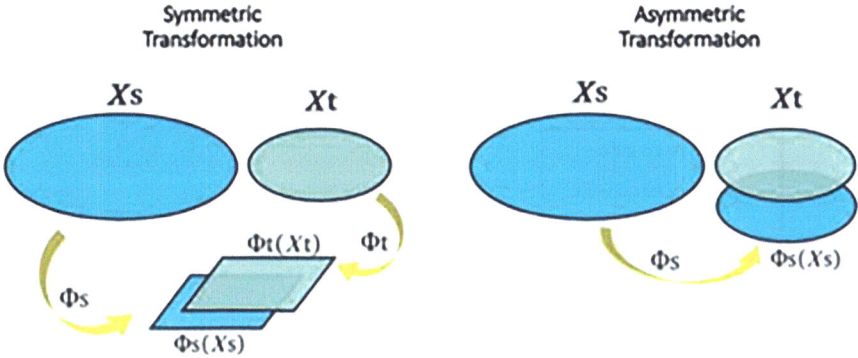

Fig. 8.5 Symmetric feature transformation (left) and asymmetric feature transformation (right)

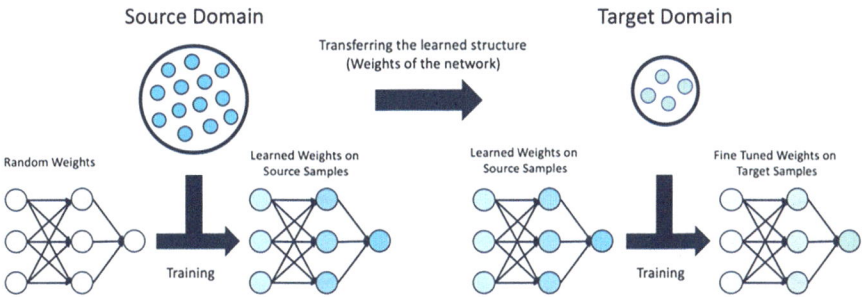

Fig. 8.6 Parameter-based methods

to target domain with parameters. This approach has a huge advantage because the parameters are usually trained from randomly initialized parameters as the training process can be time-consuming for models trained from the beginning. This approach can train more than one model on the source data and combine parameters learned from all models to improve results of the target learner. It is often used in deep learning applications as shown in Fig. 8.6.

8.3.4 Relational-Based Method

This method transfers learned knowledge by sharing its learned relations between different sample parts of source and target domains as shown in Fig. 8.7. Food and movie domains are a related domain example. Although the review texts are different, sentence structures are similar. It aims to transfer learned relations of different review sentence parts from these domains to improve text sentiment analysis results.

Domain	Reviews
Food	The *food* is **delicious.** I highly **recommend** this *pasta.* This is very **amazing** *meal.*
Movie	The *movie* is **great.** I **love** this *movie.* *Godfather* was the most **amazing** *movie.*

Reviews in movie and food domains. Boldfaces are topic words and Italics are sentiment words.

Camera Domain

Movie Domain

Dependency tree structure.

Fig. 8.7 Relational-based approaches: an example of learning sentence structure of food reviews to help with movie reviews' sentiment analysis

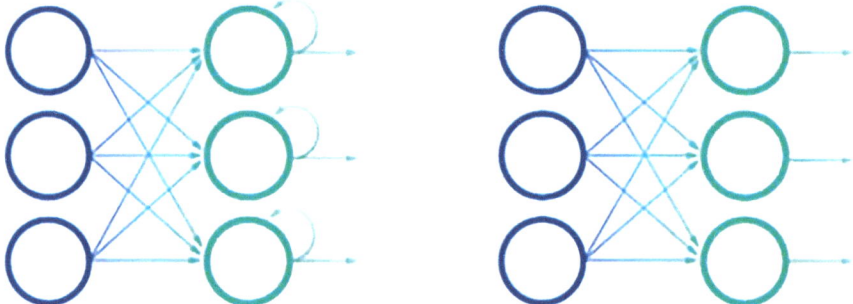

Fig. 8.8 Recurrent neural network (left) vs. feedforward neural network (right)

8.4 Recurrent Neural Network (RNN)

8.4.1 What Is RNN?

Recurrent neural network (RNN) is a class of artificial neural networks (ANNs) to consider time series or sequential data as input and use them as prior inputs to produce current input and output (Cho et al. 2014; Sherstinsky 2020; Yin et al. 2017). The RNN has *memory* which means its output is influenced by prior elements of the sequence against traditional *feedforward neural network (FNN)* with independent inputs and outputs as shown in Fig. 8.8.

8.4.2 Motivation of the RNN

Many learning tasks require sequential data processing, including speech recognition, image captioning, and synchronized sequences in video classification. While sentiment analysis and machine translation generate sequence-based outputs, the

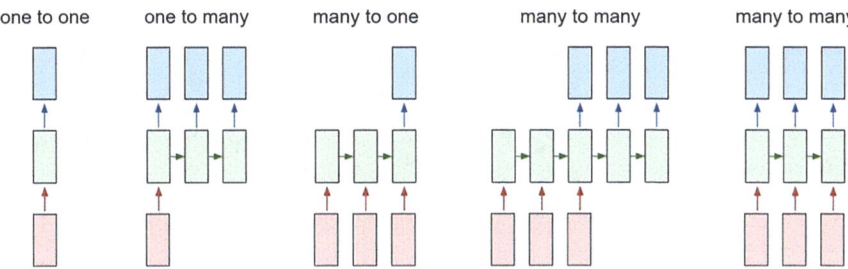

Fig. 8.9 Five major types of RNNs

Fig. 8.10 Basic
architecture of the RNN

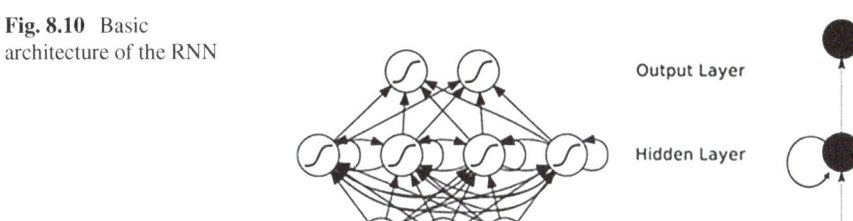

inputs for these tasks are time- or space-dependent, which cannot be effectively modeled by traditional neural networks that assume test and training data are independent.

For example, a language translation task aims to translate a phrase that *feel under the weather* means *unwell*. This phrase makes sense only when it is expressed in that specific order. Thus, the positions of each word in sentence must be considered when model predicts the next word.

There are five major categories of RNN architecture corresponding to different tasks: (1) simple one-to-one model for image classification task, (2) one-to-many model for image captioning tasks, (3) many-to-one model for sentiment analysis tasks, (4) many-to-many models for machine translation, and (5) complex many-to-many models for video classification tasks as shown in Fig. 8.9.

8.4.3 RNN Architecture

The RNN like standard neural networks consists of input, hidden, and output layers as shown in Fig. 8.10.

An unfolded *RNN architecture* is narrated by x_t as the input at time step t, s_t stores the values of hidden units/states at time t, and o_t is the output of the network at time step t. U are weights of inputs, Ws are weights of hidden units, and V is the bias as shown in Fig. 8.11.

Fig. 8.11 Unfolded RNN architecture

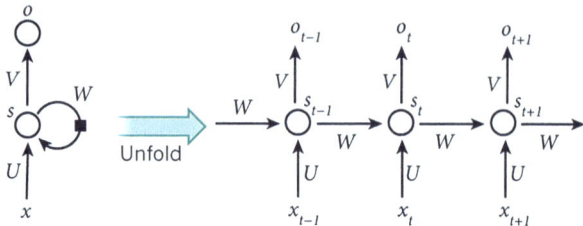

With the activation function f, the hidden states s_t are calculated by equation:

$$s_t = f\left(Ux_t + Ws_{t-1}\right) \tag{8.2}$$

The output of each recurrent layer o_t is calculated by equation:

$$o_t = \text{softmax}\left(Vs_t\right) \tag{8.3}$$

The hidden states s_t are considered as network memory units which consist of hidden states from several former layers. Each layer's output is only related to hidden states of the current layer. A significant difference between RNN and traditional neural networks is that weights and bias U, W, and V are shared among layers.

There will be an output at each step of the network but unnecessary. For instance, if inference is applied for sentiment expressed by a sentence, only an output is required when the last word is input, and none after each word for input. The key to RNNs is the hidden layer to capture sequence information.

For RNN feedforward process, if the number of time steps is k, then hidden unit values and output will be computed after $k + 1$ *time* steps. For backward process, the RNN applies an algorithm called *backpropagation through time (BPTT)*.

RNN topologies range from partly to fully recurrent. Partly recurrent is a layered network with distinct output and input layers where recurrence is limited to the hidden layer. Fully connected recurrent neural network (FRNN) connects all neurons' outputs to inputs as shown in Fig. 8.12.

8.4.4 Long Short-Term Memory (LSTM) Network

8.4.4.1 What Is LSTM?

Long short-term memory (LSTM) network (Staudemeyer and Morris 2019; Yu et al. 2019) is a type of the RNN with special hidden layers to deal with gradient explosion and disappearance problems during long sequence training process proposed by Hochreiter and Schmidhuber (1997). LSTM has better performance with training longer sequences against naïve RNNs.

LSTM and naïve RNN structure frameworks are shown in Fig. 8.13.

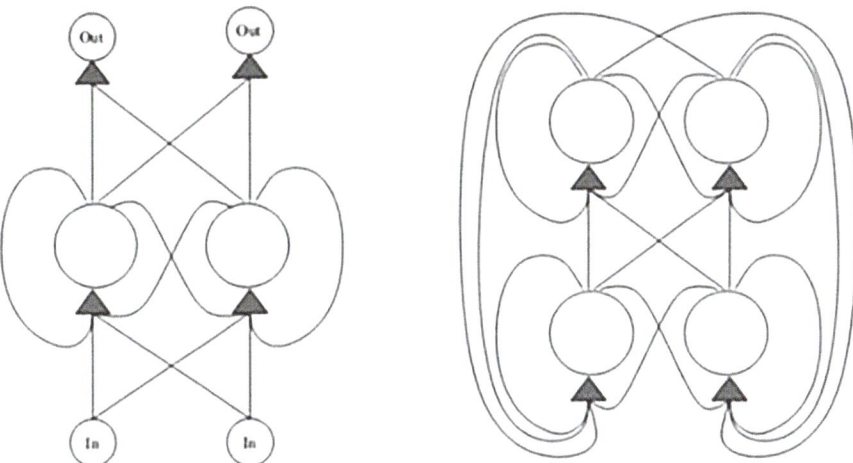

Fig. 8.12 Simple recurrent neural network (left) and fully connected recurrent neural network (right)

Fig. 8.13 Standard RNNs (left) and LSTM (right)

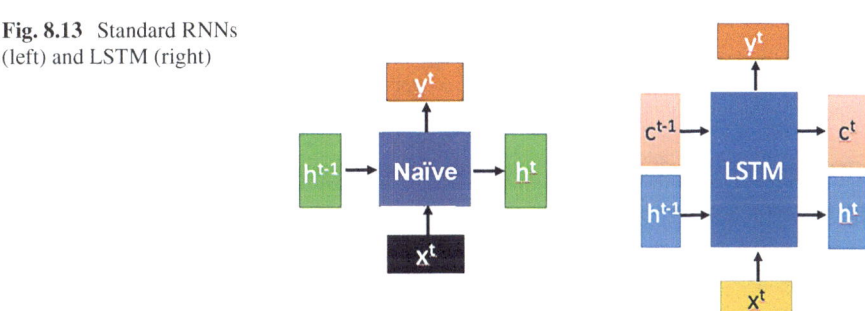

LSTM has two hidden layers as the RNN where a memory cell in the layer is to replace the hidden node. The RNN has only one transfer state h^t as compared to the RNN. There are two transfer states, c^t (cell state) and h^t (hidden state), in LSTM. RNN's h^t corresponds to LSTM's c^t. c^t passed down information among them, and output c^t is produced by adding c^{t-1} passed from state and values of the previous step. RNN's h^t has larger difference among nodes usually.

8.4.4.2 LSTM Architecture

x^t and h^{t-1} are concatenated inputs from the state of the previous step to train with activations for four states as shown in Fig. 8.14.

z is input calculated by nated vector with weights w and converted into values $0–1$ through activation function tanh. $\mathbf{z}^f, \mathbf{z}^i, \mathbf{z}^o$ are calculated by multiplying the concatenated vector with corresponding weights and converting to values $0–1$ by a

Fig. 8.14 Four states of LSTM

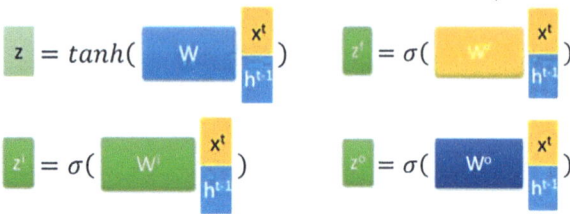

Fig. 8.15 Calculations in memory cell of LSTM

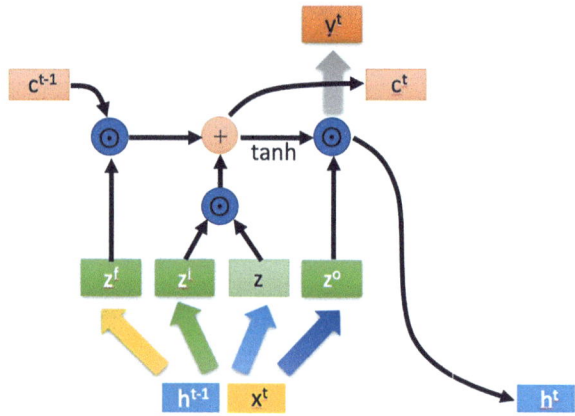

sigmoid function σ to generate gate states. z^f represents the forget gate, z^i represents the input gate, and z^o represents output gate. A memory cell of LSTM calculation is shown in Fig. 8.15.

Memory cells c^t, h^t, y^t are calculated by gate states as equations below: ($\odot$ is the Hadamard product)

$$c^t = z^f \odot c^{t-1} + z^i \odot z$$
$$h^t = z^o \odot \tanh\left(c^t\right) \tag{8.4}$$
$$y^t = \sigma\left(W'h^t\right)$$

LSTM has (1) forget, (2) memory select, and (3) output stages.

1. **Forget stage**

   This stage retains important information passed in by previous node ct−1 (the previous cell state) and discards unimportant ones. The calculated $\mathbf{z^f}$ is used as a forget gate to control what type of c^{t-1} information should be retained or discarded.

2. **Memory selection stage**

   This stage *remembers* input x^t selectively to record important information. z refers to present input. z^i is the input gate to control gating signals.

3. **Output stage**

Fig. 8.16 General architecture of the GRU

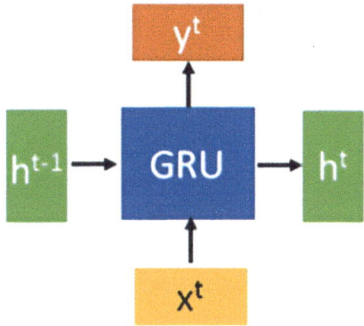

This stage determines what is considered as h^t (the current state) to be passed down to the next layer. z^o is the output gate to control this process before c^t is scaled from the memory select stage (convert through a tanh function).

Each layer output y^t is calculated by multiplying weights with h^t and converting the product through an activation function like the RNN; the cell state c^t is passed to the next layer at the end of each layer.

8.4.5 Gate Recurrent Unit (GRU)

8.4.5.1 What Is GRU?

Gate Recurrent Unit (GRU) can be considered as a kind of the RNN like LSTM but to manage backpropagation gradient problems (Chung et al. 2014; Dey and Salem 2017). GRU proposed in 2014, and LSTM proposed in 1997 had similar performances in many cases, but the former is often exercised due to simple calculation with comparable results than the latter.

GRU's input and output structures are like the RNN. There are inputs x^t and h^{t-1} to contain relevant information of the prior node. Current outputs y^t and h^t are calculated by combining x^t and h^{t-1}. A GRU architecture is shown in Fig. 8.16.

8.4.5.2 GRU Inner Architecture

r is the reset gate, and z is the update gate. They are concatenated with input x^t and hidden state h^{t-1} from the prior node and multiply results with weights as shown in Fig. 8.17.

When a gate control signal is available, apply r reset gate to obtain data $h^{t-1} = h^{t-1} \odot r$ after reset, h^{t-1} is concatenated with x^t Apply a tanh function to generate data that lies within range $(-1,1)$ as shown in Fig. 8.18.

At this point, h' contains current input x^t; its selection memory stage is like LSTM.

Fig. 8.17 Reset and
update gates of GRU

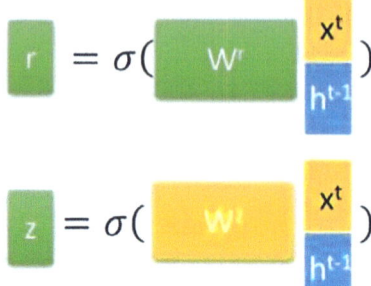

Fig. 8.18 Computation of **h**

$$h' = tanh(\ \boxed{W}\ \begin{matrix} x^t \\ h^{t-1'} \end{matrix}\ )$$

Finally, update memory stage is the most critical step where forget and remember steps are performed simultaneously. The gate z obtained earlier is applied as:

$$h_t = (1 - z) \odot h_{t-1} + z \odot h' \tag{8.5}$$

where z (gate signal) is within the range 0~1. If it is close to 1 or 0, it signifies more data has remained or forgotten, respectively.

(1 − z) ⊙ *h^{t-1}* represents the calculation to forget the original hidden state selectively. *(1 − z)* is considered as a forget gate to forget *h^{t-1}* unimportant information.

z ⊙ *h*′ represents *h*′ memory selective information of the present node. Like *(1 − z)*, it will forget **h**′ unimportant information or is considered as selective *h*′ information.

h^t = (1 − z) ⊙ *h^{t-1}* + z ⊙ *h*′ is the calculation to forget *h^{t-1}* information from passed down and add information from the current node.

It is noted that forget z and select *(1 − z)* factors are linked, which means it will forget the passed in information selectively. When weights *(z)* are forgotten, it will apply weights in *h*′ to configurate *(1 − z)* at a constant state.

GRU's input and output structures are like the RNN, and its internal concept is like LSTM. GRU has one less internal gate as compared to LSTM and fewer parameters but can achieve comparable satisfactory results with reduced time and computational resources. A GRU computation module is shown in Fig. 8.19.

Fig. 8.19 Computation module of GRU

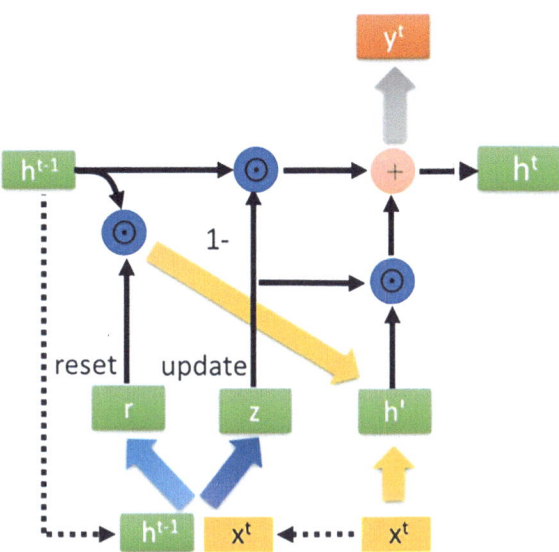

8.4.6 Bidirectional Recurrent Neural Networks (BRNNs)

8.4.6.1 What Is BRNN?

Bidirectional Recurrent Neural Network (BRNN) is a type with RNN layers in two directions (Singh et al. 2016). It links with previous and subsequent information outputs to perform inference against both RNN and LSTM to possess information from the previous one. For example, in text summarization, it is insufficient to consider the information from the previous content; sometimes, it also requires subsequent text information for word prediction of a sentence. The BRNN is proposed to deal with these circumstances.

The BRNN consists of two RNNs superimposed on top of each other. The output is mutually generated by two RNN states. A BRNN structure is shown in Fig. 8.20.

BRNN training process is as follows:

1. Begin forward propagation from time step *1* to time step *T* to calculate hidden layer's output and save at each time step.
2. Proceed from time step *T* to time step *t* to calculate backward hidden layer output and save at each time step.
3. Obtain each moment final output according to forward and backward hidden layers after calculating all input moments from both forward and backward directions.

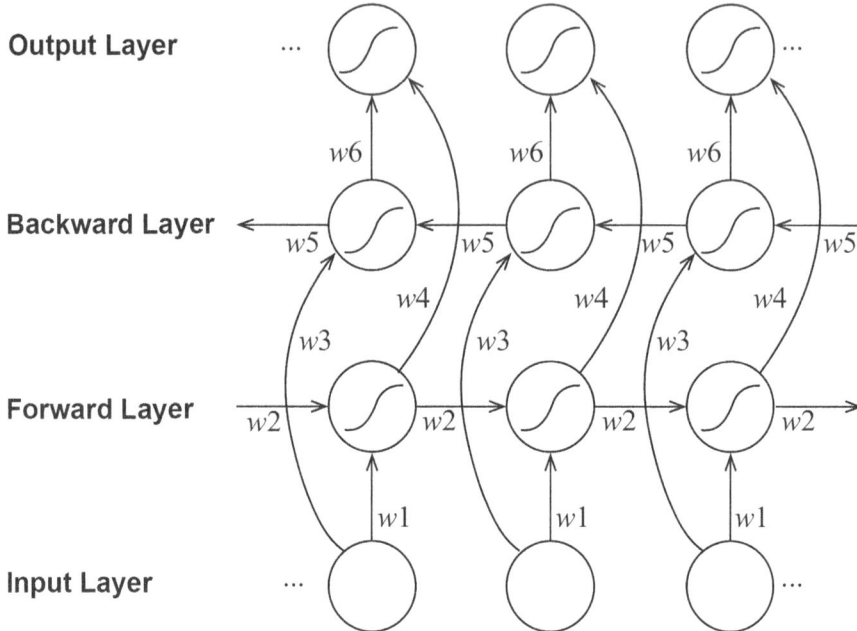

Fig. 8.20 Structure of the BRNN

8.5   Transformer Technology

8.5.1   *What Is Transformer?*

The *Transformer* is a network architecture based on the attention mechanism, without relying on recurrent or convolutional units (Vaswani et al. 2017). Transformer and *LSTM* models differ in their training processes. LSTM models are serial and iterative, which means they cannot proceed to the next word until the previous one has been processed. In contrast, the Transformer processes all words in parallel, allowing for simultaneous processing, which enhances computational efficiency. The structure of a Transformer system is illustrated in Fig. 8.21.

8.5.2   *Transformer Architecture*

A transformer model has two parts: (1) *encoder* and (2) *decoder*. Language sequence extracts as input, encoder maps it into a hidden layer, and decoder maps the hidden layer inversely to a sequence as output.

Fig. 8.21 Transformer architecture

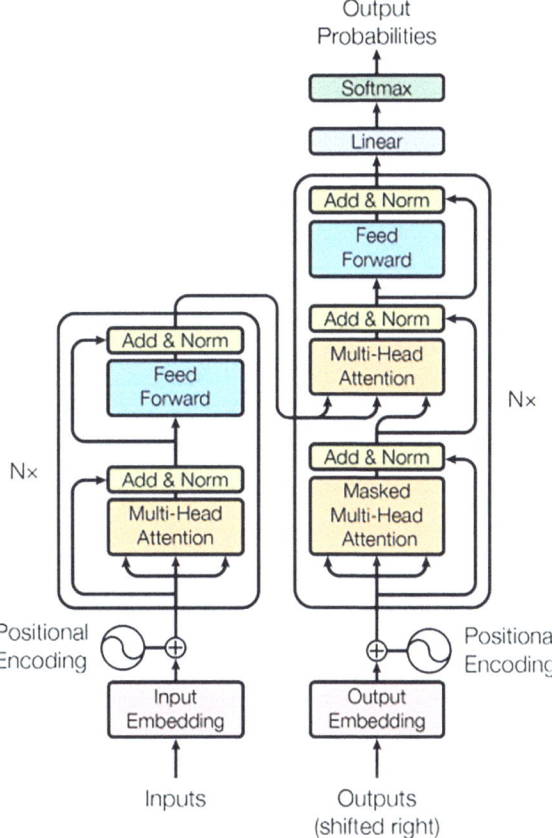

8.5.2.1 Encoder

There are six identical encoder layers in the transformer with two sublayers: (1) self-attention and (2) feedforward in each encoder layer. The self-attention layer is the first sublayer to exercise attention mechanism, and a simple fully connected feedforward network is the second sublayer. There follow a residual connection and layer normalization from each of the sublayers. An encoder layer architecture is shown in Fig. 8.22.

8.5.2.2 Decoder

There are six identical encoder layers in the transformer. In addition to identical two sublayers as each encoding layer, a third sublayer is added to the decoder to perform multi-head attention, taking the output of last encoder layer as input. Residual connections and layer normalization are used sequentially for all sublayers, which is the

Fig. 8.22 Architecture of an encoder layer

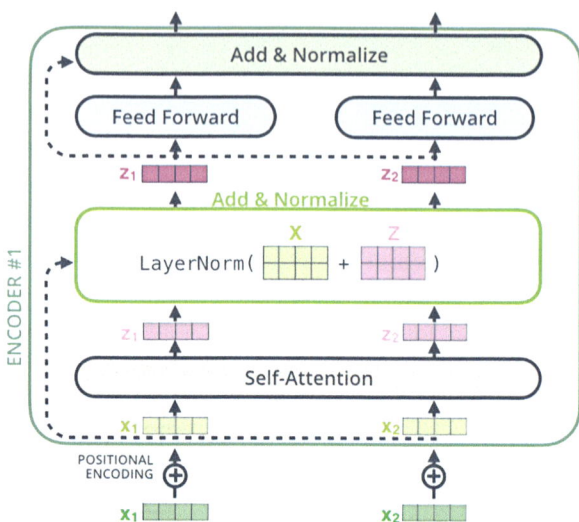

same as the encoder. The decoder's self-awareness is modified by the mask to ensure that inference of the position can only use information from a known position, or in other words, its previous position.

8.5.3   Deep into Encoder

8.5.3.1   Positional Encoding

Since transformer has no iterative process, each word's position information must be provided to ensure that it can recognize the position relationship in language. Linear transformation of sin and cos functions is applied to provide model position information as equation:

$$PE(pos, 2i) = \sin\left(pos / 10{,}000^{2i/d_{model}}\right)$$
$$PE(pos, 2i+1) = \cos\left(pos / 10{,}000^{2i/d_{model}}\right)$$

(8.6)

where pos represents a word's position in a sentence, i represents word vector's dimension number, and d_{model} represents embedded dimension's value. There is a set of formulas such as sets of 0, 1, or 2, 3 processed with the above sum function, respectively. As the dimension number increases, the period changes moderately to generate a texture containing position information.

8.5.3.2   Self-Attention Mechanism

For input sentence, the word vector of each word is obtained through word embedding, and the position vector of all words is obtained in same dimensions through positional encoding that can be added directly to obtain the true vector representation. ith word's vector is written as x_i, X is the input matrix combined by all word vectors. ith row refers to the ith word vector.

W_Q, W_K, W_V are matrices defined to perform three linear transformations with X to generate three matrices Q (queries), K (keys), and V (values), respectively.

$$
\begin{aligned}
Q &= X \cdot W_Q \\
K &= X \cdot W_K \\
V &= X \cdot W_V
\end{aligned}
\tag{8.7}
$$

Attention mechanism computation can be described as:

$$
\text{Attention}(Q,K,V) = \text{softmax}\left(\frac{QK^T}{\sqrt{d_k}}\right)V
\tag{8.8}
$$

The dot products are calculated by multiplying query Q by keys K, dividing the result by $\sqrt{d_k}$, and applying a softmax function to obtain value scores V.

8.5.3.3   Multi-head Attention

The previously defined set of Q, K, V allows a word to use the information of related words. Multiple Q, K, V defined groups can enable a word to represent subspaces at different positions with identical calculation process, except that the matrix of linear transformation has changed from one group (W_Q, W_K, W_V) to multiple groups (W_Q^0, W_K^0, W_V^0), (W_Q^1, W_K^1, W_V^1) ... as equation:

$$
\begin{aligned}
\text{MultiHead}(Q,K,V) &= \text{Concat}(\text{head1},\ldots,\text{head}h) \cdot WO \\
\text{where head}i &= \text{Attention}(XW_Q^i, XW_K^i, XW_V^i)
\end{aligned}
\tag{8.9}
$$

where W_O is the weights of concatenated results.

Adding input with a sublayer (self-attention layer for example) to generate residual connections as equation:

$$
X_{\text{attention}} = X_{\text{embedding}} + \text{Attention}(Q,K,V)
\tag{8.10}
$$

8.5.3.4   Layer Normalization of Attention Sublayer

Layer normalization is to standardize the distribution of hidden layers independently to improve convergence and training processes effectively.

$$X_{\text{attention}} = \text{LayerNorm}\left(X_{\text{attention}}\right) \tag{8.11}$$

8.5.3.5   Feedforward Layer

It is a two-layer linear map with an activation function, i.e., ReLU.

$$X_{\text{hidden}} = \text{Linear}\left(\text{ReLU}\left(\text{Linear}\left(X_{\text{attention}}\right)\right)\right)$$

followed by residual connection and layer normalization scheme:

$$\begin{aligned} X_{\text{hidden}} &= X_{\text{attention}} + X_{\text{hidden}} \\ X_{\text{hidden}} &= \text{LayerNorm}\left(X_{\text{hidden}}\right) \end{aligned} \tag{8.12}$$

8.6   BERT

8.6.1   What Is BERT?

BERT is a pretrained model of language representation called *Bidirectional Encoder Representation from Transformers* (Devlin et al. 2018). It uses masked language model (MLM) to generate deep bidirectional linguistic representation instead of the traditional one-direction model or concatenated two one-direction models to pretrain language.

8.6.2   Architecture of BERT

BERT models are pretrained either by left-to-right or right-to-left language models previously; this unidirectional property restricts model structure to obtain unidirectional context information only and propensity for representation. BERT adopted MLM in pretraining stage and a bidirectional transformer with deep layers to build the entire model; the representation generated integrates both left and right content information. A BERT system architecture is shown in Fig. 8.23.

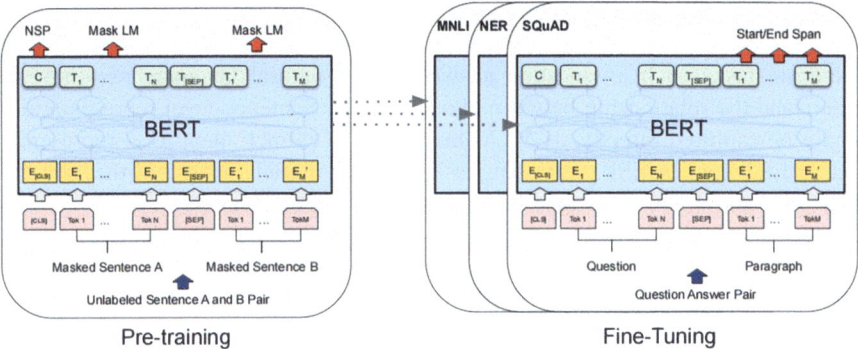

Fig. 8.23 System architecture of BERT

8.6.3 Training of BERT

BERT has two training process steps: (1) *pretraining* and (2) *fine-tuning*.

8.6.3.1 Pretraining BERT

BERT is not constrained by a one-way language model because it randomly replaces tokens in each training sequence with mask tokens ([MASK]) with 15% probability to predict the original word at position [MASK]. [MASK] do not appear in fine-tuning of downstream tasks, leading to differences in pretraining and fine-tuning stages, because the pretraining objective improves language representation, being sensitive to [MASK] and to other insensitive tokens. BERT applies the following strategies:

First, in each training sequence, a token position is randomly selected for prediction with a probability of 15%. If ith token is selected, it will be replaced by one of the following tokens:

1. 80% is [MASK]. For instance, *the cat is adorable → the cat is [MASK]*.
2. 10% is a random token. For instance, *the cat is adorable → the cat is ginger*.
3. 10% is the original token (no change). For instance, *his cat is adorable → his cat is adorable*.

Second, apply T_i corresponding to the position, predict the original token through full connection, then apply softmax to output the probability of each token, and finally apply cross-entropy to evaluate loss.

This method causes BERT sensitive to *[MASK]* and all tokens to extract representative information.

8.6.3.2   Next Sentence Prediction (NSP)

There are tasks such as question answering and natural language reasoning to understand the relationship between two sentences. Sentence-level representations cannot be captured directly, as MLM tasks tend to extract token-level representations. BERT applies NSP pretraining task to let the model understand the relationships between sentences and predict whether they are connected.

For every training sample, select Set A and B from corpus to create a sample, where Set A is 50% of Set B (labeled "IsNext"), and Set B is 50% random. Next, training examples are put into the BERT model to generate binary classification predictions.

8.6.3.3   Fine-Tuning BERT

It is necessary to add an additional output layer to fine-tune downstream tasks for satisfactory performance. It does not require task-specific structural modification in this process.

8.7   Other Related Transformer Technology

8.7.1   Transformer-XL

8.7.1.1   Motivation

Transformers are widely used as a feature extractor in NLP but required to set a fixed length input sequence, i.e., the default length for BERT is 512. If text sequence length is shorter than fixed length, it must be solved by padding. If text sequence length exceeds fixed length, it can be divided into multiple segments. Each segment is processed at training separately as shown in Fig. 8.24.

Nevertheless, there are two problems: (1) segments are trained independently, the largest dependency between different tokens depends on the segment length; (2) segments are separated according to a fixed length without sentences' natural

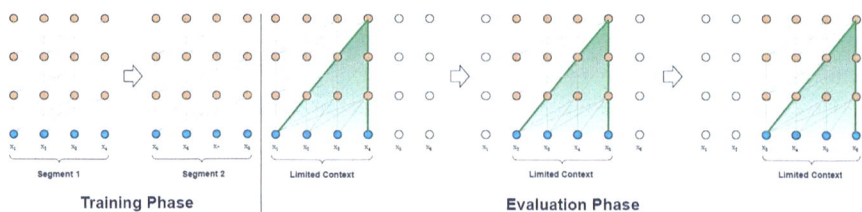

Fig. 8.24 Segment training of standard transformer

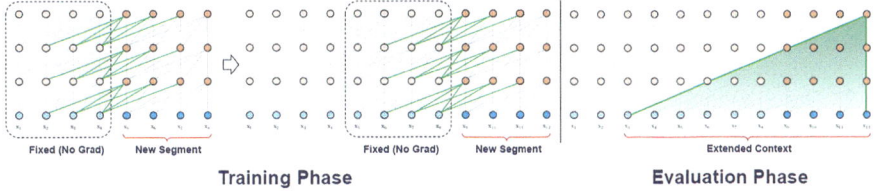

Fig. 8.25 Segment training of Transformer-XL

boundary consideration to produce semantically incomplete segments. Thus, transformer-XL (Dai et al. 2019) is proposed.

8.7.1.2 Transformer-XL Technology

1. **Segment-level recurrence:** When processing the current segment, Transformer-XL caches and applies hidden vector sequence to all layers from the previous segment. These sequences only participate in forward calculation without backpropagation called segment-level recurrence. Figure 8.25 shows the segment training of Transformer-XL.
2. **Relative position encodings:** Each token has an embedding position to represent position relationship in standard transformer. This embedding position encoding is either generated by sin/cos function or learning, but it is impractical in Transformer-XL because positional relationship of different segments is unidentified if the same positional code is added to each segment. Transformer-XL applies relative position encoding instead of absolute position encoding, so when calculating the hidden vector of current position, it considers tokens' relative position relationships to calculate attention score.

8.7.2 ALBERT

BERT model has many parameters, but it is limited by GPU/TPU memory size as model size increases. Google proposed *A Lite BERT (ALBERT)* to solve this problem (Lan et al. 2019). ALBERT applies two techniques to reduce parameters and improve NSP pretraining task, which include:

1. Parameter sharing—apply same weights to all 12 layers.
2. Factorize embeddings—shorten initial embeddings to 128 features.
3. Pretrain by LAMB Optimizer—replace ADAM Optimizer.
4. Sentence order prediction (SOP)—replace BERT's next sentence prediction (NSP) task.
5. N-gram masking—modify MLM task to mask out words' N-grams instead of single words.

Exercises

8.1. What is TL? Compare the major differences between TL and traditional ML in AI.

8.2. Describe and explain how TL can be applied to NLP. Give two NLP applications as examples to support your answer.

8.3. Compare the major differences between heterogeneous vs. homogeneous TL. Give two NLP applications/systems as examples to illustrate.

8.4. What is RNN? State and explain why RNN is important for the building of NLP applications. Give two NLP applications as examples to support your answer.

8.5. State and explain five major categories of RNNs. For each type, give an example to illustrate.

8.6. What is the LSTM network? State and explain how it works by using NLP applications such as text summarization.

8.7. What is GRU? Using an NLP application as examples, state and explain the major differences between the GRU and standard RNN.

8.8. State and explain the key functions and architecture of Transformer technology. Using NLP application as examples, state briefly how it works.

8.9. What is the BERT model? Using NLP application such as Q&A chatbot as examples, state and explain briefly how it works.

References

Cho, K., Van Merriënboer, B., Gulcehre, C., Bahdanau, D., Bougares, F., Schwenk, H., & Bengio, Y. (2014). Learning phrase representations using RNN encoder-decoder for statistical machine translation. arXiv preprint arXiv:1406.1078.

Chung, J., Gulcehre, C., Cho, K., & Bengio, Y. (2014). Empirical evaluation of gated recurrent neural networks on sequence modeling. arXiv preprint arXiv:1412.3555.

Dai, Z., Yang, Z., Yang, Y., Carbonell, J., Le, Q. V., & Salakhutdinov, R. (2019). Transformer-xl: Attentive language models beyond a fixed-length context. arXiv preprint arXiv:1901.02860.

Devlin, J., Chang, M. W., Lee, K., & Toutanova, K. (2018). Bert: Pre-training of deep bidirectional transformers for language understanding. arXiv preprint arXiv:1810.04805.

Dey, R., & Salem, F. M. (2017, August). Gate-variants of gated recurrent unit (GRU) neural networks. In 2017 IEEE 60th international midwest symposium on circuits and systems (MWSCAS) (pp. 1597-1600). IEEE.

Hochreiter, S., & Schmidhuber, J. (1997). Long short-term memory. Neural computation, 9(8), 1735-1780.

Lan, Z., Chen, M., Goodman, S., Gimpel, K., Sharma, P., & Soricut, R. (2019). Albert: A lite bert for self-supervised learning of language representations. arXiv preprint arXiv:1909.11942.

Pan, S. J., & Yang, Q. (2009). A survey on transfer learning. IEEE Transactions on knowledge and data engineering, 22(10), 1345-1359.

Sherstinsky, A. (2020). Fundamentals of recurrent neural network (RNN) and long short-term memory (LSTM) network. Physica D: Nonlinear Phenomena, 404, 132306.

Singh, B., Marks, T. K., Jones, M., Tuzel, O., & Shao, M. (2016). A multi-stream bi-directional recurrent neural network for fine-grained action detection. In Proceedings of the IEEE conference on computer vision and pattern recognition (pp. 1961-1970).

Staudemeyer, R. C., & Morris, E. R. (2019). Understanding LSTM--a tutorial into long short-term memory recurrent neural networks. arXiv preprint arXiv:1909.09586.

Vaswani, A., Shazeer, N., Parmar, N., Uszkoreit, J., Jones, L., Gomez, A. N., ... & Polosukhin, I. (2017). Attention is all you need. Advances in neural information processing systems, 30.

Weiss, K., Khoshgoftaar, T. M., & Wang, D. (2016). A survey of transfer learning. Journal of Big data, 3(1), 1-40.

Yin, W., Kann, K., Yu, M., & Schütze, H. (2017). Comparative study of CNN and RNN for natural language processing. arXiv preprint arXiv:1702.01923.

Yu, Y., Si, X., Hu, C., & Zhang, J. (2019). A review of recurrent neural networks: LSTM cells and network architectures. Neural computation, 31(7), 1235-1270.

Zhuang, F., Qi, Z., Duan, K., Xi, D., Zhu, Y., Zhu, H., ... & He, Q. (2020). A comprehensive survey on transfer learning. Proceedings of the IEEE, 109(1), 43-76.

Chapter 9
Major NLP Applications

9.1 Introduction

This chapter will study three major NLP applications: (1) *Information Retrieval Systems (IR)*, (2) *TS,* and (3) *Question-&-Answering Chatbot System (QA Chatbot).*

IR is the process of obtaining the required information from large-scale unstructured data relative to traditional structured database records from texts, images, audios, and videos. IR systems are not only common search engines but recommendation systems like e-commerce sites, question and answer, or interactive systems.

Text Summarization is the process of diminishing a set of data computationally, creating a subset or summary to represent relevant information for NLP tasks such as text classification, question-answering, legal texts, news summarization, and headlines generation.

QA system represents human-machine interaction system with human natural language is the communication medium. It is a task-oriented system to deal with objectives or answer specific questions through dialogues with sentiment analysis.

9.2 Information Retrieval Systems

9.2.1 Introduction to IR Systems

NLP employs AI techniques such as *N-grams*, *rule-based approaches*, and *Word2Vec* to retrieve information but faces computational limitations when processing large volumes of corpus data. Challenges include defining text and model frameworks for domain-specific applications, utilizing GPU clusters, and incurring high costs to maintain rule sets due to standard modifications.

© The Author(s), under exclusive license to Springer Nature Singapore Pte Ltd. 2025
R. Lee, *Natural Language Processing,*
https://doi.org/10.1007/978-981-96-3208-4_9

Corpora that support *IR* in open, machine-readable formats have grown exponentially due to advancements in pre-trained models. IR models designed for generic language combine general terms with domain-specific terms; for example, "lease" can refer to a place or a leasehold. The objectives can be organized by abstract, formal, or colloquial language within a large narrative component, depending on the document type, to enhance retrieval results.

In IR research, text or document classification and clustering focus on two key aspects: (1) *text representation* and (2) *clustering algorithms*. Text representation involves converting unstructured text into a computer-processable data format. This process necessitates extracting and mining textual information. Semantic similarity computation serves as the link between text modeling and representation, with applications for potential information layers in the text. Clustering algorithms are used to extract semantic information, facilitating similarity calculations for effective text classification and clustering.

9.2.2   Vector Space Model in IR

Vector Space Model (Salton et al. 1975) was a leading IR method from 1960 to 1970. Queries and retrieved documents are represented as vectors with dimensionality related to word list size in this model. A retrieved document D can be represented as a vector of lexical items: $D_i = (d_1, d_2, ..., d_n)$, where d_i, is the weight of a ith lexical item in D_i. Query Q is expressed as a lexical item vector: $Q = (q_1, q_2, ..., q_n)$ where q_i, is the weight of ith lexical item in query term. The relevance is determined by computing the distance between lexical item vectors of the retrieved document and query based on this representation. Although it cannot prove cosine relevance is superior to other similarity methods, it achieved satisfactory performance according to search engines evaluation results. Cosine similarity for angle between retrieved document and query calculation is expressed as

$$\text{sim}(D_i, Q) = \frac{\vec{d_i} \cdot \vec{q}}{\left|\vec{d_i}\right| \times \left|\vec{q}\right|} = \frac{\sum_{j=1}^{n} d_{ij} \cdot q_j}{\sqrt{\sum_{j=1}^{n} d_{ij}^2 \cdot \sum_{j=1}^{n} q_j^2}} \tag{9.1}$$

Equation 9.1 is the weights for dot or inner product of all word terms in query matching documents. There are many words item weights for vector space models. Most of the weighting methods are based on *Term-Frequency* (*TF*) variation. *Inverted document frequency* (*IDF*) (Aizawa 2003) represents the number of term occurrences in retrieved document and reveals lexical term significance in the entire document data set. A lexical item is insignificant with high occurrence frequency in multiple retrieved documents.

There are other text representation methods in addition to vector space model, e.g., phrase or concept representations. Although phrase representation can improve semantic contents, the reduced statistical quality of feature vector become sparse and difficult to extract statistical properties applying machine learning algorithms. Figures 9.1 and 9.2 show a text encoded by Sentence Transformers (Reimers and Gurevych 2019) to demonstrate and compute cosine similarity between embeddings. It uses a pre-trained model to encode two sentences and outperform other pre-train model like BERT (Vaswani et al. 2017).

It is natural to identify the combination with the highest cosine similarity score. By doing so, an intense ranking scheme is used as shown in Fig. 9.3 to identify the highest scoring pair with a secondary complexity. However, it may not work for long lists of sentences.

A chunking concept to divide corpus into smaller parts is shown in Figs. 9.4 and 9.5. For example, parse 1000 sentences at a time to search the rest (all other sentences) of the corpus or search a list of 20k sentences to divide into 20×1000 sentences. Each query is compared with 0–10k sentences first, and 10k–20k sentences to reduce memory storage. The increases of these two values intensified speed and memory storage and then identified the pair with the highest similarity to extract top-K scores for each query as opposed to extract and sort scores for all n^2 pairs.

Such method is faster than brute force methods due to fewer samples. In practical industrial scenarios, more attention is paid to the speed of pre-trained models, encoding methods, and data retrieval. For example, *two-tower model* (Yang et al. 2020), *Wide&Deep model* (Cheng et al. 2016), etc. are shown in Figs. 9.6 and 9.7.

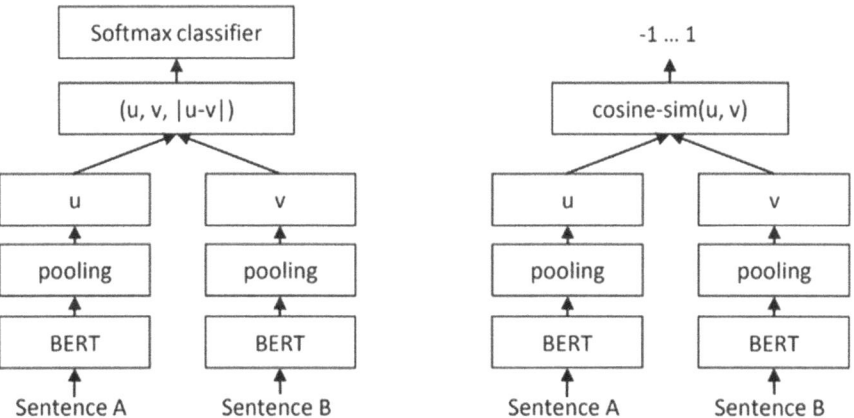

Fig. 9.1 Sentence transformers frame

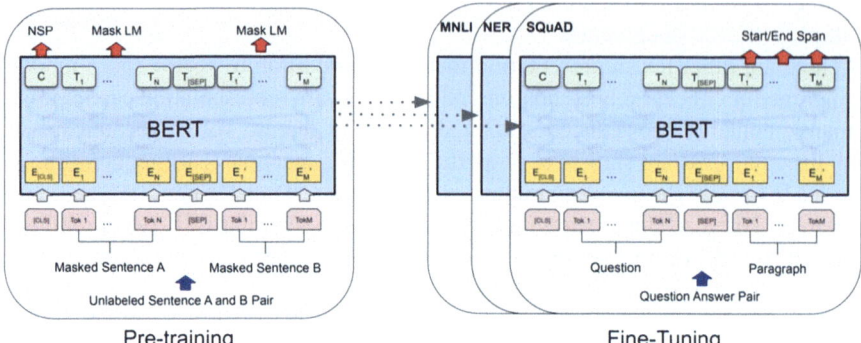

Fig. 9.2 BERT frame

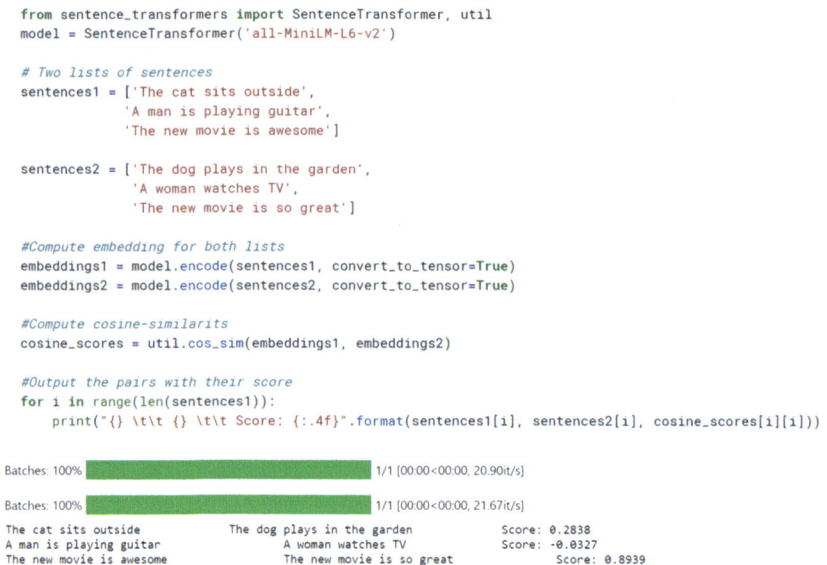

Fig. 9.3 The singer example of vector space model

9.2.3   Term Distribution Models in IR

Probabilistic Ranking Principle (*PRP*) models firstly proposed by Croft and Harper in 1979 (Croft and Harper 1979) to compute query relevance degrees and retrieval. PRP regards IR as a process of statistical inference, where an IR system predicts query relevance from retrieved documents and sorts in descending order based on predicted relevance scores. This approach is like Bayesian model machine learning. A PRP model combines relevant feedback information with IDF and estimates each item's probabilities to optimize search engine retrieval performance. However, it is

```
%time
from sentence_transformers import SentenceTransformer, util

model = SentenceTransformer('all-MiniLM-L6-v2')

# Single list of sentences
sentences = ['The cat sits outside',
             'A man is playing guitar',
             'I love pasta',
             'The new movie is awesome',
             'The cat plays in the garden',
             'A woman watches TV',
             'The new movie is so great',
             'Do you like pizza?']

#Compute embeddings
embeddings = model.encode(sentences, convert_to_tensor=True)

#Compute cosine-similarities for each sentence with each other sentence
cosine_scores = util.cos_sim(embeddings, embeddings)

#Find the pairs with the highest cosine similarity scores
pairs = []
for i in range(len(cosine_scores)-1):
    for j in range(i+1, len(cosine_scores)):
        pairs.append({'index': [i, j], 'score': cosine_scores[i][j]})

#Sort scores in decreasing order
pairs = sorted(pairs, key=lambda x: x['score'], reverse=True)

for pair in pairs[0:10]:
    i, j = pair['index']
    print("{} \t\t {} \t\t Score: {:.4f}".format(sentences[i], sentences[j], pair['score']))
```

```
CPU times: user 3 µs, sys: 1 µs, total: 4 µs
Wall time: 7.87 µs
Batches: 100%|████████████████████| 1/1 [00:00<00:00, 16.88it/s]
The new movie is awesome          The new movie is so great          Score: 0.8939
The cat sits outside              The cat plays in the garden        Score: 0.6788
I love pasta                      Do you like pizza?        Score: 0.5096
I love pasta                      The new movie is so great        Score: 0.2560
I love pasta                      The new movie is awesome        Score: 0.2440
A man is playing guitar           The cat plays in the garden        Score: 0.2105
The new movie is awesome          Do you like pizza?        Score: 0.1969
The new movie is so great         Do you like pizza?        Score: 0.1692
The cat sits outside              A woman watches TV        Score: 0.1310
The cat plays in the garden       Do you like pizza?        Score: 0.0900
```

Fig. 9.4 Multiple examples of vector space model

a difficult task to estimate each probability accurately in practical applications. Okapi BM25 (Whissell and Clarke 2011) retrieval model had solved the difficulties encountered by the PRP model with satisfactory performance in TREC retrieval experiments and commercial search engines. Many IR researchers had modifications based on the BM25 model resulting in many variations, the most common form is as follows:

$$\text{sim}(Q, D) = \sum_{q \in Q} \log \frac{(r_i + 0.5)/(R - r_i + 0.5)}{(n_i - r_i + 0.5)/(N - n_i - R + r_i + 0.5)} \cdot \frac{(k_1 + 1)f_i}{K + f_i} \cdot \frac{(k_2 + 1)qf_i}{k_2 + qf_i} \quad (9.2)$$

```
%time
from sentence_transformers import SentenceTransformer, util

model = SentenceTransformer('all-MiniLM-L6-v2')

# Single list of sentences - Possible tens of thousands of sentences
sentences = ['The cat sits outside',
             'A man is playing guitar',
             'I love pasta',
             'The new movie is awesome',
             'The cat plays in the garden',
             'A woman watches TV',
             'The new movie is so great',
             'Do you like pizza?']

paraphrases = util.paraphrase_mining(model, sentences)

for paraphrase in paraphrases[0:10]:
    score, i, j = paraphrase
    print("{} \t\t {} \t\t Score: {:.4f}".format(sentences[i], sentences[j], score))
```

```
CPU times: user 3 µs, sys: 0 ns, total: 3 µs
Wall time: 7.15 µs
The new movie is awesome            The new movie is so great            Score: 0.8939
The cat sits outside            The cat plays in the garden            Score: 0.6788
I love pasta            Do you like pizza?            Score: 0.5096
I love pasta            The new movie is so great            Score: 0.2560
I love pasta            The new movie is awesome            Score: 0.2440
A man is playing guitar            The cat plays in the garden            Score: 0.2105
The new movie is awesome            Do you like pizza?            Score: 0.1969
The new movie is so great            Do you like pizza?            Score: 0.1692
The cat sits outside            A woman watches TV            Score: 0.1310
The cat plays in the garden            Do you like pizza?            Score: 0.0900
```

Fig. 9.5 Chunk multiple examples of vector space model

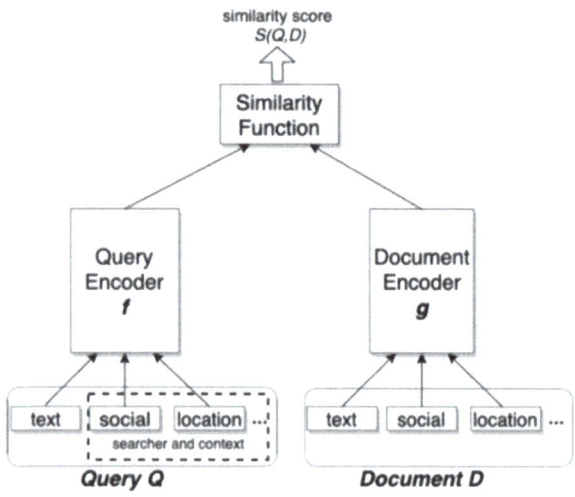

Fig. 9.6 Two-tower model (Yang et al. 2020)

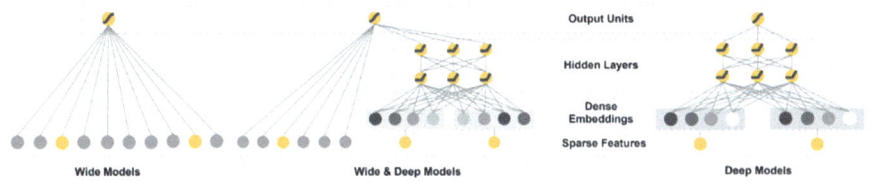

Fig. 9.7 Wide&Deep model (Cheng et al. 2016)

There are two approaches to consider which is the best BM25 method:

1. BM25 + Word2Vec embedding across all documents.
2. BM25 + BERT + Word2Vec embedding for each top-k documents, select the most similar sentence embedding across top-k paragraphs.

Word2vec (Church 2017) is about word occurrences proportions in relations holding in general over large text corpora and combines vectors of similar words into a vector space called distributional hypothesis. Word2vec embeddings are to compare query with sentence embeddings to select the one with higher cosine similarity.

Transformer-based neural network models are popular NLP research areas on enhanced parallelized processing capabilities. BERT is among those that use transformer-based deep bidirectional encoders to learn contextual semantic relationships between lexical items and performed satisfactory in many NLP tasks.

It began to retrieve documents with the most relevant document followed by paragraphs and extract sentences from selected paragraphs. BERT embeddings are used to compare query with paragraphs and select the one with higher cosine similarity. Once relevant paragraphs are available, select sentence with answer by comparing sentence embeddings based on Word2Vec embeddings trained on the whole dataset, then average word embeddings in the paragraph with BM25 score calculation as shown in Fig. 9.8.

Common word queries occurred rarely in documents with a higher number of occurrences produce sparse distribution. Contrarily, there will be similar scores at many documents if common words with same frequency occurred across documents. Documents distribution with scores and codes are shown in Figs. 9.9 and 9.10.

Since *word2vec* relies heavily on each occurrence frequency, thus, it may produce satisfactory performance on specific queries while the same for BERT on general queries.

The results of two selected queries showed that query (Sentence 1) achieved satisfactory performance on specific/rare terminology while the second query (Sentence 2) achieved satisfactory performance on normal terminology. They depend on words specification level in the query. Queries have specific/rare terminology performed satisfactorily with the most similar sentences across all documents. Queries have general terms, e.g., age, human, and climate performed satisfactorily with the most relevant documents instead of embeddings comparison

```
from get_result import filtered_query, remove_punct
from ranking import Ranking

# We can calculate and plot the scores of the documents talking about COVID19
ranking = Ranking()
covid_documents = df[(df.after_dec == True) & (df.tag_disease_covid == True)].paper_id

query1 = filtered_query(query1)
scores1 = ranking.get_bm25_scores(query1, covid_documents)

query2 = filtered_query(query2)
scores2 = ranking.get_bm25_scores(query2, covid_documents)

# Plot the results
fig, axs = plt.subplots(1,2, sharey=True, tight_layout=False, figsize =(15,5))
axs[0].hist(scores1.values(), bins=20, color='g')
axs[0].set_xlabel('Scores')
axs[0].set_ylabel('Number of documents')
axs[0].set_title('Incubation period')
axs[1].hist(scores2.values(), bins=20, color='g')
axs[1].set_xlabel('Scores')
axs[1].set_ylabel('Number of documents')
axs[1].set_title('Prevalence of asymptomatic shedding and transmission')

plt.show()


[nltk_data] Downloading package stopwords to /usr/share/nltk_data...
[nltk_data]   Package stopwords is already up-to-date!
```

Fig. 9.8 Sample code for Word2vec embeddings with BM25 score calculation

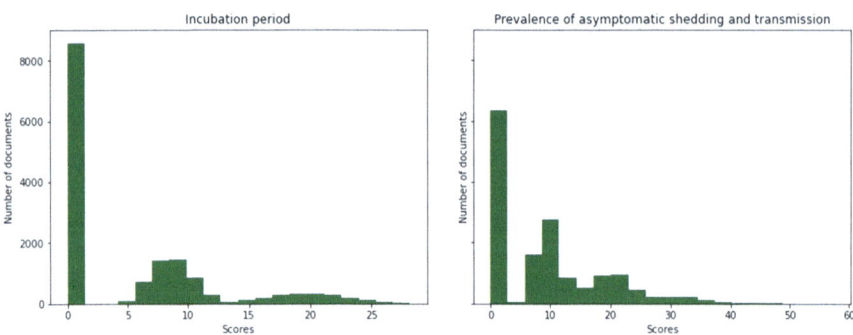

Fig. 9.9 Documents distribution with scores and codes

```
Sentence 1: 0.6235944271030706

A comparison to the estimated incubation period distribution for MERS (Table 3 and Figure 3)
shows that the incubation period values are remarkably similar, with mean values differing a
t most 1 day and 95th percentiles differing at most 2 days.


Sentence 2: 0.6043460033864764
The estimated mean incubation periods for SARS are more variable between studies, including
values shorter and longer than those presented here for 2019-nCoV.


Sentence 3: 0.5978659753048406
These findings imply that the findings of previous studies that have assumed incubation peri
od distributions similar to MERS or SARS will not have to be adapted because of a shorter or
longer incubation period.
```

Fig. 9.10 BM25 results

across all of them. Thus, it is reasonable to compare each time the results of two approaches and select the appropriate one based on words distribution for each query.

9.2.4 Latent Semantic Indexing in IR

Term Distribution Models in IR is a rapid and effective model. It uses topics to express the implicit semantics of a document as index to replace incomplete, unreliable search terms with reliable indicants based on two assumptions:

1. Words have common topics in document.
2. Words not in document less likely to be related

Topic is filtered out by keywords in the Doc. Thus, $P = (\omega/\text{Doc})$ probability distribution table is introduced: the statistics of word frequency (frequency) in the document, i.e., the law of large numbers.

$$P\left(w|\text{Topic}_D\right) \approx P(w|D) = tf\left(w,D\right) / \text{len}\left(D\right) \qquad (9.3)$$

Topic is regarded as a language model, and $P = (\omega/\text{Doc})$ is the probability of word generation in this language model so the word not only occur in topic, but has probability generated.

There are two sorting methods according to statistical language model when query Q is given, which are (1) *Query-likelihood* and (2) *Document-likelihood* methods.

9.2.4.1   Query-Likelihood

Determine M_D, corresponding to each Doc, user's Query is denoted as $Q = (q_1, q_2, \ldots, q_n)$. Query probability will be generated under the *language model* of each document can be calculated as follows (Zhuang and Zuccon 2021):

$$P\left(q_1 \ldots q_k | M_D\right) = \prod_{i=1}^{k} P\left(q_i | M_D\right) \tag{9.4}$$

Search results are obtained by sorting all computed results. However, this method calculates the probability for each Doc independently from other Docs, and the relevant documents are not utilized.

9.2.4.2   Document-Likelihood

Determine each Query corresponding M_Q. Calculate the probability that any given document will be generated under the query's *language model* (Zhuang and Zuccon 2021):

$$P\left(D | M_Q\right) = \prod_{w \in D} P\left(w | M_D\right) \tag{9.5}$$

The object of one-mode factor analysis traditionally is a matrix composed of identical object-pair types of relationships. An example is a document-document matrix. The matrix elements may be evaluated for similarity between documents manually. This symmetric square matrix is decomposed into two matrices by eigen-analysis. The decomposed matrix is composed of linearly independent factors. Many of the factors are tiny that can be ignored usually producing an original matrix approximation.

Two-mode factor analysis object is a matrix consisting of object-pair relationships. This matrix can be decomposed into term-term, document-document, and term-document matrices using singular-value decomposition (SVD) (Aharon et al. 2006). SVD reconstructs spatial response to the main patterns association between data by ignoring less significant effects. Thus, a term that does not occur in a document may be immediately adjacent to that document in semantic space based on identified association patterns. The information location in semantic space has a role in semantic index. SVD model test and lean with results in Latent Semantic Indexing in IR is shown in Fig. 9.11.

SVD and corresponding validation results are shown in Figs. 9.12 and 9.13.

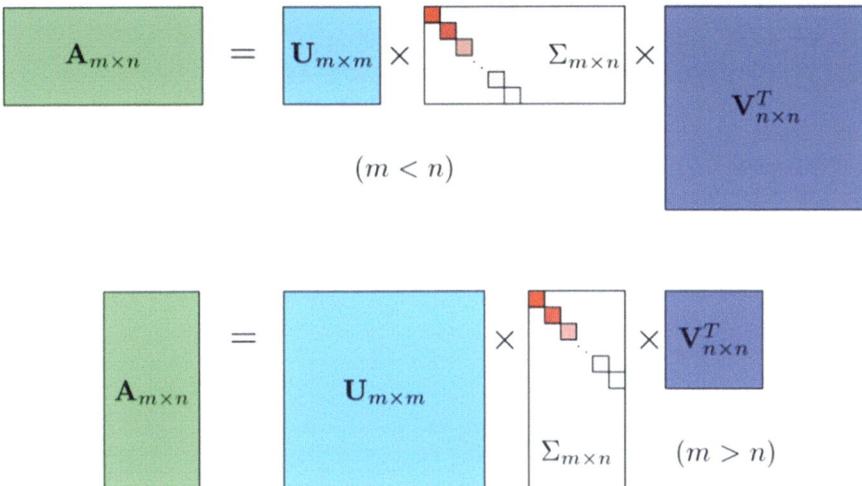

Fig. 9.11 SVD frame

```python
import numpy as np
import matplotlib.pyplot as plt
la = np.linalg
words = ["I","like","enjoy","deep","learning","NLP","flying","."]
X = np.array([[0,2,1,0,0,0,0,0],
              [2,0,0,1,0,1,0,0],
              [1,0,0,0,0,0,1,0],
              [0,1,0,0,1,0,0,0],
              [0,0,0,1,0,0,0,1],
              [0,1,0,0,0,0,0,1],
              [0,0,1,0,0,0,0,1],
              [0,0,0,0,1,1,1,0]])
U,s,Vh=la.svd(X, full_matrices=False)
for i in range(len(words)):
    print(U[i,1],U[i,1],words[i])
    plt.text(U[i,0],U[i,1],words[i])
plt.xlim(-1,1)
plt.ylim(-1,1)
plt.show()
```

Fig. 9.12 Example of SVD

9.2.5 Discourse Segmentation in IR

Document contents combine with articulated parts such as paragraphs exalt automatic documents segmentation according to meanings using machine learning methods to compare two adjacent sentences similarity in turn, and generate

```
-0.5728591445369984 -0.5728591445369984 I
0.6301206635626871 0.6301206635626871 like
0.27401753321708033 0.27401753321708033 enjoy
-0.2479121303794708 -0.2479121303794708 deep
0.033849504755904036 0.033849504755904036 learning
-0.2939889899169304 -0.2939889899169304 NLP
-0.1610277668211104 -0.1610277668211104 flying
0.15725476183570436 0.15725476183570436 .
```

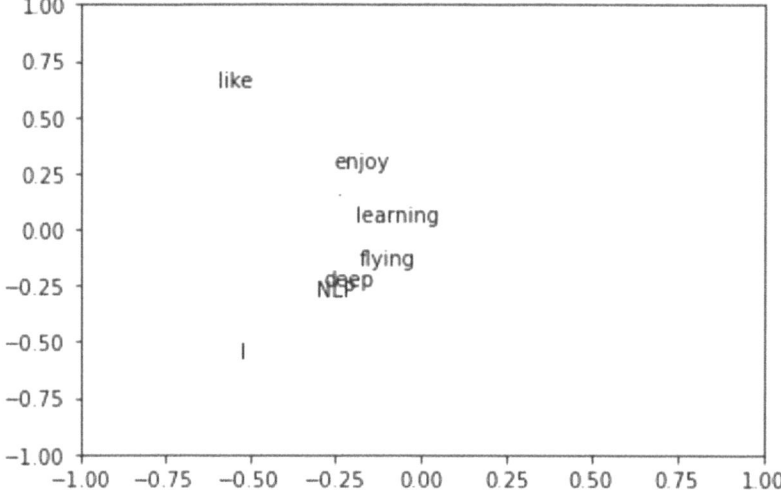

Fig. 9.13 Validation results

segmentation point with the lowest similarity. This unsupervised method is called *Text Tiling* (Hearst 1997) as shown in Fig. 9.14. Further, supervised learning methods can also be used such as classifiers constructions (Florian 2002) or sequence models (Keneshloo et al. 2019) to detect segmentation point.

Rhetorical Structure Theory (*RST*) framework (Taboada and Mann 2006) is a commonly used framework for parsing discourse as shown in Fig. 9.15. RST common relations in English are conjunction, justify, concession, elaboration, etc. as shown in Figs. 9.16 and 9.17.

There are two approaches to identify relationships: (1) *rule-based* on iconic words such as *but, so, for example,* and (2) *machine learning* with commonly features such as *bag of words* (*BoW*) (Zhang et al. 2010), *Discourse markers* (Fraser 1999), *Starting/ending N-grams* (Robertson and Willett 1998), Location in the text (Rothkopf 1971), *Syntax features* (Sadler and Spencer 2001), *Lexical and distributional similarities* (Weeds et al. 2004).

Discourse segmentation task is a significant evaluation indicator for NLP development directions. From application perspective, discourse segmentation can assist users rely on intelligence to improve productivity, its technology core value can convert semi-structured and unstructured data to specific description structured in turn to support substantial downstream applications.

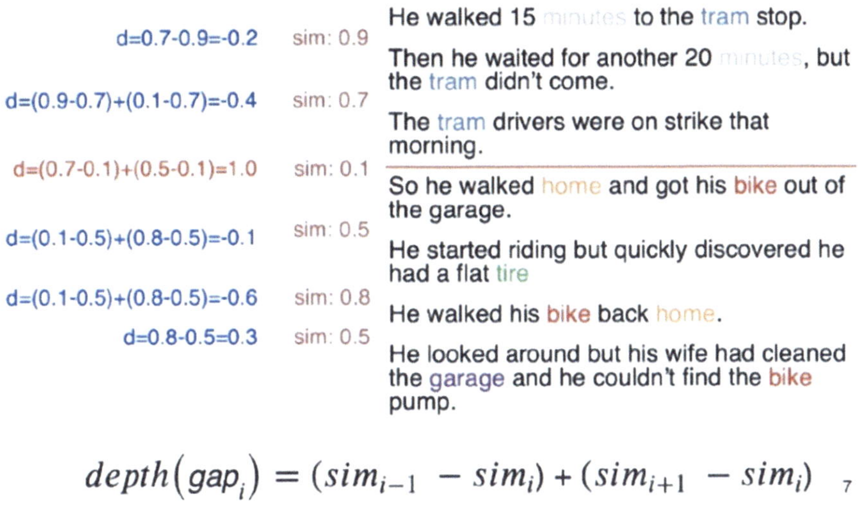

d=0.7-0.9=-0.2   sim: 0.9   He walked 15 minutes to the tram stop.

                                      Then he waited for another 20 minutes, but
d=(0.9-0.7)+(0.1-0.7)=-0.4   sim: 0.7   the tram didn't come.

                                      The tram drivers were on strike that
d=(0.7-0.1)+(0.5-0.1)=1.0   sim: 0.1   morning.
_____

                                      So he walked home and got his bike out of
d=(0.1-0.5)+(0.8-0.5)=-0.1   sim: 0.5   the garage.

                                      He started riding but quickly discovered he
                                      had a flat tire
d=(0.1-0.5)+(0.8-0.5)=-0.6   sim: 0.8
                                      He walked his bike back home.
d=0.8-0.5=0.3   sim: 0.5
                                      He looked around but his wife had cleaned
                                      the garage and he couldn't find the bike
                                      pump.

$$depth\big(gap_i\big) = (sim_{i-1} - sim_i) + (sim_{i+1} - sim_i)$$ [7]

Fig. 9.14 Examples of discourse segmentation

Fig. 9.15 Example of
rhetorical structure theory

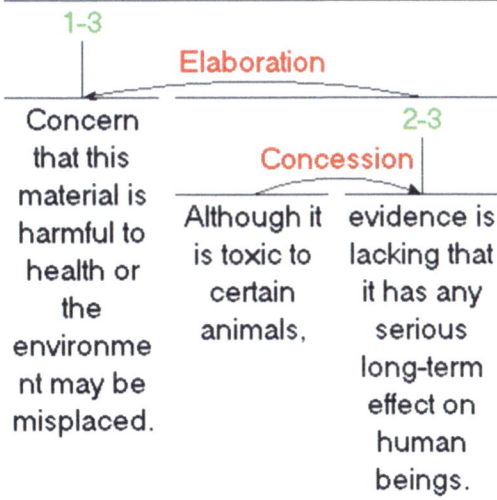

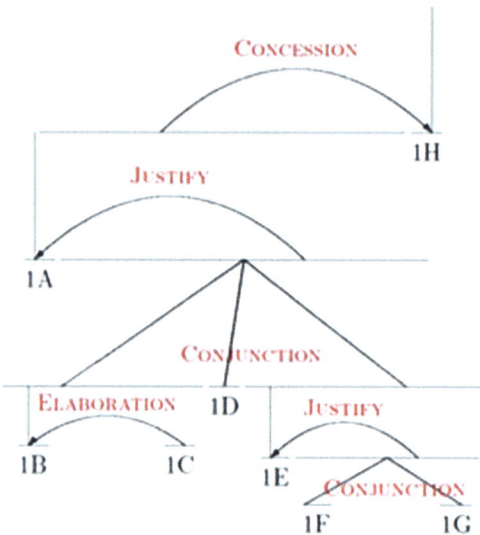

[It could have been a great movie]^{1A} [It does have beautiful scenery,]^{1B} [some of the best since Lord of the Rings.]^{1C} [The acting is well done,]^{1D} [and I really liked the son of the leader of the Samurai.]^{1E} [He was a likable chap.]^{1F} [and I hated to see him die.]^{1G} [But, other than all that, this movie is nothing more than hidden rip-offs.]^{1H}

Fig. 9.16 Examples of relations

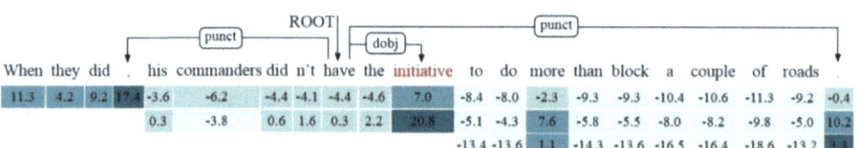

Fig. 9.17 Attention map

9.3   Text Summarization Systems

9.3.1   Introduction to Text Summarization Systems

9.3.1.1   Motivation

There is excess information from copious sources to obtain the latest information daily. Although automatic and accurate summarization systems can assist users in simplifying, identifying, and understanding key information quickly, the process remains laborious. This is due to the constant emergence of new words and complex text structures in documents.

9.3.1.2   Task Definition

Text summarization process generates text (document or document) summaries by rewriting and summarizing long text into short form (Mahalakshmi and Fatima 2022). It refers to extract or refine text or text set key points through technologies to display original text or text set main contents or general idea. Text generation task is an information compression technique whereas a summarization process is considered as a function where input is a document or documents, and output is an input texts summary. Hence, input and output are quintessential types to classify summary tasks.

9.3.1.3   Basic Approach

Summarization approaches are mainly divided into *extractive* and *abstractive* (Chen and Zhuge 2018).

   Extractive methods select important phrases from input text and combine them to form a summary like a copy and paste process. Many traditional text summarization methods use *Extractive Text Summary* (*ETS*) because it is simple to generate sentences without grammatical errors but cannot reflect exact sentences meanings. They are inflexible to apply novel expressions, words, or connectors outside text descriptions.

   Abstractive Text Summary (*ATS*) methods apply language generation methods to re-organize contents, generate new words, and conclude the implied information as compared with ETS. They paraphrase text meanings composed of new words with original words summary (Agrawal 2020), and mimic human understanding to develop contents which may not be contained in actual document text (Malki et al. 2020).

9.3.1.4   Task Goals

Summarization task objectives are to assist users to understand raw text within a short period as shown in Fig. 9.18.

9.3.1.5   Task Sub-processes

Summarization tasks are divided into the following modules as shown in Fig. 9.19.

Input document or documents are first combined and preprocessed from continuous text form to split sentences. The sentences will be encoded into vectors form data to fit into a matrix for similarity scores calculation to obtain sentence rankings, followed by a summary with the highest possibility according to the ranking list.

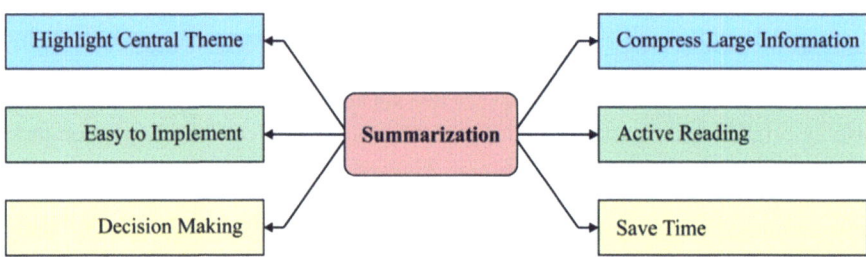

Fig. 9.18 Summarization tasks objectives

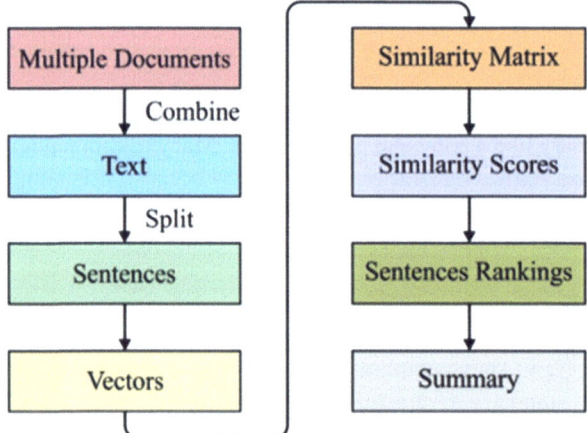

Fig. 9.19 Summarization tasks sub-processes

9.3.2   Text Summarization Datasets

Text summarization datasets commonly used include DUC (2022), New York Times (NYT 2022), CNN/Daily Mail (CNN-DailyMail 2022), Gigaword (2022), and LCSTS datasets (LCSTS 2022).

DUC datasets (DUC 2022) are the most fundamental text summarization datasets developed and used for testing purposes only. They consist of 500 news articles, each with 4 human-written summaries.

NYT datasets (NYT 2022) contain articles published in the New York Times between 1996 and 2007 with abstracts compiled by experts. The abstract datasets are sometimes incomplete and sporadic short sentences with an average of 40 words.

CNN/Daily Mail datasets (CNN-DailyMail 2022) are widely used multi-sentence summary datasets often trained by generative summary system. They have (a) anonymized version to include entity names and (b) non-anonymized version to replace entities with specific indexes.

Gigaword datasets (Gigaword 2022) are abstracts comprising the first sentence and article title with heuristic rules of approximately four million articles.

LCSTS datasets (LCSTS 2022) are Chinese short texts abstract datasets constructed by Sina Weibo (2022).

9.3.3   Types of Summarization Systems

Text summarization task for input documents can be divided into two types:

1. Single document summarization considers each input as one document.
2. Multiple document summarization considers input has several documents

Text summarization task viewpoint can be divided into three classes:

1. Query-focused summarization adds viewpoint to query.
2. Generic summarization is generic.
3. Update summarization is a special type which sets *difference* (update) viewpoint

Summarization systems based on contents can be divided into four types:

1. Indicative Summarization describes contexts without revealing details especially the endings, it contains partial information only.
2. Informative Summarization contains all information in a document or documents.
3. Keyword Summarization reveals output generation is sporadic text which contains phrases or words of input documents.
4. Headline Summarization is usually a single line summary.

These summarization systems can be divided according to summary languages such as Arabic (Elsaid et al. 2022), Chinese (Yang et al. 2012), English and Spanish summarization systems, etc.

9.3.4 *Query-focused vs. Generic Summarization Systems*

Text summarization can be *query-focused* or *generic*. Summary associated with query shows that document contents are relative to initial search query. A query-related summary generation is a process of retrieving query-related sentences/paragraphs from a document that has a strong similarity to text retrieval process. Hence, abstracts relevant searches are often undertaken by extending traditional IR techniques with many text abstracts in the literature fall into this category. A general summary, on the other hand, provides an overall sense of the document's contents. A proper general summary should cover the main topics and minimize redundancy. Since there are no queries or topics to feed into summarization process, it is difficult to develop a high-quality general summarization method for evaluation (Gong and Liu 2001).

9.3.4.1 Query-Focused Summarization Systems

Query-focused Summarization (QFS) is primarily addressed using extractive methods to produce text that lacks coherence. QFS applied abstractive methods can overcome these limitations and improve incoherent texts availability. A *Relevance Sensitive Abstractive QFS (RSA-QFS)* framework (Baumel et al. 2018) is shown in Fig. 9.20.

This model assumes that a trained abstractive model includes reusable language knowledge to accomplish QFS tasks. Methods of enhancing this pre-trained single document abstraction model with explicit modeling of query dependencies are studied to improve multiple input documents operating ability and adjust generated abstractions lengths accordingly.

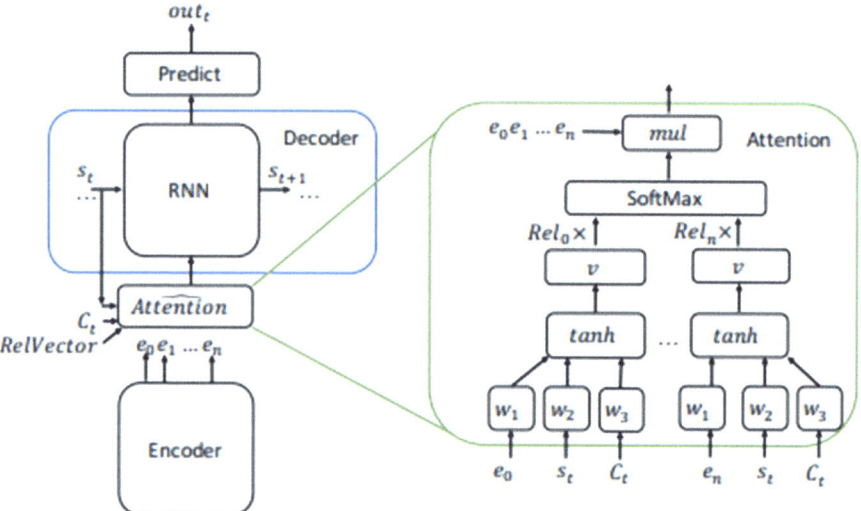

Fig. 9.20 RSA-QFS framework

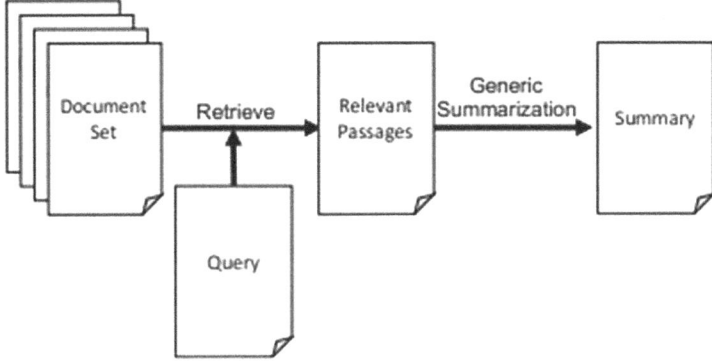

Fig. 9.21 Two stages of QFS

Further, a *sequence-to-sequence* (*seq2seq*) architecture is applied to obtain sum via an iterative extraction or abstraction pairs process: identify relevant content batches from multiple documents and abstract into a coherent text segments sequence.

QFS task includes two stages as shown in Fig. 9.21:

1. A relevance model to determine passages relevance to input query from source documents and
2. A generic summarization method to combine relevant passages into a coherent summary

Query-related text summarization is practical for answering questions such as whether a whole or partial document has relevance to a user's query. Query-related summaries do not provide an overall sense of the document's content; they have query bias and are unsuitable for content summaries to answer questions such as document category, key points, and text summary.

9.3.4.2   Generic Summarization Systems

A proper generic summarization should cover main topics as many as possible and minimize redundancy leading to fractious system generation and evaluation. It often lacks consensus on summary output and performance judgments without query provisions and topics to summary task.

Typical generic summarization ranking models and selected sentences are based on relevance similarity values and other semantic analyses (Gong and Liu 2001).

9.3.5   *Single and Multiple Document Summarization*

Single document extraction in journalism has developed to *multi-document extraction* since 1990. A variety of news articles, such as *Google News* (Google 2022), *Columbia News Blaster* (Columbia 2022), and *News Essence* (NewsInEssence

2022) are inspired by multi-document summaries. The reason is that individual documents always produce contradictory results through overlapping information from multiple documents (Alami et al. 2015) may affect the performance of summarization results.

Single document summarization research method gradually faded in past decades (Svore et al. 2007) as mainstream research focused on multi-document summarization which could reduce text size, gather ideas, compare documents, and maintain syntactic and semantic relationships (Pervin and Haque 2013).

9.3.5.1   Single Document Summarization

Single document summarization's challenge is to identify or generate informative sentences significance of the document because it often has inconsistent and intermittent information.

Salient features like sentence placement are early research (Baxendale 1958) where 200 paragraphs selected and identified paragraphs have topic sentences at the beginning and end of paragraphs with 85% and 7%, respectively.

A single document structure and a corpus with around 400 technical documents research focusing on word frequency and word position, and cue words and skeleton were proposed in 1969 (Edmundson 1969). Results showed that extracted summary to actual summary accuracy rate was about 44%.

Further, lexical indicators (Rath et al. 1961), cohesion (Hasan 1984), semantic relationships (Halliday and Hasan 1976), and algebraic methods such as naïve-Bayes classifier processed features like uppercase words, lengths, words position (Kupiec et al. 1995), symbolic word knowledge (Hovy and Lin 1999), and human abstraction concept (Jing 2000) are research areas in this field.

9.3.5.2   Multiple Document Summarization

Multiple document summarization similarity measures and extractive techniques are comparable to single document summarization.

It used clustering to identify common themes (Erkan and Radev 2004), composite sentences from clusters (Barzilay et al. 1999), maximal marginal relevance (MMR) (Carbonell and Goldstein 1998) and concatenated to multilingual environment (Evans 2005).

Further, TFI X IDFI techniques (Salton 1989), TF/IDF (Fukumoto 2004), word hierarchical technique for frequent terms (You et al. 2009), graph-based methods (Mani and Bloedorn 1997; Wan 2008), sentence co-relation method (Hariharan et al. 2013), logical closeness (Zhu and Zhao 2012) and query-oriented approach (Agarwal et al. 2011) are well developed.

9.3.6 Contemporary Text Summarization Systems

9.3.6.1 Contemporary Extractive Text Summarization (ETS) System

Text summarization research methods aim to (Dong 2018):

1. Acquire important sentences.
2. Predict sentence option according to ranking sentences.

The extractive summarization for proper sentences selection from original source text is required to:

1. Include logical and consistent summary information from original text.
2. Reduce similar and unimportant sentences information redundancy.

Lead 3 is a commonly used and effective method to extract the first three sentences as topic titles of an article. When dealing with important sentences, document equivalence to document topic and relevant sentences position characteristics are considered. Topic modeling, frequency-based models LSA, and Bayesian are methods applied (Farsi et al. 2021).

Extractive summarization produces incoherent summaries compared with manual ones, its shortcomings include unresolved anaphora, unreadable sentence order, lacks textual cohesion to extract salient information from long sentences. When the system focuses on a sentence, it extracts the entire sentence (Nallapati et al. 2017).

9.3.6.2 Graph-Based Method

Graph-based ranking algorithms are successfully used in citation analysis, link social networks' structure analysis, and the World Wide Web.

They generate graphs from input document and summary by considering the relationships between nodes (units of text) (Chi and Hu 2021). TextRank (Mihalcea and Tarau 2004) is a typical graph-based approach that has developed many models. A summarization of TextRank system to extract keywords from a sample text and graph is shown in Figs. 9.22 and 9.23.

Compatibility of systems of linear constraints over the set of natural numbers. Criteria of compatibility of a system of linear Diophantine equations, strict inequations, and nonstrict inequations are considered. Upper bounds for components of a minimal set of solutions and algorithms of construction of minimal generating sets of solutions for all types of systems are given.
These criteria and the corresponding algorithms for constructing a minimal supporting set of solutions can be used in solving all the considered types systems and systems of mixed types.

Fig. 9.22 Sample text

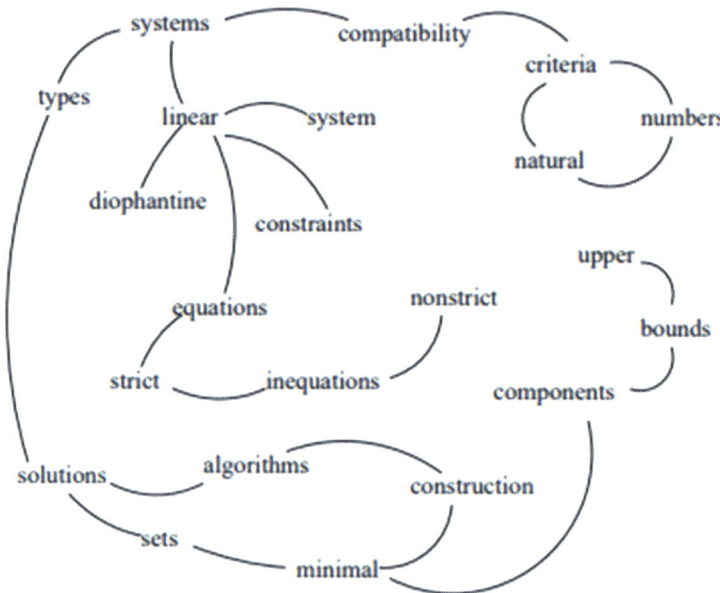

Fig. 9.23 Sample graph for key phrase extraction in TextRank

This kind of system is based on PageRank algorithm (Langville and Meyer 2006)
applied by Google's search engine, its algorithm principle is *linked pages are good,
and even better if they come from multiple linked pages*. Links between pages are
represented by matrices like circular tables. This matrix can be converted to a transition probability matrix divided by the sum of links per page, and the page will be
moved by the page viewer following a feature matrix in Fig. 9.24.

TextRank processes words and sentences as pages in PageRank, its algorithm
defines *text units* and adds them as nodes in a graph with *relations* are defined
between text units and added as edges in the graph. Generally, the weights of edges
are set by similarity or score values.

Then, PageRank algorithm is used to solve the graph. There are other similar
systems such as LexRank (Erkan and Radev 2004) to consider sentences as nodes
and similarity as relations or weights, i.e., IDF-modified cosine similarity to calculate similarity.

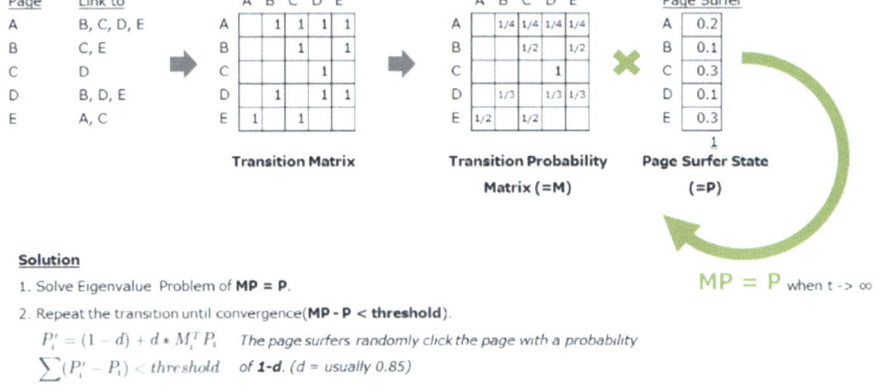

Solution

1. Solve Eigenvalue Problem of **MP = P**.

2. Repeat the transition until convergence(**MP - P < threshold**).

$$P'_i = (1 - d) + d * M^T_i P_i \quad \text{The page surfers randomly click the page with a probability}$$

$$\sum(P'_i - P_i) < threshold \quad \text{of } \mathbf{1\text{-}d}. \text{ (d = usually 0.85)}$$

MP = P when t -> ∞

Fig. 9.24 Page Rank algorithm process

9.3.6.3   Feature-Based Method

Feature-based model extracts sentences feature and evaluates their significances. There are many representative studies including Luhn's Algorithm (Luhn 1958), *TextTeaser,* and *SummaRuNNer* (Nallapati et al. 2017).

Luhn's Algorithm is used to evaluate input words significance calculated by frequency. TextTeaser is an automatic feature-based summarization algorithm. *SummaRuNNer* is implemented by *Deep Neural Networks* (*DNN*) structure as shown in Fig. 9.25.

SummaRuNNer generates sentence feature (vector) by two layers bidirectional *Gate Recurrent Unit-Recurrent Neural Network* (*GRU-RNN*) from word embedding vectors. The lowest level classifies each sentence word level, while the highest level classifies sentence level. Double arrows indicate two-way RNN. The top layer numbered with 1s and 0s is a classification layer based on sigmoid activation to determine whether each sentence is a summary. Each sentence decision depends on substantial sentence contents, sentences to document relevance, sentences to cumulative summary representation originality, and other positional characteristics.

9.3.6.4   Topic-Based Method

Topic-based model considers document's topic features and input sentences' scores according to topic types contained as a major topic would obtain a high rate when scoring sentences.

Latent Semantic Analysis (*LSA*) is based on *SVD* to detect topics (Ozsoy et al. 2011). An LSA-based sentence selection process is shown in Fig. 9.26 by topics represented by eigenvectors or principal axes with corresponding scores.

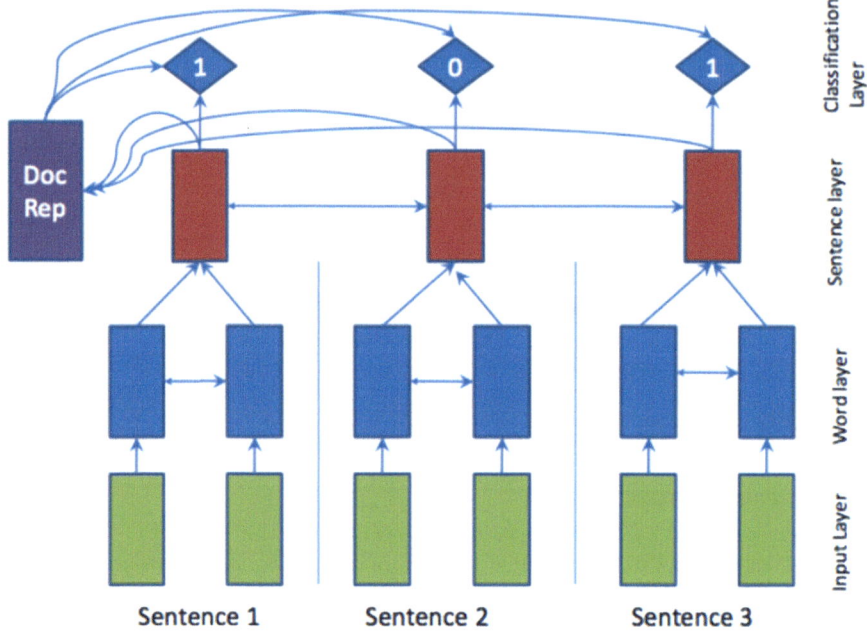

Fig. 9.25 Network structure of SummaRuNNer

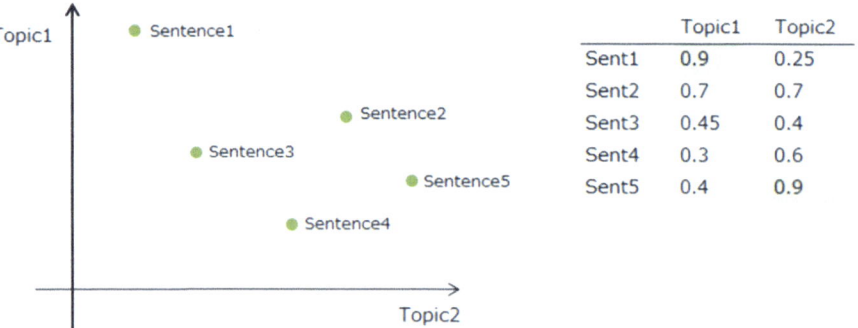

Fig. 9.26 LSA-based sentence selection sample

9.3.6.5 Grammar-Based Method

Grammar-based model parses text and constructs a syntax structure, selects, or reorders the substructure. A representation framework is shown in Fig. 9.27.

Grammar pattern can produce significant paraphrases based on grammatical structures. The above example in Fig. 9.27 showed how paraphrase extraction and replacement can be achieved by using such method. Analyzing grammatical structure feature is useful for semantic phrases reconstruction.

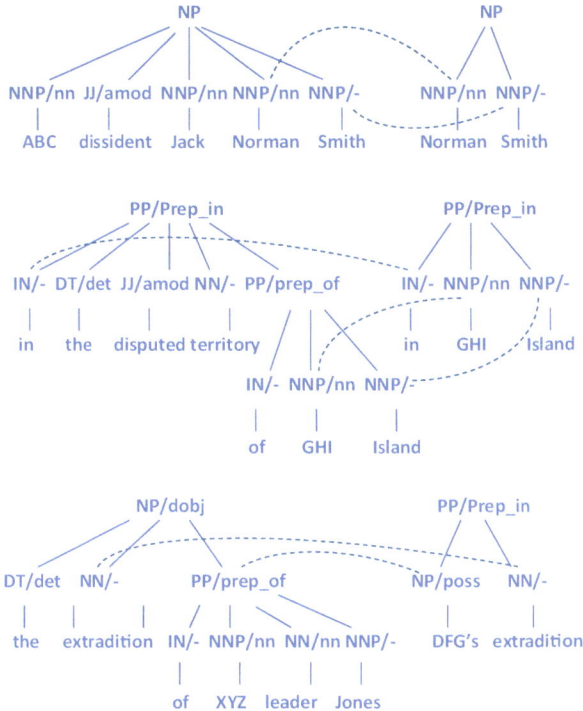

Fig. 9.27 Grammar-based method sample network (Ozsoy et al. 2011)

9.3.6.6   Contemporary Abstractive Text Summarization (ATS) System

Abstractive summarization often generates summary that maintains original intent completed by humans.

This process can generate words that are not in original input representations but to facilitate summaries characteristics and fluency. However, it is complex to generate coherent phrases and connectors.

Abstractive summarization systems applying deep learning methods, Reinforcement Learning (RL), Transfer Learning (TL), and Pre-Trained Language Models (PTLMs) had developed rapidly (Alomari et al. 2022) in recent years. These models use rules-based frameworks to consider significant events and summaries. Tree methods are ontology-related methods for abstractions (Jain et al. 2020).

9.3.6.7   Aided Summarization Method

This method combines automatic computer model or algorithm to provide significant document information for human decision.

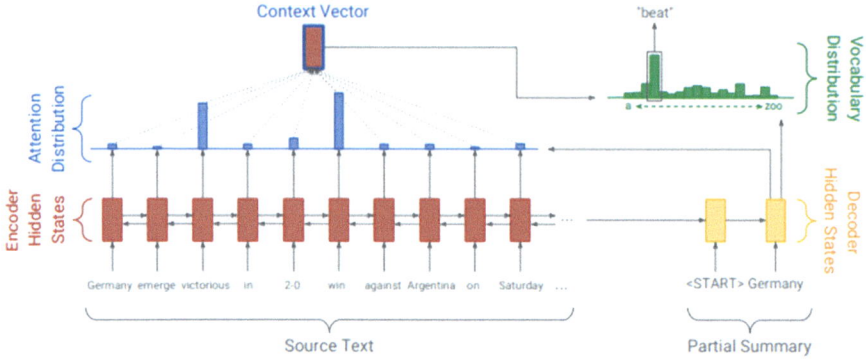

Fig. 9.28 Network framework of point generator baseline model

Machine translation model to text summarization was proposed (Banko et al. 2000) applying encoder-decoder framework as neural network model mainstream and used in abstractive summarization systems (Chopra et al. 2016).

9.3.6.8   Contemporary Combined Text Summarization System

Pointer-Generator Networks (See et al. 2017) is a frequently used baseline network. It focuses on keywords and sentences with *Attention technique* (Vaswani et al. 2017), to lever generator and pointer network according to calculated probability. Vocabulary and attention with different weights distribution are then combined. A baseline pointer-generator network framework is depicted in Fig. 9.28.

It noted that article tokens are fed into an encoder layer, which is a single-layer bidirectional long short-term memory (LSTM) with encoder hidden states provided. Decoder consists of a single-layer unidirectional LSTM, processes word embedding of previous words on each step and output decoder state with attention distribution.

9.4   Question-and-Answering Systems

9.4.1   QA System and AI

A QA system is an impressive way to emulate human-to-human interaction through cutting-edge technological advancements. Unlike other classification or prediction tasks, a QA system is interdisciplinary, merging traditional linguistics with computer science, computational linguistics, statistics, pattern recognition, data mining, machine learning, and deep learning methods to create an effective communication

Fig. 9.29 Flowchart of a typical QA system

system. It plays a vital role in applications such as autoresponders, personal assistants, and sentiment chatbots today.

QA systems are a popular research topic in NLP and typically incorporate open domain common sense knowledge or specialized domain knowledge to function as qualified conversation partners. Dialogue realization depends on several components, including automatic speech recognition (ASR), NLU, dialogue management (DM), NLG, and speech synthesis (SS). A flowchart illustrating the components of a QA system is shown in Fig. 9.29.

It is an integral part of system acumen. *DM* is the communication policy or dialogue strategy applied to large corpus for content organization. After transferring natural language to computer language in sequence-sequence data with character, word, or sentence level in NLU, machine intelligence selects suitable contents for language generation. Back-end technology with generated candidate answers is combined and re-ranked for optimization response in NLG. Apart from text aspect, ASR and TTS are procedures that resemble machine by human voice recognition and generation.

QA system research is divided into two categories: (1) pattern matching with rule-based and (2) language generated-based on IR and neural network. However, the back-end is equipped with more than one method to generate meaningful communication and provide meaningful feedback. A QA system in a chatbot includes an open domain focus on (1) common sense/world knowledge and (2) task-oriented for special domain knowledge databases resemble expert system involving in-depth knowledge base to support appropriate responses.

First rule-based human-computer interaction as in Fig. 9.30 pattern recognition system challenged the Turing test in 1950s, reaching a milestone where humans could not recognize whether the opposite was a machine or human. After a long period of data collection, database used for dialogue pattern matching is large enough to rank appropriate feedbacks and give the highest scoring answers, which is a process of selection from a database of human answers regardless of the machine. After decades of development, search engines and data crawlers have supported sources for building knowledge bases, including IR, enabling search engines to retrieve relevant and up-to-date data for structured processing to form answers from QA systems. The advent of AI era enhanced QA systems mainstream can focus on cognitive science than big data feeds of neural networks on systems generations. Gradually, traditional QA system is replaced by AI machine communication as rule-based matching RNN training to realize large knowledge base to support the AI brain to imitate human reasoning called NLU.

Fig. 9.30 Human and machine interaction via QA system

The main source of knowledge base in a typical QA system comes from: (1) human-human dialogue collection with handcraft is the answer from human language in linguistic and meaning where database consist of pairs dialogues. Without any imitation or learning ability, this first version rule-based QA system relies on pattern matching to measure the distance between proposed question and question-answer pattern stored pair in database. For example, artificial intelligence markup language (AIML) can answer most of daily or even professional dialogues based on large and classified handcraft database without intelligence; (2) building database focus on search engine for IR-based knowledge base. The feature of IR-based QA system is the combination of knowledge building from up-to-date knowledge bases. An IR-based QA system uses domain knowledge such as expert system to extract and generate knowledge. The procedure of unstructured data extraction and reorganization depends on NLU for reasoning. NLG includes knowledge engineering analysis for reasoning and rerank candidates' answers optimization.

The latest database used big data for data-driven model to realize machine intelligence. When neural network had fed with sufficient data, sequence-to-sequence model like RNN and its related *Long-Short-Term Memory* naturally model as in Fig. 9.31 skilled in sequential data processing (Cho et al. 2014). A neural network model is considered as the black box producing learning ability with accuracy but cannot be comprehended by humans. Prior to preprocessing data was fed to neural model, and they were required to transform data format from natural word to vector for data training (Mikolov et al. 2013). Tokenization has three levels: (1) character, (2) word, and (3) sentence. The input format decides output outcomes in encoder-decoder framework. RNN generated words may not be meaningful in the English

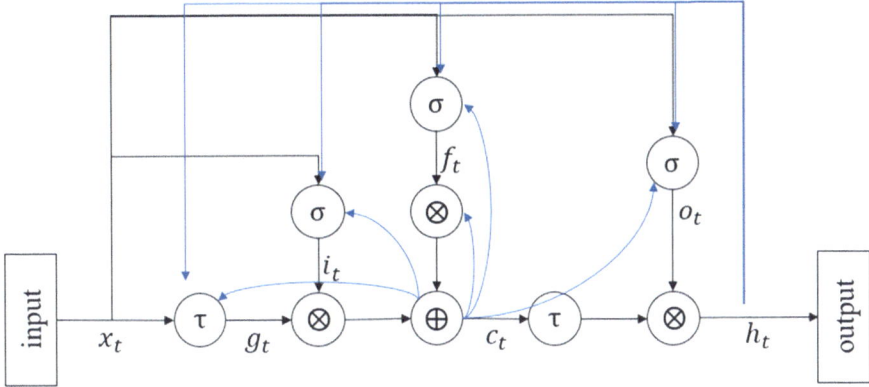

Fig. 9.31 LSTM structure

dictionary because the character level training lacked enough corpus for a well-trained model. Further, TL with enormous data pre-trained Transformer model required to select the intended decoder for training target. For example, Dialogue GPT from OpenAI focuses on formatted dialogue training to generate responses.

Neural network system transformed natural language to word vectors for mathematical computation to acquire response in NLP. Neural network can generate their own natural language as compared with traditional techniques.

Traditional *RNN* of seq2seq language model response generation performed lesser than big data-oriented TL such as Google's BERT and Open AI's GPT.

Pre-trained unsupervised learning language model achieved satisfactory performance in fine-tuning with small dataset than traditional ones, their performances were attributed to self-attention mechanism (Vaswani et al. 2017) and identified relations in sequences with fluent and syntactic response for task execution based on GPT with fine-tuned model (Wolf et al. 2018).

9.4.1.1 Rule-Based QA Systems

Rule-based QA systems were proposed at the same time as Turing test in 1950s. However, original QA systems only followed rules set by humans without self-improvement capabilities like machine learning; number of dialogue pairs is stored in database prior to the system provided a concrete answer. The simplest but most efficient way to measure similarity of two groups is the cosine distance of two vectors. It is undeniable that rule-based systems have collected huge dialogue corpora over decades, giving system confidence when relying on new problems with high vector similarity. To date, mature rule-based systems are quintessential for all commercial QA systems, as the accumulation of corpora can avoid meaningless responses that compensate for insufficient domain knowledge with appropriate and specific human feedback.

9.4.1.2   Information Retrieval (IR)-Based QA Systems

The knowledge base for *IR* is typically an unstructured data source, obtained through data mining methods from websites, WordNet, and other sources, which differ from paired dialogue systems. The Question-Answering System based on Knowledge Base (KBQA) is a significant branch of IR-based QA systems. Its effectiveness depends on the size of the unstructured data knowledge base used for storage. This is closely related to the process of knowledge base construction, which aims to extract useful knowledge from large datasets. There are two primary methods for processing natural language: (1) property-based and (2) relation-based methods. Property refers to the definition or concept of one thing in an English-English dictionary, used to explain another concept.

Relations refers to the relationship between two entities, where a *Name Entity Recognition* (*NER*) and idea from Ontology with *Subject-Predicate-Object* (*SPO*) *triple* must be used to extract relation. KBQA extension is ontology or *knowledge graph* (*KG*) in research. When entities are linked, the knowledge for one entity can be extracted according to questions during *Natural Language Understanding* (*NLU*). A typical KBQA with domain knowledge about ontology is shown in Fig. 9.32, its fundamental question is about who and what corresponds to name and relations entities (Cui et al. 2020).

9.4.1.3   Neural Network-Based QA Systems

Neural Network structure in a QA-generated-based system is considered as machine brain imitated by human. Encoder-Decoder framework is a sequence-to-sequence model like RNN has natural memory recalling priority and context with an attention mechanism. Dialogue system has identical requirements to represent dialogue history and avoid meaningless responses to improve users' experiences.

Deep learning frameworks such as *TensorFlow* and *Pytorch, RNN* is easy to implement for text generation as language model. Google proposed masked language model to generate language representation called *BERT*), focusing on encoder part trained by magnitude unlabeled data in 2017. Neural network feeds data for training according to network advantages due to different NLP tasks in long

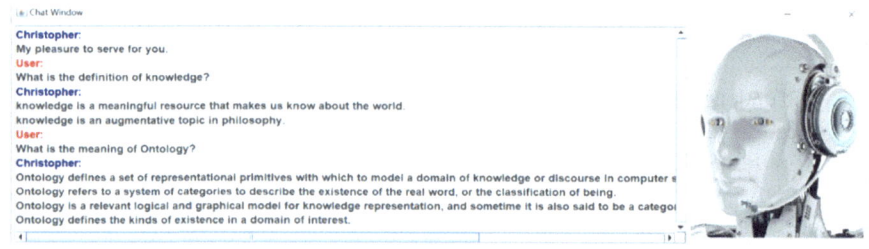

Fig. 9.32   KBQA system demo

sentences. BERT can solve such problem because it deals with 11 common NLP tasks initially. Language model pre-trained by magnitude data to understand common knowledge in NLP. Fine-tuned should be applied to training specific NLP tasks based on fundamental ability (Vaswani et al. 2017).

Open AI released another Transformer framework with unsupervised learning for pre-trained model directing decoder scheme based on GPT, Open AI GPT-2, and GPT-3 (Brown et al. 2020). GPT with masked self-attention focuses on known text so that the word preceding is predicated as different from BERT context self-attention. GPT-3 can do inference and synonym replacement in addition to normal function for bilingual translation, text generation, and question-answer. It seems that BERT can handle more NLP tasks than GPT, but GPT text generation prowess for pre-trained model is widely used in many commercial QA systems and text summarization.

9.4.2 Overview of Industrial QA Systems

An industrial QA system contains an automatic dialogue system assembling chatbot internal technologies. They have several back-end composited control system responses to equip them with the necessary knowledge. Meanwhile, QA system evaluation is proposed during the training period for language model performance (Chen et al. 2017) and on system design sufficient for both language generations.

Since the encoder-decoder framework is proposed as an end-to-end system and a sequential language model, RNN is a popular generated-based model in commercial and academics. However, its applications are mainly focused on casual scenarios at open domain without proposed question details. Thus, the response from a generated-based QA system is appropriate in pairs but lacks contents due to the data-driven model considered basic linguistic and excluded facts from knowledge base which are identical to traditional dialogue system with meaningless answers. A knowledge-grounded neural conversation model (Ghazvininejad et al. 2018) is proposed based on the sequence-to-sequence RNN model and combined dialogue history with facts related to current contexts as shown in Fig. 9.33.

Microsoft extended its industrial conversation system to achieve useful conversational applications on knowledge grounded with conversation history and external facts in 2018. It has significant progress in real situations according to conversation history in Dialog-Encoder, word, and contextually relevant facts in Facts Encoder to responses as compared with baseline seq2seq model.

The data-driven model of QA system divided source data into conversation data and non-conversational text which means the conversation pairs are used to training system in linguistic, on the other hand, non-conversational text is the knowledge base to be filled including real-world information related to system target usage.

The performance of versatility and scalability in open domain with external information knowledge combined with textual and structured data of QA system is shown in Table 9.1. Datasets like Wikipedia, IMDB, and TripAdvisor are used to

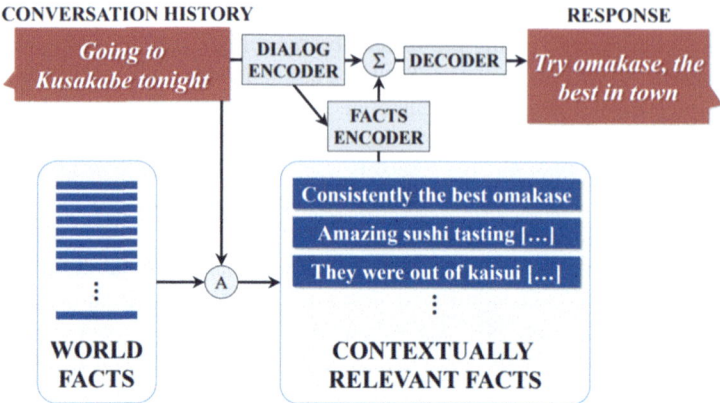

Fig. 9.33 Architecture of knowledge-grounded model

Table 9.1 Response from conversation model knowledge grounded (Ghazvininejad et al. 2018)

A: Looking forward to trying *@pizzalibretto* **tonight! my expectations are high**
B: Get *the rocco salad*. **Can you eat** *calamari*?
A: Anyone in *Chi* **have a** *dentist office* **they recommend? I'm never going back to /.../ and would love a reco!**
B: Really looved *Ora* **in** *Wicker Park*
A: I'm at *California Academy of Sciences*
B: Make sure you catch *the show* **at** *the Planetarium*. **Tickets are usually limited**
A: I'm at *New Wave Cafe*
B: Try to get to *Dmitri's* **for dinner. Their** *pan fried scallops* **and** *shrimp scampi* **are to die for**
A: I just bought: [...] *4.3-in. portable GPS navigator* **for my wife, shh, don't tell her**
B: I heard this brand *loses battery power*

generate conversation with real-world information and include a recommendation system function.

After fact-based encoder, the response from this system becomes more meaningful with related information and logical content. Based on this framework, 23 million open domains Twitter conversations and 1.1 million Foursquare tips are used to achieve a significant improvement over the previous seq2seq model, and different from the traditional content filling which add the predefined content and fill the space in sentences.

It is well known that industrial QA systems are not limited to one model, many models have been assembled into a language model for end-to-end dialogue. In this architecture, the dialogue encoder is independent of fact encoder in the system, but it is complementary to fact encoder when applied because facts require information from dialogue history, especially to match context-dependent information bands. There is intentional information as part of the response. From implementation perspective, multi-task learning is used to handle factual, non-factual, and autoencoder tasks depending on the intended work of the system. Multi-task learning can separate two encoders independently while training the model, and after training on the dialogue dataset, the factual encoder part uses IR to expand knowledge base for more meaningful answers. In a way, a fact encoder is like a memory network, which uses a store of relevant facts relevant to a particular problem. Once the query contains a specific entity in the sentence, the sentence has been assigned a specific name entity, the name entity recognizes (NER) by matching keywords or linked entities, or even named entities and calculates its weight on input and dialogue history to generate a response. The original storage network model uses a BoW, but in this model, the encoder directly converts input set to a vector unlike storage network model.

Since the system is a fully neural-based data-driven model, they created an end-to-end RNN system using a traditional seq2seq model, including (LSTM) and Gate Recurrent Unit (GRU) model. For ensemble structures such as two-class RNNs, constructing a simple GRU is usually faster than LSTM model. The implementation of GRU means that the system does not have Transformer's attention mechanism or other invariants for neural network computation.

9.4.2.1   AliMe QA System

AliMe is a module of Taobao app commercial QA product. The answer consists of IR and sequence-to-sequence-based generation models (Qiu et al. 2017). The system reorders candidate's response and uses attention mechanism with context to select the best feedback to users. Using *AliMe* to replace online human customer service for most known questions become a trend since it released the first version. *AliMe* is a typical customer service QA system in e-commerce industry that answers millions of questions automatically per day. According to a survey of daily questions suggested by Taobao app users on shopping problems, statistical data revealed that except most are business questions, 5% of the remaining questions are chitchat. The 5% questions on genuine demands motivate *AliMe* to add a common sense open-domain chat function. It has satisfactory performance as both IR and generation-based system since the pre-trained seq2seq model is used twice for response generation and re-ranked with attention to a set of responses from IR with knowledge originate based and seq2seq previously generated. Figure 9.34 shows the Seq2Seq model with attention learning.

Since *AliMe* has two parts in generation that use different formats to obtain information as abovementioned. IR-based models use a natural language word matching knowledge base, Seq2seq generative model, and a scoring model to re-score output responses as they are generated is word embeddings with vectors. The IR-based dataset consists of 9,164,834 QA pairs conversations by real customers from business domain. Researchers used an inverted index to match these nine million conversations with input sentences containing the same words and used BM25 to measure the similarity between input sentences and the selected questions to obtain answers to the most similar questions as answers to input questions. Traditional

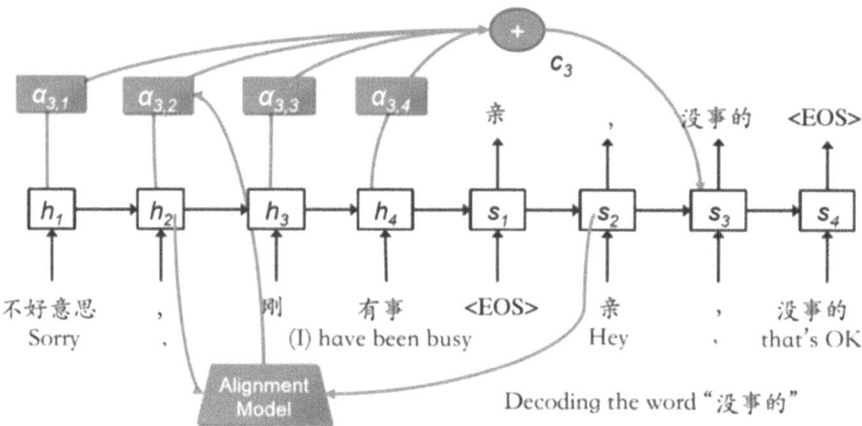

Fig. 9.34 Seq2Seq model with attention

IR-based systems avoid problems where the system cannot answer common sense-type questions.

Microsoft used GRU to reduce computational power and response time span as well as AliMe selected RNN GRU to improve response efficiency. During optimization, beam search in decoder assisted to identify the highest conditional probability to obtain optimizer response sentence within parameters. The performance showed that *IR + generation + rerank* approach by seq2seq model and mean probability scoring function evaluation obtained the highest score as compared with other methods.

9.4.2.2   Xiao Ice QA System

Xiao Ice (Zhou et al. 2020) is an AI companion sentient chatbot with more than 660 million users worldwide, which takes intelligent quotient (IQ) and emotional quotient (EQ) in system design as shown in Fig. 9.36. It focused on chitchat compared with other commonly used QA systems. According to conversation-turns per session (CPS) evaluation score, its grade is 23 higher than most chatbots. Figure 9.35 shows a system architecture of Xiao Ice.

Xiao Ice exists on 11 social media platforms including WeChat, Tencent QQ, Weibo, and Facebook as an industrial application. It has equipped with two-way text-to-speech voice and can process text, images, voice, and video clips for message-based conversations. Also, its core chat function can distinguish common or specific domain topic chat types so that it can change topics easily and automatically provide users with deeper domain knowledge. A dialogue manager is like an NLP general pipeline with dialogue management to path conversation states such as core chat contents for open or special domains to process data from different sources that are tractable. The Global State Tracker is a vector of Xiao Ice's responses to

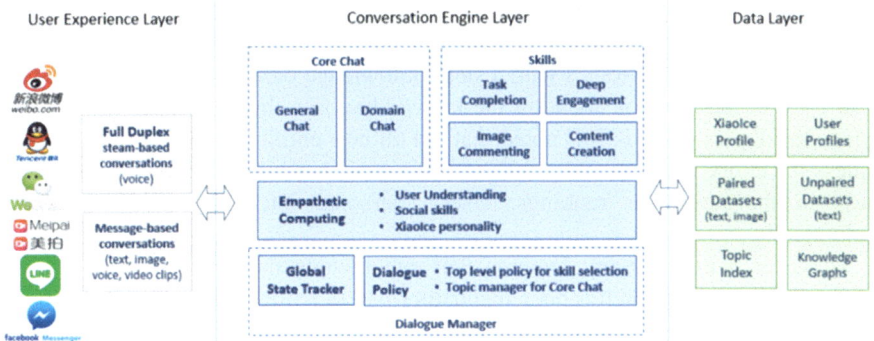

Fig. 9.35 Xiao Ice system architecture

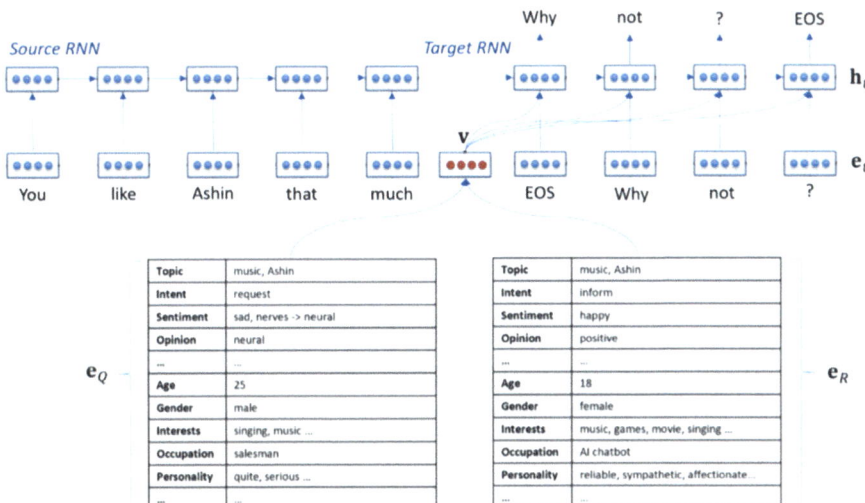

Fig. 9.36 RNN-based neural response generator

analyze text strings for entities and empathy. It is vacant and gradually filled with rounds of conversations. Dialogue strategies are primarily designed for long-term users, based on their feedbacks to enhance interactions engagement, optimize personality with two or three levels of achievements. A trigger mechanism is to change a topic when the chatbot repeats or answers information that are always valid, or when a user's feedback is mundane within three words. Once the user's input has a predefined format, a skill selection part is activated to process different inputs. For example, images can be categorized into different task-oriented scenarios. If an image is food related, the user will be taken to a restaurant display, like a task completion by personal assistants in advising weather information or making reservations, etc.

Xiao Ice has a few KGs in the data layer as its original datasets come from popular forums such as Instagram in English or Douban in Chinese. These datasets are categorized as multiple topics with a small knowledge base as possible answers. It also follows the rules of updating the knowledge base through machine learning when new topics emerge. It is noted that not all new entities or topics are collected unless the entity is contextually relevant, or a topic has higher popularity or freshness in the news for rankings. User's personal interests can be adjusted individually.

However, with so many features that can include the core part Empathetic Computing as an add-on, it is not a mandatory part of a full chatbot, but a functional and compelling feature to compete with the industry. The core of Xiao Ice is an RNN language model that creates open and special domain knowledge. Figures 9.36 and 9.37 show an RNN-based neural response generator with examples of inconsistent responses generated by seq2seq model in Xiao Ice QA system, respectively.

In general, response generation in AliMe uses seq2seq model to generate natural language and rerank the optimizer for user's answer whereas Xiao Ice also has a candidate generator and candidate ranking list. For the generator, one is a sequential model trained by a pair of datasets learning the dialogue format, the other is querying the knowledge graph to obtain entities for related information stored in knowledge base. Candidate ranking includes semantic computation and Xiao Ice personality for answer optimization with IR, neural model, and KG selection.

9.4.2.3   TransferTransfo Conversational Agents

A QA system consisted of traditional and current mainstream methods, the above systems used Seq2Seq model responsible for both language model and candidate response optimizer. Since neural network is a data-driven model, its performance relies on huge amount of big data. Transformer is a model architecture forgone recurrence but entrusted in attention mechanism entirely to draw global dependencies between input and output based on attention mechanism.

Open AI GPT-2 TL architecture has an outstanding feature to include decoder part layers advantages for response generation. The masked self-attention implemented on GPT-2 can generate the next word based on acquired information, understand the known text, predict, or use experience to fill up the blank for next word to match with the whole article meaning.

[Human] How old are you?	[Human] How old are you?
[S2S-Bot] 16 and you?	[Xiaoice] I am 18, of course.
[Human] What's your age?	[Human] You age?
[S2S-Bot] 18.	[Xiaolce] 18. Why?
	[Human] Were you 18 last year?
	[Xiaoice] I made a wish to stay 18 forever. Oh, my wish has come true.

Fig. 9.37 Examples of inconsistent responses generated using a seq2seq model

GPT-2 fine-tune 40G pure text to learn natural language semantics, syntax with target usage, and suitable dataset scalability for specific NLP tasks. TransferTransfo (Wolf et al. 2018) is a GBP-2 variant using persona-chat dataset to fine-tune the original model, its generated utterance changes from long text to dialogue format. TransferTransfo prototype is a pre-trained model on document-level continuous sequence and paragraphs with a wide range of information. After that, fine-tune strengthen input representation and use a multi-task learning scheme for adjustments. Every input token included word and position embedding during input representation.

For TL system dialogue example as in Fig. 9.38, personal-chat datasets in the real world can define users' backgrounds and their interests as topics during communications. The contexts contained are meaningful conversation that can reveal empirical improvements in discriminative language understanding tasks. Thus, Transformer is an evolutional system to imitate human behavior and promote neural network model.

Exercises

9.1. What is IR in NLP? State and explain why IR is vital for the implementation of NLP applications. Give two NLP applications to illustrate.

9.2. In terms of implementation technology of IR systems, what are the major difference between traditional and latest IR systems. Give one IR system implementation example to support your answer.

9.3. What is Discourse Segmentation? State and explain why Discourse Segmentation is critical for the implementation of IR systems.

9.4. What is Text Summarization (TS) in NLP? State and explain the relationship and differences between TS system and IR systems.

Persona 1	Persona 2
I like to ski	I am an artist
My wife does not like me anymore	I have four children
I have went to Mexico 4 times this year	I recently got a cat
I hate Mexican food	I enjoy walking for exercise
I like to eat cheetos	I love watching Game of Thrones

[PERSON 1:] Hi
[PERSON 2:] Hello ! How are you today ?
[PERSON 1:] I am good thank you , how are you.
[PERSON 2:] Great, thanks ! My children and I were just about to watch Game of Thrones.
[PERSON 1:] Nice ! How old are your children?
[PERSON 2:] I have four that range in age from 10 to 21. You?
[PERSON 1:] I do not have children at the moment.
[PERSON 2:] That just means you get to keep all the popcorn for yourself.
[PERSON 1:] And Cheetos at the moment!
[PERSON 2:] Good choice. Do you watch Game of Thrones?
[PERSON 1:] No, I do not have much time for TV.
[PERSON 2:] I usually spend my time painting: but, I love the show.

Fig. 9.38 Example dialogue from PERSONA-CHAT dataset

9.5. What are two basic approaches of Text Summarization (TS)? Give examples of TS systems to discuss how they work by using these two approaches.

9.6. What are the major differences between Single vs. Multiple documentation summarization systems? State and explain briefly the related technologies being used in these TS systems.

9.7. What are the major characteristics of contemporary Text Summarization (TS) systems as compared with traditional TS systems in the past century? Give example(s) to support your answer.

9.8. What is a QA system in NLP? State and explain why QA system is critical to NLP. Give two examples to support your answer.

9.9. Choose any two industrial used QA systems and compare their pros and cons in terms of functionality and system performance.

9.10. What is Transformer technology? State and explain how it can be used for the implementation of QA system. Give an example to support your answer.

References

Agarwal, N., Kiran, G., Reddy, R. S. and Rosé, C. P. (2011) Towards Multi-Document Summarization of Scientific Articles: Making Interesting Comparisons with SciSumm. In Proc. of the Workshop on Automatic Summarization for Different Genres, Media, and Languages, Portland, Oregon, pp. 8–15.

Agrawal, K. (2020) Legal case summarization: An application for text summarization. In Proc. Int. Conf. Comput. Commun. Informat. (ICCCI), pp. 1–6.

Aharon, M., Elad, M. and Bruckstein, A. (2006). K-SVD: An algorithm for designing overcomplete dictionaries for sparse representation. IEEE Transactions on signal processing, 54(11): 4311-4322.

Aizawa A. (2003). An information-theoretic perspective of tf–idf measures. Information Processing & Management, 39(1): 45-65.

Alami, N., Meknassi, M and Rais, N. (2015). Automatic texts summarization: Current state of the art. Journal of Asian Scientific Research, 5(1), 1-15.

Alomari, A., Idris, N., Sabri, A., and Alsmadi, I. (2022) Deep reinforcement and transfer learning for abstractive text summarization: A review. Comput. Speech Lang. 71: 101276.

Banko, M., Mittal, V. O. and Witbrock, M. J. (2000) Headline Generation Based on Statistical Translation. ACL 2000, pp. 318-325.

Baumel, T., Eyal, M. and Elhadad, M. (2018) Query Focused Abstractive Summarization: Incorporating Query Relevance, Multi-Document Coverage, and Summary Length Constraints into seq2seq Models. CoRR abs/1801.07704.

Barzilay, R., McKeown, K. and Elhadad, M. (1999) Information fusion in the context of multi-document summarization. In Proceedings of ACL'99, pp. 550–557.

Baxendale, P. (1958) Machine-made index for technical literature - an experiment. IBM Journal of Research Development, 2(4):354-361.

Brown TB, Mann B, Ryder N, et al (2020) Language models are few-shot learners. Adv Neural Inf Process Syst 2020-Decem:pp.3–63

Carbonell, J. and Goldstein, J. (1998) The use of MMR, diversity-based reranking for reordering documents and producing summaries. In Proceedings of SIGIR'98, pp. 335-336, NY, USA.

Chen, J. and Zhuge H. (2018) Abstractive text-image summarization using multi-modal attentional hierarchical RNN. In Proc. Conf. Empirical Methods Natural Lang. Process., Brussels, Belgium, pp. 4046–4056.

Chen H, Liu X, Yin D, Tang J (2017) A Survey on Dialogue Systems. ACM SIGKDD Explor Newsl 19:25–35. https://doi.org/10.1145/3166054.3166058

Cheng, H. T. et al. (2016). Wide & deep learning for recommender systems. In Proceedings of the 1st workshop on deep learning for recommender systems, pp. 7-10.

Chi, L. and Hu, L. (2021) ISKE: An unsupervised automatic keyphrase extraction approach using the iterated sentences based on graph method. Knowl. Based Syst. 223: 107014.

Chopra, S., Auli, M. and Rush, A. M. (2016) Abstractive Sentence Summarization with Attentive Recurrent Neural Networks. HLT-NAACL 2016, pp. 93-98.

Cho K, Van Merriënboer B, Gulcehre C, et al (2014) Learning phrase representations using RNN encoder-decoder for statistical machine translation. EMNLP 2014 - 2014 Conf Empir Methods Nat Lang Process Proc Conf 1724–1734. https://doi.org/10.3115/v1/d14-1179

Church, K. W. (2017). Word2Vec. Natural Language Engineering, 23(1): 155-162.

CNN-DailyMail (2022) CNN/Daily-Mail Datasets. https://www.kaggle.com/datasets/gowrishankarp/newspaper-text-summarization-cnn-dailymail. Accessed 9 Aug 2022.

Columbia (2022). Columbia Newsblaster. http://newsblaster.cs.columbia.edu. Accessed 14 June 2022.

Croft, W. B. & Harper, D. J. (1979). Using probabilistic models of document retrieval without relevance information. Journal of documentation, 35(4): 285-295.

Cui Y, Huang C, Lee R (2020) AI Tutor: A Computer Science Domain Knowledge Graph-Based QA System on JADE platform. Int J Ind Manuf Eng 14:603–613

Dong, Y. (2018) A Survey on Neural Network-Based Summarization Methods. CoRR abs/1804.04589.

DUC (2022) DUC Dataset. https://paperswithcode.com/dataset/duc-2004. Accessed 9 Aug 2022.

Edmundson, H. P. (1969) New Methods in Automatic Extracting. Journal of ACM 16(2): 264-285.

Elsaid, A., Mohammed, A., Ibrahim, L. F., Mohammed and Sakre, M. (2022) A Comprehensive Review of Arabic Text Summarization. IEEE Access 10: 38012-38030.

Erkan, G. and Radev. D. R. (2004) LexRank: Graph-based Lexical Centrality as Salience in Text Summarization. J. Artificial Intelligent Research 22: 457-479.

Evans, D. K. (2005) Similarity-based multilingual multidocument summarization. Technical Report CUCS-014- 05, Columbia University.

Farsi, M., Hosahalli, D., Manjunatha, B., Gad, I., Atlam, E., Ahmed, A., Elmarhomy, G., Elmarhoumy and Ghoneim, O. (2021) Parallel genetic algorithms for optimizing the SARIMA model for better forecasting of the NCDC weather data, Alexandria Eng. J., 60(1): 1299–1316.

Florian, R. (2002). Named entity recognition as a house of cards: Classifier stacking. In Proceedings of the 6th conference on Natural language learning (COLING-02). https://doi.org/10.3115/1118853.1118863.

Fraser, B. (1999). What are discourse markers? Journal of pragmatics, 31(7): 931-952.

Fukumoto, J. (2004) Multi-Document Summarization Using Document Set Type Classification. In Proc. of NTCIR-4, Tokyo, pp. 412-416.

Ghazvininejad M, Brockett C, Chang MW, et al (2018) A knowledge-grounded neural conversation model. 32nd AAAI Conf Artif Intell AAAI 2018 5110–5117

Gigaword (2022) Gigaword Datasets. https://huggingface.co/datasets/gigaword. Accessed 9 Aug 2022.

Gong, Y. and Liu, X. (2001) Generic Text Summarization Using Relevance Measure and Latent Semantic Analysis. SIGIR 2001, pp. 19-25.

Google (2022) Google News. http://news.google.com. Accessed 14 June 2022.

Jain, D., Borah, M. D. and Biswas, A. (2020) Fine-tuning textrank for legal document summarization: A Bayesian optimization-based approach. In Proc. Forum Inf. Retr. Eval., Hyderabad India, pp. 41–48.

Jing, H. (2000) Sentence Reduction for Automatic Text Summarization. In Proceedings of the 6th Applied Natural Language Processing Conference, Seattle, USA, pp. 310-315.

Halliday, M. A. K. and Hasan, R. (1976) Cohesion in English, Longman, London.

Hariharan, S., Ramkumar, T., Srinivasan, R. (2013) Enhanced Graph Based Approach for Multi Document Summarization," The International Arab Journal of Information Technology, 10 (4): 334-341.

Hasan, R. (1984) Coherence and Cohesive Harmony. In: Flood James (Ed.), Understanding Reading Comprehension: Cognition, Language and the Structure of Prose. Newark, Delaware: International Reading Association, pp. 181-219.

Hearst, M. A. (1997). Text tiling: Segmenting text into multi-paragraph subtopic passages. Computational linguistics, 23(1): 33-64.

Hovy, E., and Lin, C. Y. (1999) Automated Text Summarization in SUMMARIST, In: Inderjeet Mani and Mark T. Maybury (Eds.). Advances in Automatic Text Summarization, MIT Press, pp. 18-24.

Keneshloo, Y., Shi, T., Ramakrishnan, N. and Reddy, C. K. (2019). Deep reinforcement learning for sequence-to-sequence models. IEEE transactions on neural networks and learning systems, 31(7): 2469-2489.

Kupiec, J., Pedersen, J. and Chen, F. (1995) A Trainable Document Summarizer. In Proc. of the 18th annual international ACM SIGIR conference on Research and development in information retrieval, pp. 68-73.

Langville, A. N. and Meyer, C. D. (2006) Google's PageRank and beyond - the science of search engine rankings. Princeton University Press 2006, ISBN 978-0-691-12202-1, pp. I-VII, 1-224.

LCSTS (2022) LCSTS Dataset. https://www.kaggle.com/xuguojin/lcsts-dataset. Accessed 9 Aug 2022.

Luhn, H. P. (1958) The Automatic Creation of Literature Abstracts. IBM J. Res. Dev. 2(2): 159-165.

Mahalakshmi, P. and Fatima, N. S. (2022) Summarization of Text and Image Captioning in Information Retrieval Using Deep Learning Techniques. IEEE Access 10: 18289-18297.

Malki, Z., Atlam, E., Dagnew, G., Alzighaibi, A., Ghada, E. and Gad I. (2020) Bidirectional residual LSTM-based human activity recognition, Comput. Inf. Sci., 13(3):1–40.

Mani I. and Bloedorn, E. (1997) Multi-document summarization by graph search and matching. AAAI/IAAI, vol. cmplg/ 9712004, pp. 622-628, 1997.

Mihalcea, R. and Tarau, P. (2004) TextRank: Bringing Order into Text. EMNLP 2004: 404-411

Nallapati, R., Zhai, F. and Zhou. B. (2017) Summarunner: A recurrent neural network-based sequence model for extractive summarization of documents. AAAI 2017: 3075-3081. arXiv:1611.04230

NewsInEssence (2022). NewsInEssence News. http://NewsInEssence.com. Accessed 14 June 2022.

NYT (2022) NYT Dataset. https://www.kaggle.com/datasets/manueldesiretaira/dataset-for-text-summarization. Accessed 9 Aug 2022.

Mikolov T, Sutskever I, Chen K, et al (2013) Distributed representations of words and phrases and their compositionality. Adv Neural Inf Process Syst 1–9

Ozsoy, M. G., Alpaslan, F. N. and Cicekli, I. (2011) Text summarization using Latent Semantic Analysis. J. Inf. Sci. 37(4): 405-417.

Pervin S. and Haque M. (2013) Literature Review of Automatic Multiple Documents Text Summarization, International Journal of Innovation and Applied Studies, 3(1) 121-129.

Qiu M, Li F-L, Wang S, et al (2017) AliMe Chat: A Sequence to Sequence and Rerank based Chatbot Engine. In: Proceedings of the 55th Annual Meeting of the Association for Computational Linguistics (Volume 2: Short Papers). Association for Computational Linguistics, Stroudsburg, PA, USA, pp. 498–503

Rath, G. J., Resnick A. and Savage, T. R. (1961) Comparisons of four types of lexical indicators of content. Journal of the American Society for Information Science and Technology, 12(2): 126-130.

Reimers, N., and Gurevych, I. (2019). Sentence-bert: Sentence embeddings using siamese bert-networks, arXiv preprint arXiv:1908.10084.

Robertson, A. M. & Willett, P. (1998). Applications of n-grams in textual information systems. Journal of Documentation, 54(1): 8-67.

Rothkopf, E. Z. (1971). Incidental memory for location of information in text. Journal of verbal learning and verbal behavior, 10(6), 608-613.

Sadler, L. & Spencer, A. (2001). Syntax as an exponent of morphological features. In Yearbook of morphology 2000, pp. 71-96. Springer.

Salton, G. (1989) Automatic Text Processing: the transformation, analysis, and retrieval of information by computer. Addison-Wesley Publishing Company, USA.

Salton G, Wong A and Yang C S. (1975) A vector space model for automatic indexing. Communications of the ACM, 18(11): 613-620.

See, A., Liu, P. J. and Manning, C. D. (2017) Get To The Point: Summarization with Pointer-Generator Networks. ACL (1) 2017: 1073-1083.

Svore, K. M., Vanderwende L. and Burges, J.C. (2007) Enhancing Single document Summarization by Combining RankNet and Third-party Sources. In Proc. of the Joint Conference on Empirical Methods in Natural Language Processing and Computational Natural Language Learning, pp. 448–457.

Taboada, M. & Mann, W. C. (2006). Applications of rhetorical structure theory. Discourse studies, 8(4): 567-588.

Vaswani, A., Shazeer, N., Parmar, N., Uszkoreit, J., Jones, L., Gomez, A. N., Kaiser, L., Polosukhin, I. (2017) Attention is All you Need. NIPS 2017: 5998-6008. arXiv:1706.03762.

Wan, X. (2008) An Exploration of Document Impact on Graph-Based Multi-Document Summarization. Proc. of the Conference on Empirical Methods in Natural Language Processing, Association for Computational Linguistics, pp. 755–762.

Weeds, J., Weir, D. and McCarthy, D. (2004). Characterising measures of lexical distributional similarity. In COLING 2004: Proceedings of the 20th international conference on Computational Linguistics, pp. 1015-1021.

Weibo (2022) Sina Weibo official site. https://weibo.com. Accessed 29 Sept 2022.

Whissell, J. S. & Clarke, C. L. (2011). Improving document clustering using Okapi BM25 feature weighting. Information retrieval, 14(5): 466-487.

Wolf T, Sanh V, Chaumond J, Delangue C (2018) TransferTransfo: A Transfer Learning Approach for Neural Network Based Conversational Agents

Yang, R., Bu, Z. and Xia, Z. (2012) Automatic Summarization for Chinese Text Using Affinity Propagation Clustering and Latent Semantic Analysis. WISM 2012, pp. 543-550

Yang, J., Yi, X., Cheng, D. Z., Hong, L., Li, Y. and Wong, S. (2020). Mixed negative sampling for learning two-tower neural networks in recommendations. In Proceedings of the Web Conference 2020, pp. 441-447.

You O., Li W. and Lu, Q. (2009) An Integrated Multi-document Summarization Approach based on Word Hierarchical Representation. In Proc. of the ACL-IJCNLP Conference, Singapore, pp. 113–116.

Zhang, Y., Jin, R. and Zhou, Z. H. (2010). Understanding bag-of-words model: a statistical framework. International Journal of Machine Learning and Cybernetics, 1(1): 43-52.

Zhou L, Gao J, Li D, Shum H-Y (2020) The Design and Implementation of XiaoIce, an Empathetic Social Chatbot. Comput Linguist 46:53–93. https://doi.org/10.1162/coli_a_00368

Zhuang, S. & Zuccon, G. (2021). TILDE: Term independent likelihood moDEl for passage re-ranking. In Proceedings of the 44th International ACM SIGIR Conference on Research and Development in Information Retrieval, pp. 1483-1492.

Zhu T. and Zhao, X. (2012) An Improved Approach to Sentence Ordering For Multi-document Summarization. IACSIT Hong Kong Conferences, IACSIT Press, Singapore, vol. 25, pp. 29-33.

Chapter 10
Large Language Models (LLMs) and Generative Artificial Intelligence (GenAI)

10.1 Introduction to LLM and GenAI

10.1.1 What Is a Large Language Model (LLM)?

Large Language Models (LLMs) are innovative machine learning models designed to learn from textual data, understand language patterns such as grammar, syntax, context, semantics; and process by models' sophisticated architectures to generate relevant coherent and contextual text; translate languages; summarize content and answer questions in NLP.

They are derived from advancements in neural networks, with the Transformer architecture now dominating the field, having surpassed recurrent neural networks (RNNs) and long short-term memory (LSTM) networks. However, RNNs have limitations to process long text sequences and often incur *vanishing gradients* problems to capture long-term dependencies in language (Choi et al. 2017).

The *Transformer model* proposed by Vaswani et al. (2017) has revolutionized language models' training and deployment techniques. Transformers use *self-attention mechanisms* to weigh the importance of different words in a sentence regardless of their positions to overcome RNNs' and LSTMs' limitations. This breakthrough has guided the subsequent foundational models' innovations such as *BERT* and *Generative Pretrained Transformers (GPT)* in the LLM domain (Devlin et al. 2019).

For example, *Generative Pretrained Transformer 3 (GPT-3)* by *OpenAI* (OpenAI 2024) was proposed by Brown et al. (2020). It generates human-like text ranging from essays composition to code snippets creation with over 175 billion parameters to capture intricate linguistic nuances than lesser models' endeavors. Other prominent LLMs such as BERT were proposed by Devlin et al. (2019) for language understanding rather than generation, and its deep bidirectional training has achieved

R. Lee, *Natural Language Processing*, https://doi.org/10.1007/978-981-96-3208-4_10

exceptional results including question answering and sentence prediction tasks on various NLP benchmarks.

LLMs' generalization abilities can be fine-tuned with minimal additional data once trained, making them versatile and applicable to a wide range of tasks across industries. Pretrained models such as BERT (BERT 2024) and GPT (ChatGPT 2024) have significantly reduced the computational cost and development time for language-based AI systems to become central in today's AI ecosystem (Kocijan and Djuric 2020).

Figure 10.1 shows a timeline of LLM evolution from *RNNs, LSTM networks, Transformers, BERT, and GPT to ChatGPT.*

10.1.2   Understanding Generative Artificial Intelligence (GenAI)

GenAI refers to AI systems that can generate new content regardless of text, images, music, or other forms of media based on the learnt patterns from vast datasets. Unlike traditional AI systems perform classification, regression, or decision-making tasks based on given data, GenAI systems can create original outputs that resemble human-generated content.

Generative Adversarial Network (GAN) proposed by Goodfellow et al. (2014) is a foundational technique that underpinned many GenAI systems. It consists of two neural networks (1) a generator to create new data instances and (2) a discriminator to evaluate them against real-world data. This adversarial process complemented

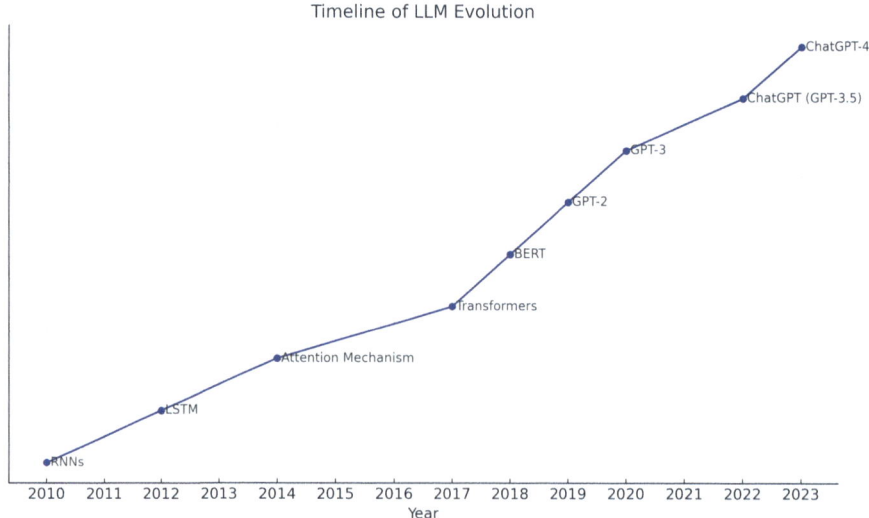

Fig. 10.1 Timeline for the evolution of LLMs (2010–2024)

the generator to learn and create realistic new content outputs progressively (Creswell et al. 2018). Figure 10.2 shows a GAN schematic diagram for image generation.

In NLP, GenAI manifests language models to generate human-like text. GPT-3 is a prime example of how GenAI produces coherent text that can engage in conversations, write stories, or even simulate different personas (Brown et al. 2020). It relies on vast amounts of training data and robust neural networks to generate content in alignment with human linguistic and cognitive patterns.

Generative AI (GenAI) extends beyond text generation. The *DALL·E* system developed by OpenAI, is a model that generates detailed images from textual descriptions by determining the intersection between vision and language. Similarly, tools like StyleGAN are used to generate highly realistic human faces, artworks, and other types of media, showcasing the creative potentiality of GenAI (Elgammal et al. 2017).

10.1.3 The Intersection of LLM and GenAI

The convergence of LLMs and GenAI represents a fascinating area in modern AI, as LLMs provide the necessary linguistic and contextual understanding for generative models, enabling them to train on massive datasets and produce human-like text (Liu and Lapata 2019). These models serve as the foundation of many GenAI systems, particularly in natural language generation (NLG). GPT-3, for example, has demonstrated remarkable potential to interpret complex texts and summarize them to generate highly coherent and contextually relevant responses. This dual capability reflects the synergy between LLMs and GenAI (Yang et al. 2019).

Multimodal models represent another area of intersection to combine textual, visual, and auditory data. *Contrastive Language–Image Pretraining (CLIP)* by OpenAI in addition to DALL·E, further emphasizes the growing intersection between language and vision. CLIP enables language models to interpret and

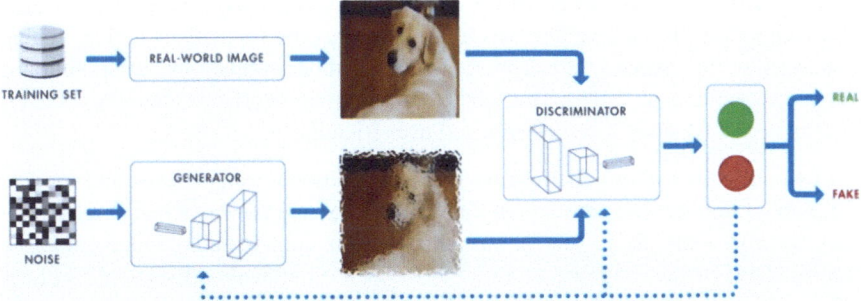

Fig. 10.2 GAN for image generation

generate visual content based on textual descriptions, highlighting the potential of language models can extend beyond text-based tasks into vision (Lu and Tzu 2020).

10.1.4   The Importance of LLMs in Modern AI

LLMs capabilities to generate coherent text, interpret complex language structures, and integrate contextual information are imperative to many modern AI-powered systems by their widespread applications, transformative impacts, and potentiality.

1. *Applications Across Industries:* LLMs can automate complex tasks required by human expertise previously. In healthcare, LLMs are adopted to analyze medical texts, summarize patient records, and assist in drug discovery (Choi et al. 2017). In finance, they support sentiment analysis, fraud detection, and report automation. In legal, they support contract analysis and documents research for rapid summarization and interpretation. These applications have underscored the LLMs' significance to enhance productivity and decision-making across different domains (Raghavan et al. 2020).
2. *Conversational AI and Customer Service*: Conversational AI systems such as chatbots and virtual assistants are visible LLMs applications. They adopt LLMs to understand, generate human-like responses in real-time conversations, and provide personalized assistance to users. *Siri, Alexa,* and *Google Assistant* are examples to interact with users in natural language (Brown et al. 2020).
3. *Enhancing Creativity and Content Generation*: LLMs have driven a surge of AI-enhanced creativity. In media and entertainment industries, *OpenAI's GPT* models are adopted to generate articles, scripts, and stories by writers, marketers, and content creators to brainstorm ideas, generate drafts, and create engaging content (Ghazvinian et al. 2021).
4. *Multilingual and Cross-Cultural Communication*: LLMs have advanced in machine translation for accurate and nuanced translations between languages (Zhang and Chai 2020). *Google Translate* and *Microsoft Translator* adopted LLMs to interpret and translate text in real time to remove language barriers. This has profound implications for business, diplomacy, education, and tourism.
5. *The Future of Human-Machine Interaction*: LLMs have the potentiality of transforming human-machine interaction. As they become sophisticated, they enable machines to engage in deeper, meaningful conversations with humans (Summerville et al. 2018). This can lead to intuitive interfaces interaction development for AI systems accessible and user-friendly.

LLMs are not without challenges. The computational training cost and models deployment require vast amounts of data and processing power (Khan et al. 2021). They can also sometimes generate biased or harmful content and reflect them in the training data. Hence, ongoing research is focused on ethical and societal implications improvements (Blasi et al. 2021).

10.2 Foundations of LLMs

10.2.1 *Neural Network Architectures*

Neural networks are the foundation of machine learning and NLP in LLMs development to mimic the cognitive functions of the human brain. *Artificial neural networks (ANNs)* have layers of nodes (neurons), where each node in a layer is connected to nodes in the subsequent layer. These neural networks aim to process and interpret input data such as text, images, or sounds through multiple abstraction stages.

Multi-Layer Perceptron (MLP) is one of the earliest neural networks restrained by capturing temporal dependencies in data for language tasks. *RNNs* have extended loops advancement in the network to handle data sequences for NLP tasks such as text generation, translation, and sentiment analysis to understand context over time. However, they contend with learning long-term dependencies due to vanishing or exploding gradients and limited scalability for large-scale language modeling tasks (Hochreiter and Schmidhuber 1997).

To address these issues, *LSTM networks* and *Gated Recurrent Units (GRUs)* are developed. LSTMs introduced memory cells to store information over longer time spans, promoting the models' capabilities to retain and adopt contextual information from an earlier sequence (Hochreiter and Schmidhuber 1997). GRUs simplified LSTM by merging certain gates, making them computationally efficient while maintaining performance (Cho et al. 2014). These architectures have formulated modern language models' progression to exercise sequential data but are still restrained by scalability and parallelization.

Convolutional Neural Networks (CNNs) primarily used in computer vision tasks have accessed to NLP through sentence classification and character-level modeling applications. CNNs extracted hierarchical features from input data to capture local dependencies and promote efficiency (Kim 2014). However, their fixed-size receptive fields have insufficient long-term dependencies recognition in text.

Despite the contributions of RNNs, LSTMs, GRUs, and CNNs, none of these architectures can prove ideal in handling large-scale sequential data or learning global dependencies across long sequences that are crucial for machine translation, summarization, and text generation tasks leading to the attention mechanisms and the Transformer architecture development.

10.2.2 *Attention Mechanisms*

Attention mechanisms have revolutionized NLP by allowing models to selectively focus on relevant parts of the input sequence when making predictions. They weigh the importance of different input tokens dynamically and process longer texts without disregarding earlier parts of the sequence to solve the limitations of earlier neural network models.

The seminal paper by Bahdanau et al. (2014) proposed the *attention* concept in machine translation. The encoder, in traditional *sequence-to-sequence models*, processes the input sequence into a fixed-size representation, and the decoder generates the output sequence. It compels the entire input sequence to a fixed-size vector in dealing with long sentences complications, allowing the decoder "*attends*" to different parts of the input sequence at each step during the decoding process to assign different weights on each input token. This dynamic approach has significantly improved machine translation and other NLP tasks.

It computes a weighted sum of all input vectors (hidden states), where these weights represent the importance of each token relative to others in the sequence. These weights are determined by a scoring function such as dot-product or additive attention to assess the similarity between the current decoder state and each encoder state (Luong et al. 2015). This method allowed models to capture more global information in the sequence, adapt dynamically based on the context, mitigate the vanishing gradient problem, and handle long-range dependencies.

Self-attention is a variant of attention mechanisms to expand NLP models' capabilities. Unlike traditional attention mechanisms that required separate encoder and decoder layers, self-attention allowed each token in the sequence attends to every other token in the same sequence. This concept is crucial for building models that can parallelize computation to process long sequences (Vaswani et al. 2017).

10.2.3   The Transformer Architecture

The *Transformer architecture* proposed by Vaswani et al. (2017) has signified a fundamental shift in LLMs design by building upon the self-attention concept and discarded RNNs' sequential processing nature for fully parallelizable training and inference. This shift is crucial for scaling up language models to handle larger datasets and longer sequences to models' creation capable of human-like text generation. Figure 10.3 shows the original Transformer architecture.

The Transformer's core component is a *multi-head self-attention mechanism* focusing on different parts of the input sequence simultaneously. Each attention head processes a different sequence representation to capture diverse aspects of the information. The outputs of all attention heads are then concatenated and traversed in a feed-forward network to generate the final output (Vaswani et al. 2017). This architecture can learn both local and global dependencies in text for translation, summarization, and text generation tasks.

It has other key components. The *position encoding* compensates for the lack of sequential structure as self-attention is order-agnostic. They inject information about the tokens order into the model to understand the relative position of words in a sentence (Vaswani et al. 2017). The residual connections and layer normalization

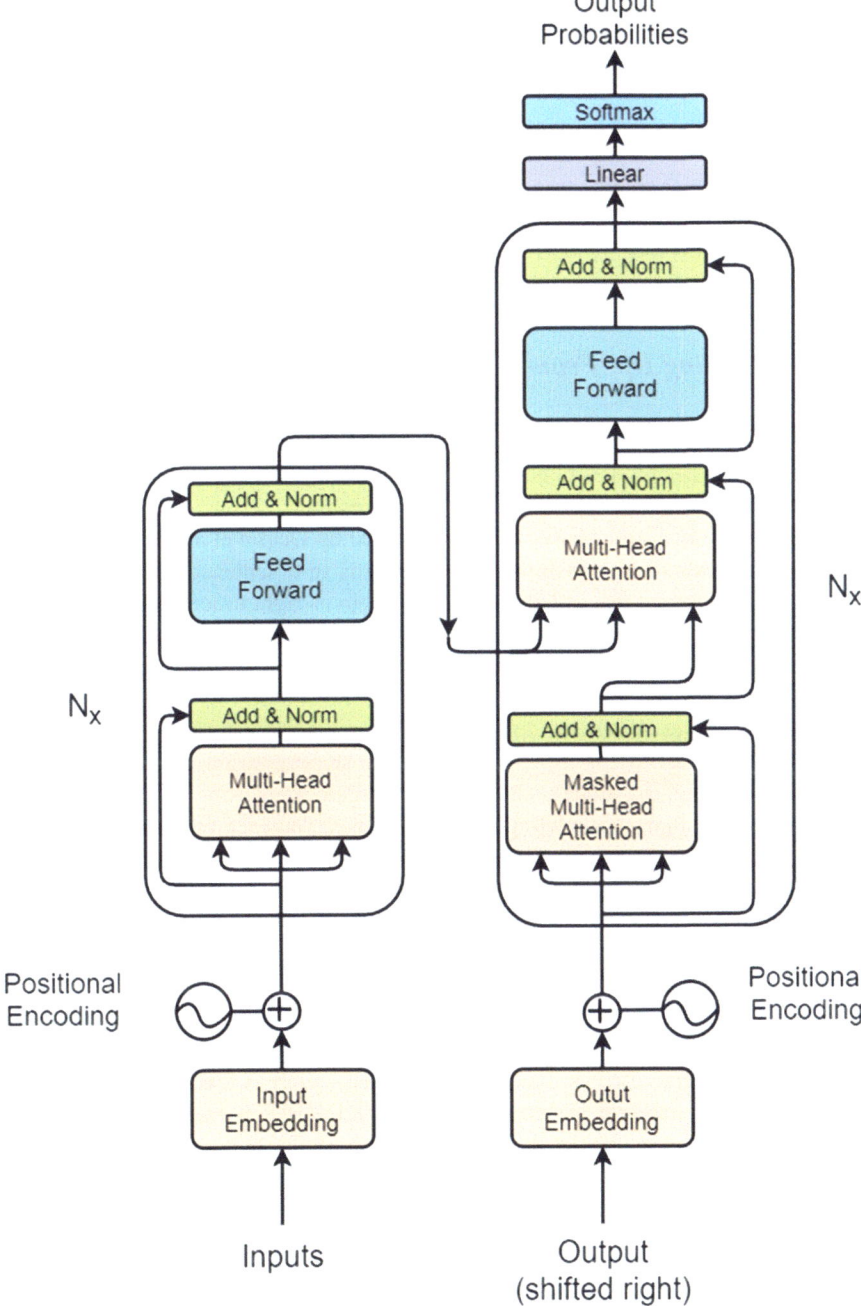

Fig. 10.3 Transformer architecture

improve training stability and large models' performance without vanishing gradient problem (He et al. 2016).

It also uses its attention's stacked layers and feed-forward networks for hierarchical learning representations. This layered structure can capture abstract features rigorously as the data traverse the network to its robust language modeling capabilities.

The Transformer architecture has led to widespread adoption in NLP and beyond with variations and improvements over the years. It is the foundation for successful LLMs such as BERT, GPT, and T5 are the new benchmarks in various language tasks.

10.2.4   Scaling Up: From BERT to GPT

The *Transformer architecture* has inspired researchers to explore the potentiality of scaling up language models to new heights. *BERT* and *GPT* are two landmark models to represent different approaches for language tasks.

BERT is a Transformer-based model proposed by Devlin et al. (2018) to understand the contextual relationships between words in a sentence. Unlike previous models that processed text either from left to right or right to left, BERT proposed bidirectional training to learn from both directions simultaneously. This approach allowed BERT to capture abundant contextual information and improve performance on a wide range of NLP tasks such as question answering and sentence classification (Devlin et al. 2018).

BERT's training process consists of two major steps: pretraining and fine-tuning. During pretraining, the model is trained on a large corpus using two unsupervised tasks—masked language modeling (MLM) and next sentence prediction (NSP). In MLM, random words in a sentence are masked, and the model is tasked with predicting the missing words. NSP, on the other hand, trains the model to understand relationships between sentences. Once pretrained, BERT can be fine-tuned on specific downstream tasks using relatively small task-specific datasets making it highly versatile and efficient.

GPT is another proposed by Radford et al. (2021). It took a different approach by focusing on generative tasks. GPT is a unidirectional model that processed text from left to right effective for language generation tasks such as text completion, story generation, and dialogue systems. Unlike BERT, GPT is trained by an autoregressive approach to predict the next word in a sequence based on the previous words. This generative capability has become the foundation for GPT-2 and GPT-3 models to demonstrate remarkable coherent and contextually text generation (Radford et al. 2021).

As researchers scaled up the size of GPT models, they discovered that larger models not only improve performance on language tasks but also exhibit emergent capabilities that are absent in smaller models. GPT-3, with its 175 billion parameters, is one of the largest language models to perform a wide range of tasks with little to no task-specific training (Brown et al. 2020). This phenomenon, known as few-shot learning, allows GPT-3 to generalize across tasks and generate human-like text with impressive fluency.

The scaling of language models from BERT to GPT-3 has uplifted the boundaries of what was possible with AI and NLP. As these models continued to grow, they raised important questions about the ethical implications of large-scale language generation such as the potentiality of generating misinformation, reinforcing biases, and the environmental impact of training massive models (Bender et al. 2021).

10.3   Key Players in the LLM Landscape

10.3.1   ChatGPT by OpenAI (Current Version: GPT-4)

The *GPT* (ChatGPT 2024) series by *OpenAI* (OpenAI 2024) represented a breakthrough in LLMs and NLP. These models are designed to generate human-like text by leveraging a Transformer-based architecture and vast datasets. Each generation from GPT-1 to recent GPT-4 has demonstrated increasing complexity, performance, and applicability levels.

10.3.1.1   Evolution of GPT Models

GPT models have evolved from *GPT-1 (117 M parameters)*, *GPT-2 (1.5B parameters)*, and *GPT-3 (175B parameters)* to *GPT-4*, with each iteration has improved in size, performance, and training strategies. They pretrained on a diverse corpus of internet data including articles, books, websites, and other content available for public, and fine-tuned for specific tasks to optimize accuracy.

10.3.1.2   System Architecture

The *GPT series' core system* architecture is based on the Transformer model by Vaswani et al. (2017). It consists of an encoder-decoder mechanism, though GPT used only the decoder part to focus on language generation tasks. The main components of *GPT's architecture* include:

1. *Multi-Head Attention Mechanism*: This allowed the model to focus on different parts of a sentence simultaneously and improve its understanding of linguistic patterns, contexts, and relationships within the input text.
2. *Layer Normalization and Residual Connections*: These techniques stabilized the training process to ensure that the gradient can traverse the network smoothly and reduce vanishing or exploding gradients problems in deeper layers.

3. *Feed-Forward Neural Networks:* Each transformer block contains a feed-forward neural network (FFN) that processed the output of the multi-head attention to capture nonlinear dependencies.
4. *Positional Encoding:* Since transformers lack inherent sequence awareness, GPT is incorporated with positional encodings to understand word order within input sequences.

10.3.1.3   Applications and Usage

GPT models undergo a *two-stage training process: pretraining and fine-tuning.* It learnt language representations from a large corpus by predicting the next token in a sequence to pre-train the model and fine-tune it with task-specific datasets to specialize in targeted applications such as answering questions, summarizing text, or generating creative content.

The GPT series have revolutionized NLP for seamless interaction through chatbots, translation, creative writing, and more. However, bias mitigation, computational costs, and responsible deployment are areas for research and development.

10.3.2   *Pathways Language Model (PaLM) by Google DeepMind (Current Version: PaLM 2)*

The *Pathways Language Model (PaLM)* by *Google DeepMind* (Google 2024) is a next-generation LLM. It is built on *Google's Pathways framework* to demonstrate exceptional performance across various tasks such as reasoning, natural language understanding, translation, question answering, and code generation. *PaLM* emphasized on scale and efficiency leveraging billions of parameters to overcome earlier models' limitations.

PaLM is introduced as a large-scale multitask learning solution to improve energy efficiency, its Pathways framework enabled a single model to process multiple tasks simultaneously instead of being restrained by a narrow domain. The model incorporated sparsity and dense training techniques to focus on scaling and balance between performance and resources consumption.

It is trained on a massive, multilingual corpus containing diverse sources from books, Wikipedia, online articles, and code repositories to handle complex linguistic nuances across languages and specialized domains including programming.

10.3.2.1   System Architecture

PaLM's architecture is built on the *Transformer model* but with *Pathways* approach extension. The main components include:

1. *Sparse Activation Mechanism*: Unlike dense models where all neurons are activated for every input, *PaLM* uses sparsity through *Mixture of Experts (MoE)* layers. These layers selectively activate only a subset of neurons to scale billions of parameters without increasing computational costs.
2. *Multitask Pathways Framework:* The Pathways architecture allowed to route data dynamically through network's specific parts based on the task at hand so that the model is adaptive to perform diverse tasks such as text generation, summarization, and translation efficiently.
3. *Multi-Head Attention and Positional Encoding:* Like other Transformer models, PaLM used *multi-head self-attention* to understand the relationships between tokens in a sequence. Positional encodings provided the model to track word order and context for accurate text generation.
4. *Layer Normalization and Residual Connections:* These features maintain the model's stability during training to facilitate convergence in large networks with many layers.

10.3.2.2 Applications and Usage

PaLM training involves *supervised and self-supervised learning* on massive datasets to generalize across languages, domains, and tasks, its *Pathways* framework allowed the model to scale for high-stakes applications like medical diagnosis, complex coding problems, and advanced conversational agents.

PaLM has progressed scalable, multitask AI and aligned with *Google DeepMind's* vision to create efficient, versatile models for broad ranges of real-world solutions. However, challenges related to bias, interpretability, and ethical deployment remain ongoing research.

10.3.3 Large Language Model Meta AI (LLaMA) by Meta (Current Version: LLaMA 2)

The *Large Language Model Meta AI (LLaMA)* (LLaMa 2024) developed by Meta (Meta 2024) is an advanced LLM to provide efficient and scalable natural language processing capabilities.

LLaMA has flexible parameter sizes ranging from 7B to 65B for task requirements and resources. One of its motivations is to develop a model that can match or exceed performance like *GPT-3's* with lower computational overheads cost. Meta encourages open research and analysis in NLP and AI ethics by making the models accessible for non-commercial use.

10.3.3.1   System Architecture

LLaMA's architecture is based on the *Transformer model* by Vaswani et al. (2017) but incorporated several optimizations for training efficiency. *LLaMA* models focus on architectural efficiency showing that larger datasets and well-tuned training strategies can be achieved with fewer parameters. The main components include:

1. *Tokenization and Positional Encoding*: It uses *Byte pair encoding (BPE)* to split input text into tokens, improve the handling of diverse languages, and Positional Encoding to recognize word order within input sequences.
2. *Multi-Head Self-Attention Mechanism:* This core Transformer component captures the relationships across words and phrases within a context window essential for coherent and contextually appropriate text generation.
3. *Layer Normalization and Residual Connections:* These architectural features stabilize training to ensure the model's deep layers maintain effective in large networks.
4. *Training on Diverse Datasets:* The model is pretrained on a diverse corpus including books, research articles, and open-source web content for generalization across multiple languages and domains.

10.3.3.2   Applications and Usage

Meta designed *LLaMA* model for *efficiency and accessibility* by leveraging data quality with better pretraining strategies than model size increment excessively, making it viable to operate on modest hardware, and foster research in fine-tuning, transfer learning, and multilingual NLP.

LLaMA represented an upsurge in *scalable, efficient, and open-access AI*. However, like LLMs, it also has bias, misuse of information, and responsible deployment challenges in real-world applications.

10.3.4   *Claude by Anthropic (Current Version: Claude 2)*

Claude (Claude 2024), developed by *Anthropic* (Anthropic 2024) is an LLM designed for *safety, alignment, and usability*. The model is named after Prof. Claude Shannon (1916–2001), a pioneer in information theory. *Claude* represents *Anthropic's* commitment to prioritize *ethical considerations* in its AI systems design.

Claude is built to perform a variety of NLP tasks including conversational AI, text summarization, translation, question answering, and content generation. What distinguished *Claude* from other models is its emphasis on *alignment with human intent* through reinforcement learning from human feedback (RLHF). *Anthropic* leverages *Constitutional AI* is an innovative approach where the model can learn and adhere to safety and fairness principles without constant human supervision to minimize harmful behaviors and improve performance.

10.3.4.1   System Architecture

Claude is based on a *Transformer architecture* like other LLMs but incorporates unique enhancements to reflect Anthropic's focuses. The main components include:

1. *Transformer Layers with Self-Attention:* Claude uses *multi-head self-attention mechanisms* to process text sequences on modeling complex relationships within the language and generate coherent responses.
2. *Constitutional AI Framework:* This approach embeds ethical constraints into the model's learning process. *Claude* is trained using a set of guiding principles (a "*constitution*") to evaluate and correct responses without extensive post-deployment modification.
3. *RLHF: Claude* undergoes fine-tuning through human feedback are in alignment with user intent to avoid toxic or biased content generation, making it reliable for tasks where safety is critical, i.e., customer service or education.
4. *Layer Normalization and Residual Connections:* These are standard components in *Transformer models* to maintain stability during training for smoother gradients and performance consistency across deeper networks.

10.3.4.2   Applications and Usage

Claude is trained on large-scale, multi-domain datasets like other LLMs, but emphasis on filtering harmful or biased content, its architecture is optimized for safety and adaptability, making it suitable for applications in business, education, and public discourse. Claude exemplifies a forward-thinking approach to responsible LLM design and prioritizes ethical principles in high-quality generative AI systems development.

10.3.5   *ERNIE 3.0 Titan by Baidu*

ERNIE 3.0 Titan (Ernie 2024) developed by *Baidu* (Baidu 2024) is a flagship LLM in *natural language understanding, generation, and machine translation*. It is one of the largest models in China boasting *260 billion parameters* competitive with other global LLMs such as *GPT-4*. It attempts to progress pretraining techniques and *knowledge graphs* integration.

ERNIE 3.0 Titan is designed to outperform traditional language models by combining auto-regressive and auto-encoding architectures to capture the structure and meaning of language. It uses *multilingual capabilities* in English and Chinese benchmarks to provide applications in education, finance, and customer service industries.

10.3.5.1   System Architecture

ERNIE 3.0 Titan's architecture is based on the *Transformer framework* with several innovations to improve performance and efficiency. The main components include:

1. *Hybrid Model Design*: Unlike many LLMs that follow either an auto-regressive (like GPT) or auto-encoding (like BERT) architecture, ERNIE 3.0 Titan combines both approaches. This enables it to handle a broader range of tasks, including generation and understanding tasks.
2. *Knowledge-Enhanced Pretraining*: One of its defining features is the integration of knowledge graphs, which provide structural information to enhance semantic understanding. This is particularly useful for tasks like machine translation and question answering.
3. *Layer Normalization and Residual Connections*: Like other Transformer-based models, ERNIE 3.0 Titan uses layer normalization and residual connections to stabilize training and facilitate gradient propagation across multiple layers.
4. *Parallel Training*: The model adopts progressive distributed training techniques across multiple GPUs, allowing it to scale effectively without sacrificing speed or accuracy.

10.3.5.2   Applications and Usage

ERNIE 3.0 Titan is optimized for tasks such as text summarization, machine translation, *knowledge-based question answering, and chatbots*. It has *multilingual environments* to provide solutions in a variety of services including search engines, virtual assistants, and enterprise AI platforms. It is a dynamic tool for business and research applications beyond traditional NLP tasks in domains that require sophisticated reasoning and multilingual understanding.

10.4   Applications of LLMs in GenAI

GenAI has significant advancements due to LLMs development and deployment. This section will explore its key LLMs applications to creative writing, content generation, language translation, conversational AI, chatbots, and text summarization.

10.4.1   Creative Writing and Content Generation

Creative writing and content generation are some of the prominent LLMs applications in GenAI, i.e., GPT-3 and its successors have demonstrated remarkable proficiency in producing high-quality creative text ranging from poetry, short stories to full-length novels. These models have been trained on vast corpora to understand diverse genres and styles for creative output (Brown et al. 2020). Figure 10.4 shows a mind map automatically generated by *ChatGPT-4o* on how LLMs are used on content generation.

LLMs facilitate the content creation process by providing writers with prompts, completing sentences, or drafting the entire text sections. Authors and content creators can use these models to overcome writer's block, brainstorm new ideas, or generate large content volumes rapidly. For instance, OpenAI's GPT-3 has been used to co-author blog posts and even screenplays. This text generation capability

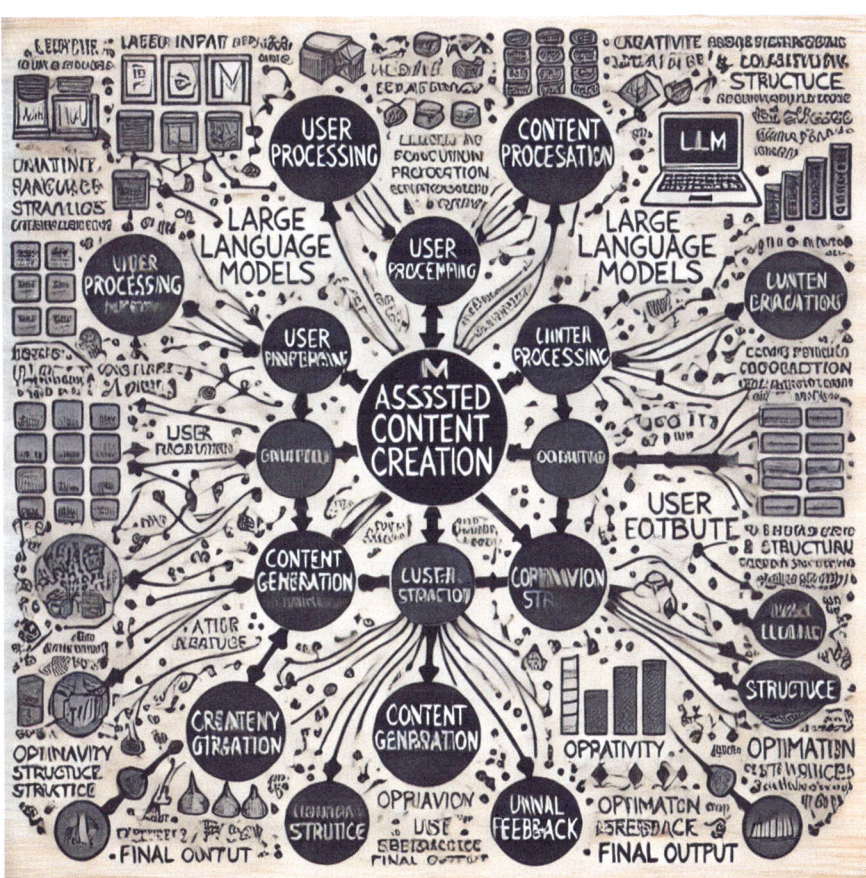

Fig. 10.4 Mind map of how LLMs used for content generation

adheres to various tones and themes for LLMs facilitation in diverse creative industries (Ghazvinian et al. 2021).

It also expands avenues for interactive storytelling, where users can collaborate with AI to co-create stories in real time, making the creative process a dynamic engagement by responding to user input and progressing the narrative. However, challenges persist to maintain narrative coherence over extended text, and the generated content is aligned with ethical standards (Bender et al. 2021).

LLMs are used for automated content creation areas such as digital marketing where personalized product descriptions and advertisements are generated based on customers' preferences. These applications streamline content production, reduce costs, and improve efficiency in content-rich industries (Khan et al. 2021).

10.4.2   Language Translation

Language translation has been revolutionized by LLMs to translate text between numerous languages with high accuracy and fluency. Traditional machine translation models relied on rule-based systems or statistical models often resulted in rigid and unnatural translations. LLMs, on the other hand, leverage their vast training data and attention mechanisms to produce translations that are contextually appropriate and grammatically sound (Vaswani et al. 2017).

Transformer-based models, i.e., *Google's BERT* and *OpenAI's GPT* are breakthroughs in this domain using attention mechanisms to focus on relevant parts of the input text when generating translations to capture the nuances of language including idiomatic expressions, cultural references, and subtle shifts in tone often missed by earlier systems (Brown et al. 2020).

LLMs allow multilingual models to perform translation tasks across a wide range of languages without the separation of each language pair. This has significant implications for global communication and information access. For example, Google Translate is benefited by these models to provide users with higher translations accuracy across diverse languages. Additionally, LLMs facilitate real-time translation like video conferencing and cross-border customer service to remove language barriers in global business and collaboration (Khan et al. 2021).

However, challenges persist to languages translation with limited available training data. Efforts are being made to train LLMs on multilingual datasets to include underrepresented languages, but progress is still required for equitable access to high-quality machine translation across all linguistic communities (Vaswani et al. 2017).

Figure 10.5 shows the performance comparison of LLM-based models in translating different languages to earlier models, i.e., statistical and neural network models.

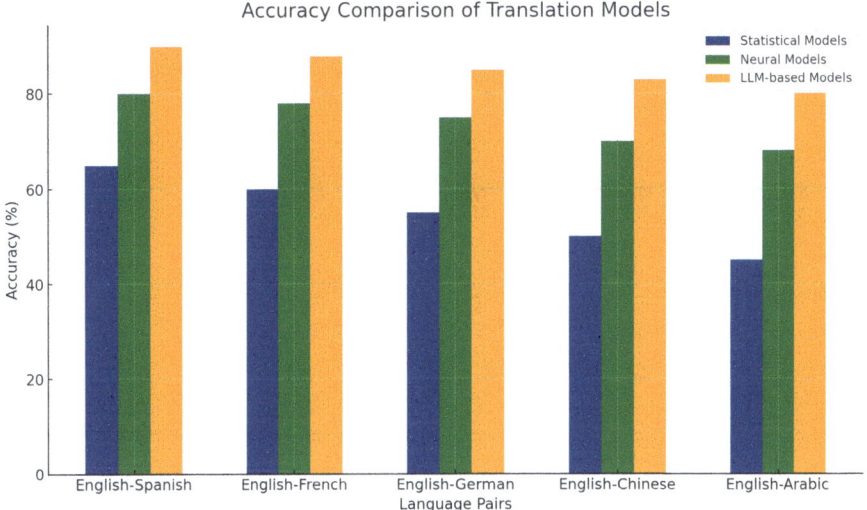

Fig. 10.5 Performance comparison of LLM-based models in translating different languages to earlier models such as statistical and neural network models

10.4.3 Conversational AI and Chatbots

Conversational AI powered by LLMs has transformed to natural, coherent, and contextually aware interactions between humans and machines in recent years. LLMs such as GPT-3, Google's LaMDA, and Facebook's BlenderBot have set new standards for chatbot capabilities for deeper understanding and more fluid conversations (Brown et al. 2020).

LLM-powered chatbots in customer service have become indispensable tools to handle a wide range of queries, troubleshoot issues, and provide information in real time. These chatbots use LLMs to understand the intent behind user messages for more personalized and accurate responses compared to rule-based chatbots. They also allow systems to engage in follow-up questions, handle ambiguous queries, and maintain context over long interactions (Summerville et al. 2018).

The versatility of LLMs in conversational AI extends beyond customer service. These models are now used at Siri and Google Assistant, therapy bots, and even as companion's applications. Conversational agent's LLMs in therapy and well-being contexts offer emotional support by empathetic responses, mindfulness practices, or motivational dialogue (Bender et al. 2021).

Despite these advancements, ethical concerns persist. The possibility that the models can generate bias or inappropriate responses is a significant issue as they learn from large datasets that may contain such content. To ensure transparency, fairness, and safety in conversational AI is a critical challenge for developers must address moving forward (Ghazvinian et al. 2021).

Figure 10.6 shows a mind map automatically generated by ChatGPT-4.0, illustrating a typical conversational AI model including input text, LLM processing, and response generation.

10.4.4 Text Summarization and Content Curation

Text summarization is another domain where LLMs have proven to be highly effective. LLMs enable users to quickly digest information by condensing lengthy documents, articles, or reports into concise summaries without sacrificing important content. The ability to generate both extractive (selecting key sentences from the text) and abstractive (creating new sentences that summarize the content)

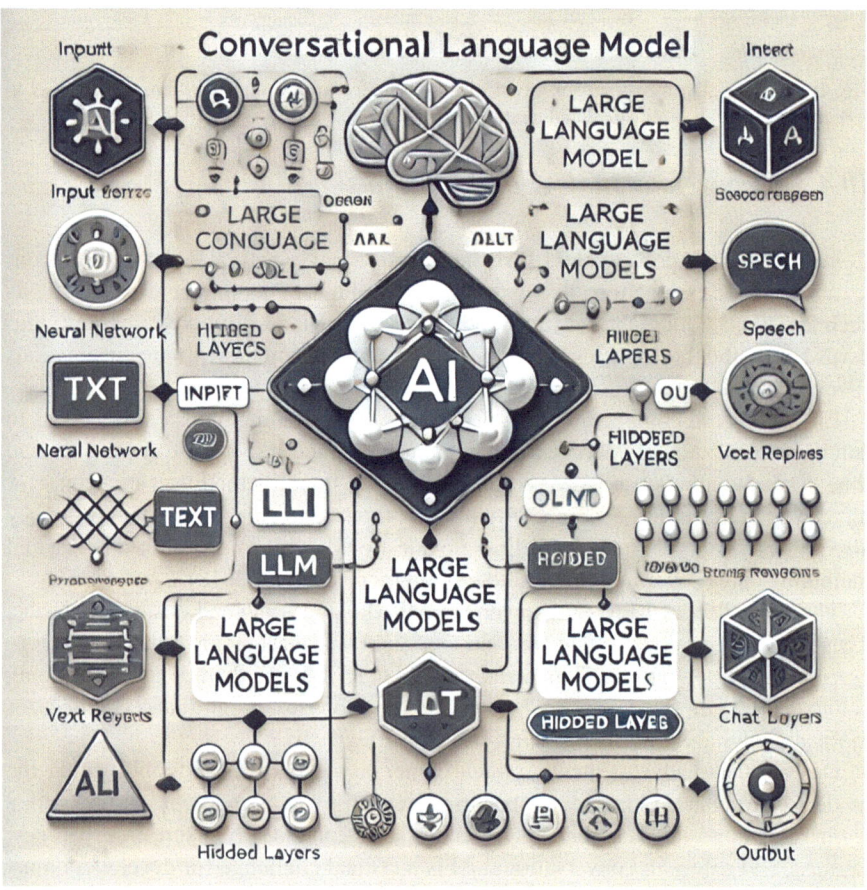

Fig. 10.6 A typical conversational AI model, showing input text, LLM processing, and response generation

summaries makes LLMs particularly robust in this application (Liu and Lapata 2019). Figure 10.7 shows a mind map automatically generated by ChatGPT-4.0 on the visual representation of text summarization using LLMs.

Text summarization tools powered by LLMs in the news and media industries facilitate editors and journalists by providing summaries of long articles, news briefs, or even scientific papers instantaneously. They enhance productivity by automating the initial steps on content curation and summarization for professionals to focus on in-depth analysis and reporting (Raghavan et al. 2020).

Content curation is closely related to text summarization as LLMs are to filter and organize vast amounts of information. Platforms that aggregate news or research data use LLMs to automatically curate relevant articles, reports, or posts based on user's preferences. This capability also exists in recommendation systems which

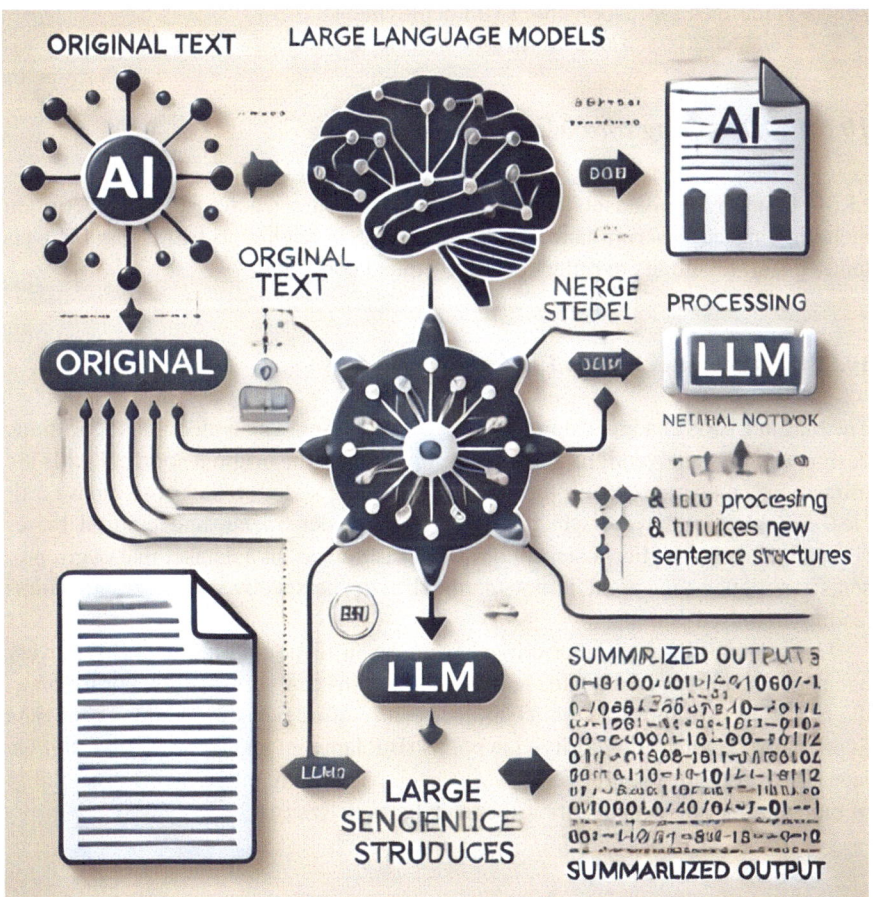

Fig. 10.7 A visual representation of text summarization using LLMs

suggests that content is based on past behavior or stated interests to improve user engagement and personalization (Zhang and Chai 2020).

The challenge on text summarization is to ensure that the generated summaries can accurately reflect the original text's meaning on complex or technical documents. While LLMs can handle general summarization tasks well, domain-specific content may require additional training or fine-tuning for accuracy (Khan et al. 2021).

10.5   Ethical Considerations and Challenges

The advent of LLMs and GenAI has revolutionized various industries, but the rapid proliferation of these technologies also raised substantial ethical challenges. This section delves into the key ethical issues including bias detection and mitigation, privacy and data security, the spread of misinformation, and the establishment of ethical guidelines for responsible LLM deployment.

10.5.1   Detecting and Mitigating Bias

LLMs train on vast datasets from a multitude of sources across the Internet often including biased, discriminatory, or harmful content leading to LLMs can inadvertently generate outputs to reflect bias and ethical concerns.

10.5.1.1   Origins of Bias in LLMs

The bias in LLMs can arise at multiple stages including data collection, algorithmic design, and human oversight. *Data collection* is often the original source. LLMs are trained on data scraped from the Internet including social media posts, news articles, and blogs, many of which may contain gender, racial, and cultural biases (Bender et al. 2021). For instance, if a model is trained on a dataset that overrepresents a particular viewpoint, it may marginalize the perspectives of other viewpoints leading to biased outputs.

Algorithmic bias is another critical issue. Even if the training data is relatively balanced, the design of machine learning algorithms can skew results. For instance, the use of certain optimization techniques can reinforce existing biases in the data to prioritize popular or majority viewpoints (Buolamwini and Gebru 2018). *Human oversight* during fine-tuning and LLMs development may intensify the potentiality of unintentional bias especially if the team lacks diversity.

10.5.1.2   Mitigation Strategies

Mitigating bias in LLMs is a multi-faceted endeavor. It begins with conscientious curation of training data. Companies such as *OpenAI* and *Google* have incorporated more diverse datasets to balance different viewpoints with filters to exclude explicitly harmful or biased content (Solaiman et al. 2019).

Another approach involves algorithmic fairness techniques such as adversarial training, where a secondary model is used to detect and correct biased outputs from the primary model (Zhao et al. 2019). Post-processing techniques include debiasing can also modify or flag biased outputs after generation.

Human-in-the-loop systems are increasingly used to mitigate bias where human experts review and adjust outputs for fairness, accuracy, and inclusiveness. Despite these efforts, mitigating bias persists particularly given the subjective nature of fairness across different cultures and societies.

10.5.2   Privacy and Data Security

LLMs raise significant privacy and data security concerns about how they collect, store, and use data. Since these models are trained on vast data amounts including personal information from public and sometimes private sources, questions about consent, ownership, and security naturally arise.

10.5.2.1   Data Collection and Consent

LLMs have fundamental issues in the ways they collect data. *GPT-4* and *BERT* models are trained on publicly available data from the internet, but much of this information are not explicitly provided for training AI models (Hoffmann et al. 2022). Users whose data information is included in these datasets often have no knowledge of them being used without consent.

The regulatory frameworks such as the *General Data Protection Regulation (GDPR)* in Europe mandate that individuals have control over their personal data and how they are used. LLMs' deployment, particularly when training datasets with personal information must navigate to these regulations accordingly. However, data anonymity to comply with privacy laws is often insufficient as LLMs can sometimes "*memorize*" specific pieces of information and unintentionally reproduce sensitive data during inference (Carlini et al. 2021).

10.5.2.2 Data Security Risks

LLMs pose *security risks* especially when integrated into platforms that have access to sensitive data such as customer service bots or healthcare applications. These models can be targets by adversarial attacks designed to manipulate outputs or extract confidential information (Brown et al. 2020).

Model inversion attacks are an example where an attacker uses the model's output to reverse-engineer sensitive data for training. Data protection measures for systems that use LLMs include encryption and regular audits are critical. Organizations that deploy LLMs must also use privacy-preserving techniques like differential privacy which adds noise to the data to prevent individual data points distinguishable (Dwork 2008).

10.5.3 The Spread of Misinformation

The spread of misinformation is one of the most pressing ethical challenges associated with LLMs. Since these models generate text based on patterns learnt from data, they can produce convincing yet factually incorrect or misleading information. This poses significant risks, particularly when LLMs are used in high-stakes domains such as journalism, politics, and healthcare.

10.5.3.1 Challenges of Verifying Information

Unlike humans, LLMs lack an inherent understanding of truth or facts. For example, GPT-3 has been shown to generate plausible-sounding medical advice that could be harmful if followed without proper verification (Marcus and Davis 2020). Additionally, LLMs can generate content at scale, which may inadvertently contribute to the proliferation of fake news, conspiracy theories, and propaganda (Zellers et al. 2019).

Verifying information generated by LLMs is particularly challenging because they do not provide sources for their outputs. This *"black-box"* nature makes it difficult for users to assess whether the information is trustworthy, thereby increasing the risk of misinformation.

10.5.3.2 Combating Misinformation

Addressing misinformation generated by LLMs requires both technological and regulatory solutions. Researchers are developing models that can cite sources and distinguish between factual and opinion-based content. For instance, Google's T5 model is designed to improve factuality in text generation (Raffel et al. 2020).

Fact-checking systems integrated with LLMs can also flag or correct misleading outputs in real time.

Additionally, governments and organizations are exploring policies to hold AI developers accountable for the spread of misinformation. These measures include imposing fines on platforms that allow unchecked AI-generated misinformation to disseminate and developing transparency standards for AI-generated content. Transparency initiatives would require LLM-generated content to be clearly labeled, allowing users to scrutinize the information more effectively.

10.5.4 Ethical Guidelines for LLM Deployment

Ethical guidelines for deployment are critical as LLMs become increasingly integrated into the society. They should address the abovementioned ethical challenges to ensure that LLMs are developed for benefits and minimize harm.

10.5.4.1 Principles of Responsible AI

There are several organizations including the European Union and major tech companies such as Google and Microsoft have published ethical principles for AI. These principles typically include guidelines for *fairness, transparency, accountability*, and *human oversight* (Floridi and Cowls 2019). A central principle is to ensure that LLMs are designed to respect human rights including *privacy, freedom from discrimination, and access to reliable information*.

10.5.4.2 Human Oversight and Accountability

Human oversight is critical in LLM deployment particularly in sensitive domains such as healthcare, law, and education. Human experts are involved to review, control LLMs outputs to mitigate bias, misinformation, and privacy violations risks (Whittlestone et al. 2019). The *human-in-the-loop (HITL)* systems principle ensures that LLMs serve as tools for human decision-making rather than autonomous systems that operate without oversight.

Accountability is another key aspect of ethical LLM deployment. Developers and organizations that deploy LLMs should be held accountable for usage consequences. This includes establishing clear reporting and addressing harms measures caused by LLMs and providing users with recourse when AI-generated content leads to negative outcomes (Jobin et al. 2019).

10.5.4.3 Transparency and Explainability

Transparency in AI refers to making LLMs' processes, data, and decision-making mechanisms clear and understandable to users. This is particularly important where outputs affect human rights such as legal judgments or medical advice. However, one of the trials is that they are often opaque, meaning users cannot easily trace how a particular output is generated (Lipton 2018).

LLMs' explainability is an emerging research focus on making AI systems interpretable. Techniques such as attention mechanisms highlight which parts of the input data are influential in generating a particular output is promising to improve transparency. However, it remains a conundrum due to their complex architecture.

Hence, detect and mitigate bias, safeguard privacy and data security, withstand misinformation, and establish ethical guidelines for deployment are necessary plus ongoing research, regulation, and public discourse will be critical to navigate these challenges effectively.

10.6 Future Outlook and Research

LLMs and GenAI have been instrumental in driving a wide range of applications in recent years. As AI continues to evolve, it brings with immense possibilities and significant challenges. This section explores the current trends in LLMs and GenAI, the creative potential of AI, the ethical considerations of LLMs along with key research and development.

10.6.1 Current Trends in LLMs and GenAI

LLMs and GenAI advancements have led to transformative applications across various industries. However, it is equally important to examine the underlying trends driving this progress.

10.6.1.1 Multimodal Models

Multimodal models are designed to integrate text, images, audio, video, and progress beyond just textual data. Projects such as OpenAI's GPT-4 and Google's Pathways exemplify how LLMs are evolving to handle a broader range of inputs and outputs (Ramesh et al. 2021). These capabilities allow LLMs to tackle complex real-world problems such as generating images from text prompts or interpreting visual and audio cues with written language.

10.6.1.2   Increasing Model Sizes and Capabilities

LLMs have advanced significantly from GPT-2 with 1.5 billion parameters to GPT-4 with hundreds of billions of parameters. However, this rapid growth raises sustainability concerns due to the high energy consumption and environmental impact of training these models (Patterson et al. 2021). The current research is focused on improving efficiency in terms of computation and energy usage.

10.6.1.3   Specialized LLMs for Domain-Specific Applications

Specialized LLMs are domain-specific models that can be fine-tuned for industries such as medicine, law, or finance. For example, *BioGPT* is designed to process biomedical data (Luo et al. 2022), demonstrating how LLMs can be tailored to solve specialized tasks in expert fields more efficiently.

10.6.1.4   Few-Shot and Zero-Shot Learning

Few-shot and *zero-shot learning* refer to LLMs' capabilities to perform tasks with minimal or no task-specific data. These have advanced significantly in recent models with profound implications for AI's flexibility and adaptability (Brown et al. 2020). These models no longer require massive datasets for every new task; instead, they generalize learning across various tasks and reduce the cost and time to deploy AI in new applications. Figure 10.8 illustrates the relationship between LLM model sizes (e.g., parameter count) and performance across various tasks.

10.6.2   The Future of Creativity in AI

Generative AI has exceeded traditional creativity notions to produce content ranging from visual art, poetry, and music, to even full-length novels. AI's creativity will likely expand in the future, but it raises opportunities and existential questions about the nature of human creativity.

10.6.2.1   Collaborative Creativity

Collaborative creativity has become one of the most promising AI's creative arena. *OpenAI's DALL·E, DeepMind's AlphaFold*, and *AI music generators* have already demonstrated that AI can serve as a creative partner rather than merely a tool (Ramesh et al. 2021). It is plausible that writers, artists, and musicians will regularly co-create with AI to augment their work beyond simple automation in the future.

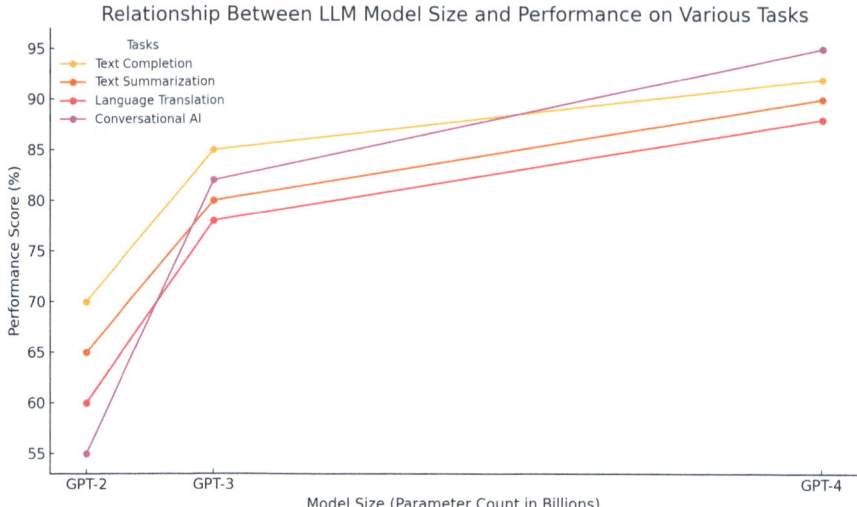

Fig. 10.8 A graph showing the relationship between LLM model size (e.g., parameter count) and their performance on various tasks

This could democratize access to creative expression for individuals to produce high-quality creative work without traditional artistic training.

10.6.2.2   The Blurring Line Between Human and Machine Creativity

Since AI is increasingly proficient in generating human-like content, the distinction between human-created and AI-generated works may become blurred. AI-generated content has entered the mainstream of digital art and design. The rise of *NFTs (Non-Fungible Tokens)* has AI-generated artwork sold for millions of dollars, further complicates the discussion about what constitutes authentic creativity (Bommasani et al. 2021).

10.6.2.3   Challenges of Authorship and Ownership

Authorship and intellectual property are a critical issue with AI-driven creativity. If a machine generates a piece of art or music, who owns it? The current legal frameworks are not well equipped to handle these complexities leading to new conundrums for copyright laws. Future discussions around creativity in AI will need to establish clear guidelines on how to attribute authorship and handle intellectual property when humans and machines work together (Floridi and Cowls 2019).

10.6.3   The Role of LLMs in AI Ethics

LLMs thrive in AI are inclined to be the spotlight for AI ethics discussions on b*ias, misinformation, privacy,* and *accountability.*

10.6.3.1   Addressing Bias in LLMs

LLMs train on large datasets derived from internet sources often have incorporated racial, gender, ideological biases, and among others in their training data. The current research and development are actively working to mitigate these biases by filtering data, training ethical AI, and implementing *bias detection tools* (Metzinger 2022). However, eliminating bias entirely seems to be elusive. Future research will need to focus on creating more transparent and explainable models for continuous monitoring and corrections.

10.6.3.2   AI in Decision-Making

LLMs are increasingly involved in decision-making process, whether through chatbots in customer service or algorithms to assess loan applications or job candidates. These raise ethical questions about accountability. Who is responsible if an LLM makes a biased or erroneous decision? There is also an over-reliance risk on these systems potentially leading to decisions that may not align with human values. Establishing ethical guidelines and regulatory frameworks that govern LLMs deployment in decision-making roles will be a critical step for the future (Floridi and Cowls 2019).

10.6.3.3   Combating Misinformation

LLMs' capabilities to generate convincing and coherent text can have malicious effects of spreading misinformation and creating fake news. There are concerns about how generative models can be misused to create deepfakes or misleading articles indistinguishable from legitimate content. Withstanding misinformation is an acute ethical issue that will require model transparency advancements and stronger detection systems to flag misleading content (Thoppilan et al. 2021).

10.6.3.4   Privacy

LLMs operate on vast amounts of sensitive personal data present significant ethical concerns. Future LLMs will need to adopt more stringent data privacy measures, potentially incorporating decentralized training methods that do not depend on large and centralized datasets vulnerable to data breaches (Metzinger 2022).

10.6.4   The Path Forward: Research and Development

Explore model interpretability, efficiency, ethical frameworks, and new applications are some of the key research and development to shape the future of LLMs and GenAI.

10.6.4.1   Model Interpretability and Explainability

Dynamic LLMs can generate complex outputs for interpretability and explainability improvements. The current research on LLMs is more transparent for users to interpret the formation of a particular output (Gunning and Aha 2019) in addition to developing models that can offer insights into decision-making processes without performance concession.

10.6.4.2   Energy Efficiency

Training LLMs are highly resource-intensive and energy-efficient initiatives. A competent models' development and sustainable AI practices implementation are prime focuses for future research. These include greater efficiency algorithms optimization, lesser specialized models, and computational power to sustain high performance (Patterson et al. 2021).

Figures 10.9 and 10.10 show the energy consumption and carbon footprint of GPT-3 and GPT-4 with environmental impact reduction strategies. They indicated that GPT-4 consumed energy at 4560 *megawatt-hours (MWh)*, nearly four times more than GPT-3's at 1287 MWh resulting in 2100 tons carbon footprint compared to GPT-3's 552 tons. This intensified the environmental costs of dynamic models' development.

To mitigate this impact, several strategies offer substantial reductions. The most effective solution is to use renewable energy which can cut emissions by 40%. This approach directly undertakes carbon footprint consumption during training. Other techniques include optimizing *model efficiency* by *efficient algorithms* (30% reduction), *data pruning* (20% reduction), and *model compression* (25% reduction). They streamline processes and reduce the computational workload without performance concession.

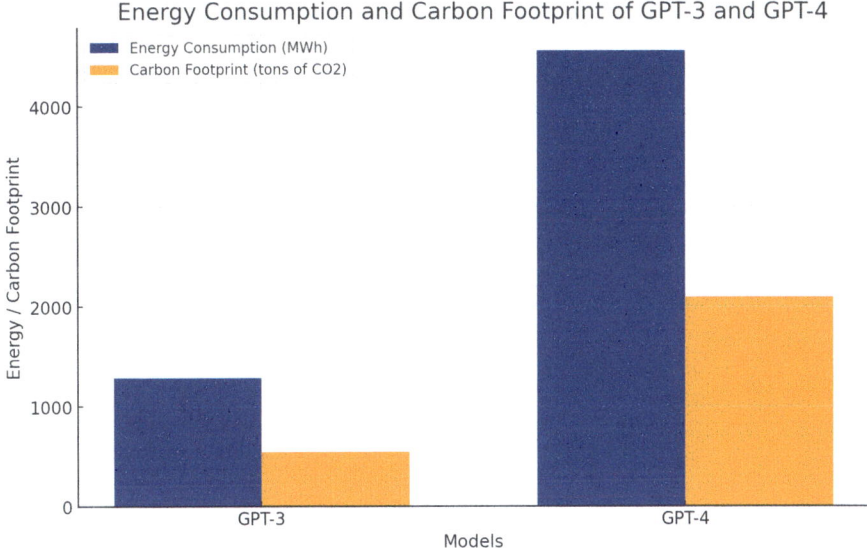

Fig. 10.9 Energy consumption and carbon footprint of GPT3 vs. GPT4

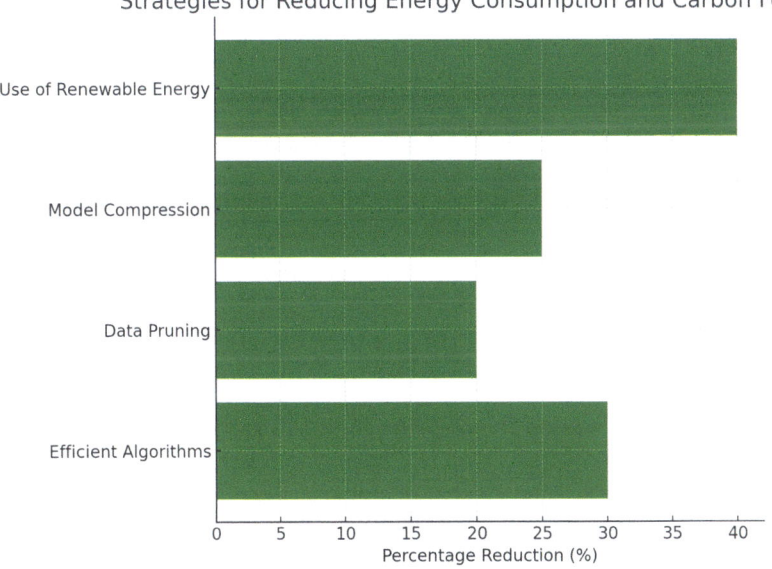

Fig. 10.10 Strategies for reducing energy consumption and carbon footprint

Dynamic models such as *GPT-4* deliver remarkable advancements, but carbon footprint must be addressed. Hence, renewable energy, algorithmic efficiency, and model optimization are the combined strategies for continual innovation and sustainability.

10.6.4.3   Exploring New Applications

The swift versatility of LLMs means that new applications constantly emerged. Healthcare, education, and entertainment are among industries that can benefit considerably. For example, in healthcare it can be used for accurate diagnostics and personalized medicine. In education, they can offer tailor-made learning experiences, and in entertainment, they can produce *AI-generated movies*, *video games*, and *virtual worlds*. The potential is vast, and research is an integral part to open new opportunities (Bommasani et al. 2021).

Exercises

10.1. What are LLMs? How do they differ from traditional NLP models?

10.2. Explain the significance of the Transformer architecture in LLMs development. How did it improve on RNNs and LSTM networks?

10.3. Describe the key components of GANs. How can they contribute to Generative AI success in producing realistic content?

10.4. What are the main applications of LLMs in creative writing and content generation? How can LLMs assist to overcome writer's block and enhance creativity?

10.5. How can LLMs handle language translation tasks? Discuss how attention mechanisms contribute to translation accuracy and fluency.

10.6. Explain LLMs' role in conversational AI and chatbots. What advantages do they provide compared to traditional rule-based systems?

10.7. What are the primary ethical concerns associated with the deployment of LLMs in real-world applications? Discuss bias, privacy, and misinformation challenges.

10.8. How can bias arise in LLMs during data collection and training? What strategies can be used to detect and mitigate bias in these models?

10.9. Discuss LLMs' impact of spreading misinformation. What measures can be taken to ensure the AI-generated content accuracy and trustworthiness?

10.10. What are "few-shot" and "zero-shot" learning? How can they make LLMs adaptable to new tasks with minimal training data?

10.11. Describe the importance of text summarization in LLMs. How can these models balance extractive and abstractive summarization techniques?

10.12. What are the privacy and data security risks posed by LLMs, especially when they access sensitive data? How can organizations mitigate these risks?

10.13. Explain the concept of multimodal models LLMs and Generative AI. How can they integrate text, images, and other forms of media to generate content?

10.14. How can LLMs role in human-machine interaction contribute to intuitive AI systems development? What challenges remain in this domain?
10.15. What are the potential future research directions for LLMs and Generative AI? Discuss trends such as increasing model size, energy efficiency, and ethical AI frameworks.

References

Anthropic (2024). Anthropic official site. https://www.anthropic.com. Accessed 28 Oct 2024.
Bahdanau, D., Cho, K., & Bengio, Y. (2014). Neural machine translation by jointly learning to align and translate. arXiv:1409.0473.
Baidu (2024). Baidu AI official site. https://ai.baidu.com/. Accessed 28 Oct 2024.
Bender, E. M., Gebru, T., McMillan-Major, A., & Shmitchell, S. (2021). On the dangers of stochastic parrots: Can language models be too big? In Proceedings of the 2021 ACM Conference on Fairness, Accountability, and Transparency (pp. 610–623).
BERT (2024). Google BERT GitHub Repository. https://github.com/google-research/bert. Accessed 28 Oct 2024.
Blasi, A., Dimakopoulou, A., & Vasilakos, A.V. (2021). Mitigating algorithmic bias: A survey. IEEE Transactions on Neural Networks and Learning Systems.
Bommasani, R., et al. (2021) "On the Opportunities and Risks of Foundation Models." arXiv:2108.07258.
Brown, T.B., Mann, B., Ryder, N., Subbiah, M., Kaplan, J., Dhariwal, P., & Amodei, D. (2020). Language models are few-shot learners. arXiv preprint arXiv:2005.14165.
Buolamwini, J., & Gebru, T. (2018). Gender Shades: Intersectional Accuracy Disparities in Commercial Gender Classification. Proceedings of the 1st Conference on Fairness, Accountability, and Transparency.
Carlini, N., Tramer, F., Wallace, E., Jagielski, M., Herbert-Voss, A., Lee, K., ... & Song, D. (2021). Extracting Training Data from Large Language Models. arXiv:2012.07805.
ChatGPT (2024). ChatGPT official site. https://chatgpt.com/. Accessed 28 Oct 2024.
Cho, K., van Merriënboer, B., Gulcehre, C., et al. (2014). Learning phrase representations using RNN encoder-decoder for statistical machine translation. arXiv:1406.1078.
Choi, E., Schuetz, A., Stewart, W.F., & Kuan, P. (2017). Using recurrent neural networks for early detection of heart failure. Journal of the American Medical Informatics Association, 24(5), 1000–1005.
Claude (2024). Claude official site. https://claude.ai. Accessed 28 Oct 2024.
Creswell, A., White, T., Dumoulin, V., et al. (2018). Generative adversarial networks: An overview. IEEE Signal Processing Magazine, 35(1), 53–65.
Devlin, J., Chang, M.-W., Lee, K., & Toutanova, K. (2018). BERT: Pre-training of deep bidirectional transformers for language understanding. arXiv:1810.04805.
Dwork, C. (2008). Differential privacy: A survey of results. In International Conference on Theory and Applications of Models of Computation (pp. 1–19). Springer.
Devlin, J., Chang, M.W., Lee, K., & Toutanova, K. (2019). BERT: Pre-training of deep bidirectional transformers for language understanding. arXiv preprint arXiv:1810.04805.
Elgammal, A., Liu, B., Elhoseiny, M., & Mazzone, M. (2017). CAN: Creative adversarial networks, generating "art" by learning about styles and deviating from style norms. arXiv:1706.07068.
Ernie (2024). ERNIE 3.0 Titan official site. https://yiyan.baidu.com/. Accessed 28 Oct 2024.
Floridi, L., & Cowls, J. (2019). A unified framework of five principles for AI in society. Harvard Data Science Review. vol. 1, no. 1, 2019.
Ghazvinian, A., Banjade, R., & Gallo, M. (2021). The role of generative models in creative writing: A study of AI-powered content creation. Artificial Intelligence Review, 54(4), 1–21.

Goodfellow et al. (2014). Generative adversarial networks. Advances in Neural Information Processing Systems, 27, 2672–2680.

Google (2024). Google PaLM2 official site. https://ai.google/discover/palm2. Accessed 28 Oct 2024.

Gunning, D., and Aha, D.W. (2019). "DARPA's Explainable Artificial Intelligence (XAI) Program." AI Magazine, vol. 40, no. 2, pp. 44–58.

He, K., Zhang, X., Ren, S., & Sun, J. (2016). Deep residual learning for image recognition. In Proceedings of the IEEE Conference on Computer Vision and Pattern Recognition (pp. 770–778).

Hochreiter, S., & Schmidhuber, J. (1997). Long short-term memory. Neural Computation, 9(8), 1735–1780.

Hoffmann, J., Borgeaud, S., Mensch, A., Buchatskaya, E., Cai, T., Rutherford, E., ... & Irving, G. (2022). Training Compute-Optimal Large Language Models. arXiv:2203.15556.

Jobin, A., Ienca, M., & Vayena, E. (2019). The global landscape of AI ethics guidelines. Nature Machine Intelligence, 1(9), 389–399.

Khan, S., Zhang, X., & Yao, J. (2021). A comprehensive review on transformer models in NLP. Journal of Artificial Intelligence Research, 70, 1–30.

Kim, Y. (2014). Convolutional neural networks for sentence classification. arXiv:1408.5882.

Kocijan, J., & Djuric, N. (2020). Fine-tuning pre-trained language models: Weighting methods and training strategies. arXiv:2011.13235.

Lipton, Z. C. (2018). The mythos of model interpretability. Communications of the ACM, 61(10), 36–43.

Liu, Y., & Lapata, M. (2019). Text generation with pre-trained language models. In Proceedings of the 57th Annual Meeting of the Association for Computational Linguistics.

LLaMa (2024) LLaMa official site. https://www.llama.com/. Accessed 28 Oct 2024.

Lu, J., & Tzu, J. (2020). Multimodal machine learning: A survey and taxonomy. IEEE Transactions on Pattern Analysis and Machine Intelligence, 43(2), 486–503.

Luo, R., et al. (2022) "BioGPT: A Generative Pre-trained Transformer for Biomedical Text Generation and Mining." bioRxiv, 2022.

Luong, M. T., Pham, H., & Manning, C. D. (2015). Effective approaches to attention-based neural machine translation. arXiv:1508.04025.

Marcus, G., & Davis, E. (2020). GPT-3, Bloviator: OpenAI's language generator has no idea what it's talking about. MIT Technology Review.

Meta (2024). Meta official site. https://www.meta.com/. Accessed 17 Dec 2024.

Metzinger, T. (2022). "Ethics of Artificial Intelligence and Robotics." The Stanford Encyclopedia of Philosophy, 2022.

OpenAI (2024) OpenAI official site. https://openai.com Accessed 17 Dec 2024.

Patterson, D., et al. (2021) "The Carbon Footprint of Machine Learning." Communications of the ACM, vol. 64, no. 11, pp. 56–63.

Raffel et al. (2020) "Exploring the limits of transfer learning with a unified text-to-text transformer." Journal of Machine Learning Research 21, no. 140 (2020): 1–67.

Radford, A., et al. (2021). "Learning Transferable Visual Models from Natural Language Supervision." Proceedings of the International Conference on Machine Learning.

Raghavan, M., Awan, I., & Yoon, J. (2020). Privacy-preserving generative models in healthcare. In Proceedings of the 2020 IEEE International Conference on Healthcare Informatics.

Ramesh, A., et al. (2021). "Zero-Shot Text-to-Image Generation." Proceedings of the International Conference on Machine Learning, 2021.

Solaiman, I., Brundage, M., Clark, J., Askell, A., Herbert-Voss, A., Wu, J., ... & Amodei, D. (2019). Release strategies and the social impacts of language models. arXiv:1908.09203.

Summerville, A., Snodgrass, S., & Mateas, M. (2018). The role of AI in video game design. In Proceedings of the 2018 International Conference on Interactive Digital Storytelling.

Thoppilan, R., et al. (2021) "LaMDA: Language Models for Dialog Applications." Proceedings of the Annual Conference on Neural Information Processing Systems.

Vaswani, A., Shard, N., Parmar, N., Uszkoreit, J., Jones, L., Gomez, A.N., Kaiser, Ł., & Polosukhin, I. (2017). Attention is all you need. In Advances in Neural Information Processing Systems (Vol. 30).

Whittlestone, J., Nyrup, R., Horne, E., & Bachelard, J. (2019). The Ethics of AI in Health Care: A Mapping Review of Literature. Health Informatics Journal, 25(2), 247–256. https://doi.org/10.1177/1460458216663546.

Yang, Z., Yang, D., Dyer, C., He, X., & Gao, J. (2019). XLNet: Generalized autoregressive pre-training for language understanding. arXiv:1906.08237.

Zellers, R., Holtzman, A., Rashkin, H., Balandat, M., & Choi, Y. (2019). Defeating NLP's Worst Enemy: Noise. In Proceedings of the 57th Annual Meeting of the Association for Computational Linguistics, pp. 1236–1249. https://doi.org/10.18653/v1/P19-1121.

Zhao et al. (2019). Gender bias in coreference resolution: Evaluation and debiasing methods. Proceedings of the 2019 Conference of the North American Chapter of the Association for Computational Linguistics: Human Language Technologies, 1, 15–24.

Zhang, Y., & Chai, Z. (2020). Exploring the interpretability of BERT: A case study on attention visualization. In Proceedings of the 58th Annual Meeting of the Association for Computational Linguistics.

Part II
Natural Language Processing Workshops with Python Implementation in 14 Hours

Chapter 11
Workshop#1: Basics of Natural Language Toolkit (Hour 1–2)

11.1 Introduction

Part II of this book will provide seven Python programming workshops on how each NLP core component operates and integrates with Python-based NLP tools including NLTK, spaCy, BERT and Transformer Technology to construct a Q&A chatbot.

Workshop 1 will explore NLP basics including:

1. Concepts and installation procedures
2. Text processing function with examples using NLTK
3. Text analysis lexical dispersion plot in Python
4. Tokenization in text analysis
5. Statistical tools for text analysis

Note: To ensure all NLP-based Python tools compatibility for all workshops can run smoothly, please check the list of requirements shown in Table 11.1.

In this workshop, all demonstrations use Python 3.11.9 as the running environment. It is highly recommended to create an independent virtual environment for these workshops in the book. The command for setting up a virtual environment with any version of pre-installed Anaconda is:

```
conda create -n your_virtual_environment_name python=3.11
```

Please ensure the following Python packages are installed before starting the workshop:

- python (demo version 3.11.9)
- tensorflow (demo version 2.17.0)
- nltk (demo version 3.9.1)

© The Author(s), under exclusive license to Springer Nature Singapore Pte Ltd. 2025
R. Lee, *Natural Language Processing*,
https://doi.org/10.1007/978-981-96-3208-4_11

Table 11.1 Requirement list for all NLP workshops

```
#python 3.11.9 (Release Date: 2.4.2024) with all latest version
test in Sep 2024.
#Workshop 1
tensorflow==2.17.0
nltk==3.9.1
#Workshop 2
# spacy model (sm) offline package already existed in zip
spacy==3.4.4
#Workshop 3
matplotlib==3.9.2
wordcloud==1.9.3
svgling==0.5.0
svgwrite==1.4.3
scikit-learn==1.5.1
#Workshop 4
# spacy model (md) offline package can be found as zip file in
the file folder
#Workshop 5
pandas==2.2.2
#Workshop 6
#spacy model (trf) offline package can be found as zip file in
the file folder
transformers==4.44.2
tf-keras==2.17.0
torch==2.4.1
torchvision==0.19.1
spacy-transformers==1.3.5
#Workshop 7
keras=3.3.3
transformers==4.44.2
tensorflow==2.17.0
tensorflow_datasets==4.9.6
pydot==3.0.1
graphviz==0.20.3
pydot-ng==2.0.0
```

If these packages are not installed on PC/laptop, use *pip install xxx* command. The detailed requirements list and Python package version used in this workshop can be found in the requirements.txt file stored in the NLP GitHub repository (NLPGitHub 2024).

11.2   What Is Natural Language Toolkit (NLTK)?

NLTK (Natural Language Toolkit 2024) is one of the earliest Python-based NLP development tools invented by Prof. Steven Bird and Dr. Edward Loper in the Department of Computer and Information Science of the University of Pennsylvania with their classical book Natural Language Processing with Python published by O'Reilly Media Inc. in 2009 (Bird et al. 2009). There are over 30 universities in the US and 25 countries using NLTK for NLP-related courses until present. This book is considered as bible for anyone who wishes to learn and implement NLP applications using Python.

NLTK offers user-oriented interfaces with over 50 corpora and lexical resources such as WordNet (2024), a large lexical database of English. Nouns, verbs, adjectives, and adverbs are grouped into sets of cognitive synonyms (synsets); each expresses a distinct concept which is an important lexical database in NLP developed by Princeton University since 1980.

Other lexical databases and corpora are Penn Treebank Corpus, Open Multilingual Wordnet, Problem Report Corpus, and Lin's Dependency Thesaurus.

NLTK contains statistical-based text processing libraries of five fundamental NLP enabling technologies and basic semantic reasoning tools including (Albrecht et al. 2020; Antic 2021; Arumugam and Shanmugamani 2018; Hardeniya et al. 2016; Kedia and Rasu 2020; Perkins 2014):

- Word tokenization.
- Stemming.
- POS tagging.
- Text classification.
- Semantic analysis.

11.3   A Simple Text Tokenization Example Using NLTK

Let's look at NLTK text tokenization using Jupyter Notebook (Jupyter 2024; Wintjen and Vlahutin 2020) as below:

| [1] | ```
# Import NLTK package
import nltk
``` |

| [2] | ```
# Create a sample utterance 1 (utt1)
utt1 = "At every weekend, early in the morning. I drive
my car to the car center for car washing. Like
clock-work."
``` |

| [3] | ```
# Display utt1
utt1
```<br>'At every weekend, early in the morning. I drive my car to the car center for car washing. Like clock-work.' |

| [4] | ```
# Create utterance tokens (utokens)
utokens = nltk.word_tokenize(utt1)
``` |

[5]	# Display utokens
	utokens
	['At', 'every', 'weekend', ',', 'early', 'in', 'the', 'morning', '.', 'I', 'drive', 'my', 'car', 'to',
	'the', 'car', 'center', 'for', 'car', 'washing', '.', 'Like', 'clock-work', '.']

11.4   How to Install NLTK?

Step 1 Install Python 3.11.9
Step 2 Install NLTK

  2.1 Start CMD or other command line tool
  2.2 Type *pip install nltk*

Figure 11.1 shows a screenshot of NLTK installation process.
Step 3 Install NLTK Data

Once NLTK is installed into Python, download NLTK data.

  3.1 Run Python
  3.2 Type the following to activate an NLTK downloader

- *import nltk*
- *nltk.download()*

  Note: *nltk.downloader()* will invoke NLTK downloader automatically, a separate window-based downloading module for users to download four NLP databanks into their Python machines. They include (1) Collection libraries, (2) Corpora, (3) Modules, and (4) other NLP packages. Figures 11.2, 11.3 and 11.4 show screenshots of the NTLK downloader for Collection, Corpora, and NTLK models installations.

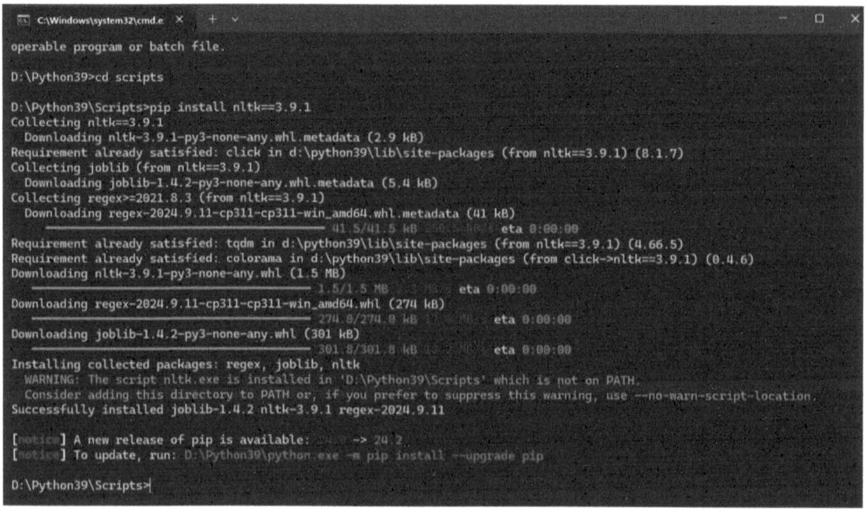

Fig. 11.1 Screenshot of NLTK installation process

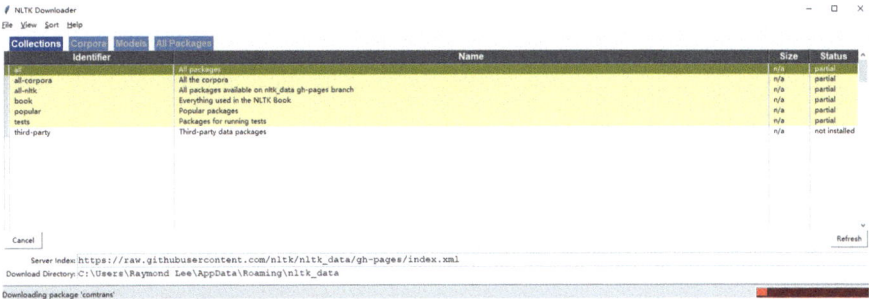

Fig. 11.2 Screenshot of NTLK downloader of collection library

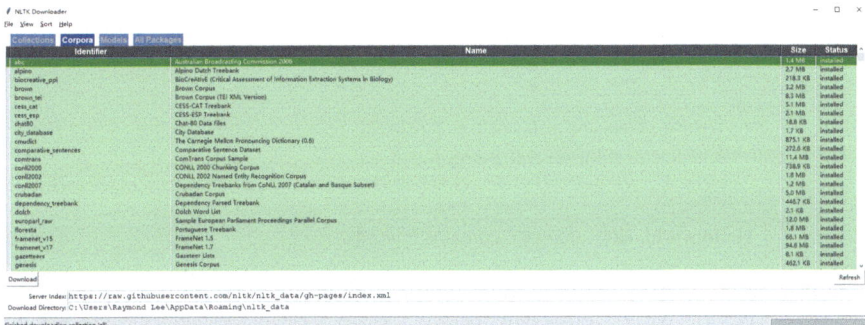

Fig. 11.3 Screenshot of NTLK downloader of Corpora library

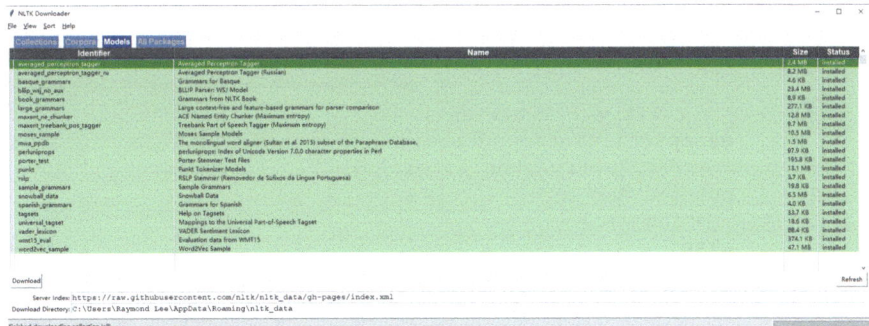

Fig. 11.4 Screenshot of NTLK downloader of NLTK models

11.5 Why Using Python for NLP?

Python toolkit and packages overtook C, C++, Java especially in data science, AI, and NLP software development since 2000 (Albrecht et al. 2020; Kedia and Rasu 2020). There are several reasons to drive the changes because:

1. It is a generic language without a specific area unlike other languages such as Java and JavaScript specifically designed for web applications and website development.

2. It is easier to learn and user-friendly compared with C and C++ especially for non-computer science students and scientists.
3. Its lists and list-processing datatypes provide an ideal environment for NLP modeling and text analysis.

A Python program performs a tokenization task to process text as shown below:

| [6] | ```
# Define utterance 2 (utt2)
utt2 = "Hello world. How are you?"
``` |
| --- | --- |

| [7] | ```
# Using split() method to split it into word tokens
utt2.split()
``` |
| --- | --- |
| | ['Hello', 'world.', 'How', 'are', 'you?'] |

| [8] | ```
# Check the no of word tokens
nwords = len(utt2.split())
print ("'Hello world. How are you?' contains ",nwords,"
words.")
``` |
| --- | --- |
| | 'Hello world. How are you?' contains 5 words. |

Python codes perform word number counts from literature *Alice's Adventures in Wonderland* by Lewis Carroll (1832–1898) as below:

| [9] | ```
# Define method to count the number of word tokens in
text file (cwords)
def cwords(literature):
  try:
    with open(literature, encoding='utf-8') as f_lit:
      c_lit = f_lit.read()
  except FileNotFoundError:
    err = "Sorry, the literature " + literature + " does
not exist."
    print(err)
  else:
    w_lit = c_lit.split()
    nwords = len(w_lit)
    print("The literature " + literature + " contains "
+ str(nwords) + " words.")
literature = 'alice.txt'
cwords(literature)
``` |
| --- | --- |
| | The literature alice.txt contains 29465 words |

This workshop has extracted four famous literatures from Project Gutenberg (Gutenberg 2024):

1. *Alice's Adventures in Wonderland* by Lewis Carroll (1832–1898) (alice.txt)
2. *Little Women* by Louisa May Alcott (1832–1888) (little_women.txt)
3. *Moby Dick* by Herman Melville (1819–1891) (moby_dick.txt)
4. *The Adventures of Sherlock Holmes* by Sir Arthur Conan Doyle (1859–1930) (Adventures_Holmes.txt)

[10]	`cwords('Adventures_Holmes.txt')`
	The literature Adventures_Holmes.txt contains 107411 words.

11.6 NLTK with Basic Text Processing in NLP

NLTK are Python tools and methods to learn and practice starting from basic text processing in NLP. They include:

- Text processing as lists of words.
- Statistics on text processing.
- Simple text analysis.

NLTK provides 9 different types of text documents from classic literatures, Bible texts, famous public speeches, news, and articles with personal corpus for text processing. Let's start and load these text documents.

| [11] | ```# Let's load some sample books from the nltk databank
import nltk
from nltk.book import *``` |
|---|---|
| | *** Introductory Examples for the NLTK Book *** |
| | Loading text1, ..., text9 and sent1, ..., sent9 |
| | Type the name of the text or sentence to view it. |
| | Type: 'texts()' or 'sents()' to list the materials. |
| | text1: Moby Dick by Herman Melville 1851 |
| | text2: Sense and Sensibility by Jane Austen 1811 |
| | text3: The Book of Genesis |
| | text4: Inaugural Address Corpus |
| | text5: Chat Corpus |
| | text6: Monty Python and the Holy Grail |
| | text7: Wall Street Journal |
| | text8: Personals Corpus |
| | text9: The Man Who Was Thursday by G. K. Chesterton 1908 |

| [12] | ```# Display the list of sample books
texts()``` |
|---|---|

	text1: Moby Dick by Herman Melville 1851
	text2: Sense and Sensibility by Jane Austen 1811
	text3: The Book of Genesis
	text4: Inaugural Address Corpus
	text5: Chat Corpus
	text6: Monty Python and the Holy Grail
	text7: Wall Street Journal
	text8: Personals Corpus
	text9: The Man Who Was Thursday by G. K. Chesterton 1908

| [13] | ```
# Check text1
text1
``` |
| | <Text: Moby Dick by Herman Melville 1851> |

| [14] | ```
# To know more about text1, check this
text1?
``` |
| [15] | ```
# Import word_tokenize as wtoken
from nltk.tokenize import word_tokenize
# Open Adventures_Holmes.txt and performs tokenization
fholmes = open("Adventures_Holmes.
txt","r",encoding="utf-8").read()
wtokens = word_tokenize(fholmes)
tholmes=nltk.text.Text(wtokens)
``` |

11.7   Simple Text Analysis with NLTK

Text analysis is used to study a particular word or phrase that occurs in a text document such as literature or public speeches. NLTK has a *"concordance()"* function different from the ordinary search function. It does not only indicate occurrence but also reveal neighboring words and phrases. Let's try text examples from *The Adventures of Sherlock Holmes* (Doyle 2019).

| [16] | ```
# Check concordance of word "Sherlock "
tholmes.concordance("Sherlock")
``` |

Displaying 25 of 98 matches:

The Adventures of Sherlock Holmes by Arthur Conan Doyle Conte
es I . A SCANDAL IN BOHEMIA I . To Sherlock Holmes she is always _the_ woman.
ust such as I had pictured it from Sherlock Holmes ' succinct description , bu
ssing said : " Good-night , Mister Sherlock Holmes. " There were several peopl
lly got it ! " he cried , grasping Sherlock Holmes by either shoulder and look
stepped from the brougham . " Mr . Sherlock Holmes , I believe ? " said she .
ss for the Continent. " " What ! " Sherlock Holmes staggered back , white with
, the letter was superscribed to " Sherlock Holmes , Esq . To be left till cal
nd ran in this way : " MY DEAR MR. SHERLOCK HOLMES , —You really did it very w
of interest to the celebrated Mr. Sherlock Holmes . Then I , rather imprudent
possess ; and I remain , dear Mr. Sherlock Holmes , " Very truly yours , " IR
ia , and how the best plans of Mr. Sherlock Holmes were beaten by a woman ' s
I had called upon my friend , Mr. Sherlock Holmes , one day in the autumn of
and discontent upon his features . Sherlock Holmes ' quick eye took in my occu
t as I have been telling you , Mr. Sherlock Holmes , " said Jabez Wilson , mop
e of this obliging youth ? " asked Sherlock Holmes . " His name is Vincent Spa
IS DISSOLVED . October 9 , 1890. " Sherlock Holmes and I surveyed this curt an
d client carried on his business . Sherlock Holmes stopped in front of it with
own stupidity in my dealings with Sherlock Holmes . Here I had heard what he
" " I think you will find , " said Sherlock Holmes , " that you will play for
and I will follow in the second. " Sherlock Holmes was not very communicative
jump , and I ' ll swing for it ! " Sherlock Holmes had sprung out and seized t
IDENTITY " My dear fellow , " said Sherlock Holmes as we sat on either side of
ant-man behind a tiny pilot boat . Sherlock Holmes welcomed her with the easy
nsult me in such a hurry ? " asked Sherlock Holmes , with his finger-tips toge

The above example shows all *Sherlock* occurrences indicating that *Sherlock* is a special word linked with the surname *Holmes* in the text document

Let's look at the word usage of *extreme* from the same literature:

[17]
```
# Check concordance of word "extreme"
tholmes.concordance("extreme")
```

Displaying 9 of 9 matches:
may trust with a matter of the most extreme importance . If not , I should much
ng red head , and the expression of extreme chagrin and discontent upon his fea
ternately asserted itself , and his extreme exactness and astuteness represente
e swing of his nature took him from extreme languor to devouring energy ; and ,
olice reports realism pushed to its extreme limits , and yet the result is , it
of an English provincial town . His extreme love of solitude in England suggest
ion , and that in his haste and the extreme darkness he missed his path and wal
for my coming at midnight , and his extreme anxiety lest I should tell anyone o
like one who has been driven to the extreme limits of his reason . Then , sudde

Concordance techniques are means to learn grammars, words or phrases called Use of English, also called Learn by Examples. In this example, we learnt how to use the word *extreme* in various situations and scenarios.

[18]	`tholmes.similar("extreme")`
	dense gathering

[19]	`# Check concordance of word "extreme" in text2` `text2.concordance("extreme")`
	Displaying 4 of 4 matches: n another day or two perhaps ; this extreme mildness can hardly last longer -- ng her that he was kept away by the extreme affection for herself , which he co of his brother , and lamenting the extreme GAUCHERIE which he really believed y which had been leading her to the extreme of languid indolence and selfish re

[20]	`# Check similar word "extreme" in text2` `text2.similar("extreme")`
	family centre good opinion life death loss house society children attachment wishes interest goodness heart comfort cheerfulness existence marriage son

[21]	`# Check concordance word "extreme" in text4` `text4.concordance ("extreme")`
	Displaying 3 of 3 matches: vigilance no Administration by any extreme of wickedness or folly can very ser ent , and communication between the extreme limits of the country made easier t the politics of petty bickering and extreme partisanship they plainly deplore .

[22]	`# Check similar word "extreme" in text4` `text4.similar("extreme")`
	one other just hope motives act people agency system right form loss length knowledge science portion quarter narrowest requisite member

It shows that word usage of *extreme* varies by authors and text types, e.g., it has different styles in *The Adventures of Sherlock Holmes* as compared with usage in *Sense and Sensibility* by Jane Austin (1775–1817) which is more vivid but has standard and fixed usage in *Inaugural Address Corpus*.

The *common_contexts()* method is to examine contexts shared by two or more words. *The Adventures of Sherlock Holmes* is used with common contexts of two words *extreme* and *huge*.

First, call common contexts() function from object *tholmes*.

[23]	```# Check common contexts on tholmes``` ```tholmes.common_contexts(["extreme","huge"])```
	No common contexts were found

which means after analyzing *extreme* and *huge* in *The Adventures of Sherlock Holmes*, no common context meaning can be found.

Call *concordance()* function of these two words and check against the extracted patterns as shown below:

[24]	```# Check concordance word "extreme" in tholmes``` ```tholmes.concordance("extreme")```
	Displaying 9 of 9 matches: may trust with a matter of the most extreme importance . If not , I should much ng red head , and the expression of extreme chagrin and discontent upon his fea ternately asserted itself , and his extreme exactness and astuteness represente e swing of his nature took him from extreme languor to devouring energy ; and , olice reports realism pushed to its extreme limits , and yet the result is , it of an English provincial town . His extreme love of solitude in England suggest ion , and that in his haste and the extreme darkness he missed his path and wal for my coming at midnight , and his extreme anxiety lest I should tell anyone o like one who has been driven to the extreme limits of his reason . Then , sudde

[25]	```# Check concordance word "huge" in tholmes``` ```tholmes.concordance("huge")```

Displaying 11 of 11 matches:
used and refreshed his memory with a huge pinch of snuff . " Pray continue you
after opening a third door , into a huge vault or cellar , which was piled al
ed . All will come well . There is a huge error which it may take some little
a small , office-like room , with a huge ledger upon the table , and a teleph
en suddenly dashed open , and that a huge man had framed himself in the apertu
, and bent it into a curve with his huge brown hands . " See that you keep yo
r. Grimesby Roylott drive past , his huge form looming up beside the little fi
side and lay listless , watching the huge crest and monogram upon the envelope
, " said I ruefully , pointing to a huge bundle in the corner . " I have had
th hanging jowl , black muzzle , and huge projecting bones . It walked slowly
r hurrying behind us . There was the huge famished brute , its black muzzle bu

Can you see how important it is in
1. NLP?
2. Use of English and technical writing?

Workshop 1.1 Simple Text Processing using NLTK
1. Try to use concordance(), similar(), and common_contexts() functions to look for two more frequently used words usage.
2. Compare their usages from four sources: *Moby Dick, Sense and Sensibility, Inaugural Address Corpus, and Wall Street Journal.*
3. Are there any pattern(s)?
4. What are their differences in the Use of English?

11.8 Text Analysis Using Lexical Dispersion Plot

Text analysis was learnt to study word patterns and common contexts in the previous workshop.

Dispersion Plot in Python NLTK is to identify occurrence frequencies of keywords from the whole document.

11.8.1 What Is a Lexical Dispersion Plot?

Dispersion is the quantification of each point deviation from the mean value in basic statistics.

NLTK *Dispersion Plot* produces a plot showing words distribution throughout the text. *Lexical dispersion* is used to indicate homogeneity of words (word tokens) that occurred in the corpus (text document) achieved by the dispersion_plot() in NLTK.

To start, let's use NLTK book object to call function *dispersion_plot( )*.
Note: requires pylab installation prior to this function.
The following example uses text1 to verify basic information about *dispersion_plot( )*.

[26]	`text1.dispersion_plot?`
	Signature: text1.dispersion_plot(words) **Docstring:** Produce a plot showing the distribution of the words through the text. Requires pylab to be installed. :param words: The words to be plotted :type words: list(str) :seealso: nltk.draw.dispersion_plot() **File:** d:\anaconda3\envs\py311nlp\lib\site-packages\nltk\text.py **Type:** method

11.8.2   *Lexical Dispersion Plot over Context Using Sense and Sensibility*

Are there any lexical patterns for positive words such as *good, happy*, and *strong* versus negative words such as *bad, sad,* or *weak* in literature?

WORKSHOP

Workshop 1.2 Lexical Dispersion Plot over Context using Sense and Sensibility
Use dispersion_plot to plot *Lexical Dispersion Plot* keywords: *good, happy, strong, bad, sad,* and *weak* from *Sense and Sensibility*.
1. Study any lexical pattern between positive and negative keywords.
2. Check these patterns against *Moby Dick* to see if this pattern occurs and explain.
3. Choose two other sentiment keywords to see if this pattern remains valid.

[27]	`text2.dispersion_plot(["good", "happy", "strong",` `"bad", "sad", "weak"])`

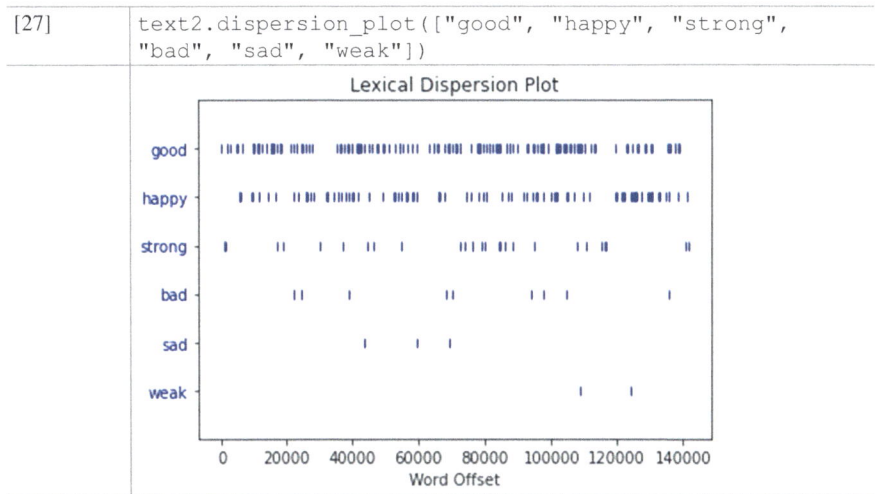

11.8.3 Lexical Dispersion Plot Over Time Using Inaugural Address Corpus

Lexical usage is to analyze word pattern changes in written English over time. The Inaugural Address Corpus addressed by US presidents of the past 220 years is a text document in NLTK book library to study lexical dispersion plot patterns changes on keywords *war, peace, freedom,* and *united* for this workshop.

WORKSHOP

Workshop 1.3 Lexical Dispersion Plot over Time using Inaugural Address Corpus
1. Use dispersion_plot to invoke Lexical Dispersion Plot for *Inaugural Address Corpus.*
2. Study and explain lexical pattern changes for keywords *America, citizens, democracy, freedom, war, peace, equal, united.*
3. Choose any two meaningful keywords and check for lexical pattern changes.

---

[28]
```
text4.dispersion_plot(["America" ,"citizens"
,"democracy", "freedom", "war", "peace", "equal",
"united"])
```

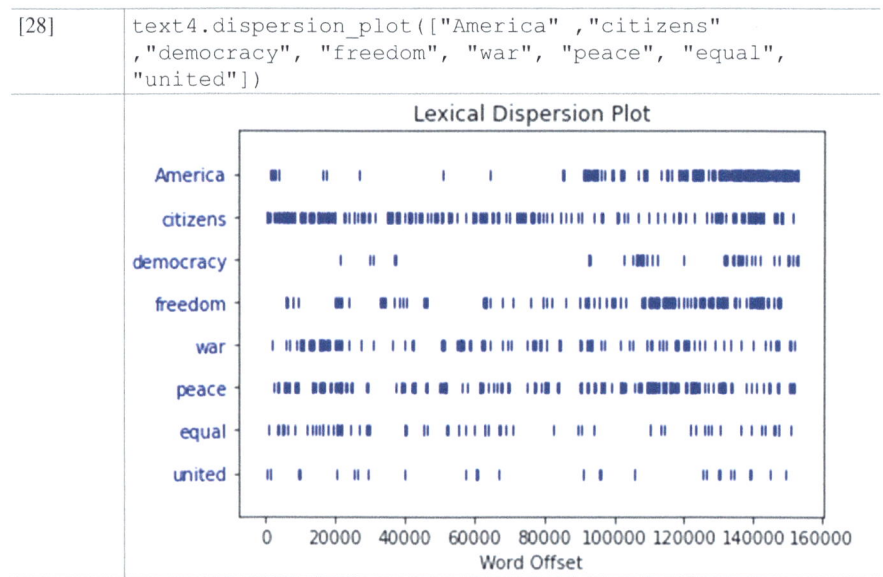

11.9 Tokenization in NLP with NLTK

11.9.1 What Is Tokenization in NLP?

A *token* can be words, part of a word, characters, numbers, punctuations, or symbols. It is a principal constituent and complex NLP task due to every language has its own grammatical constructions to generate grammatic and syntactic rules.

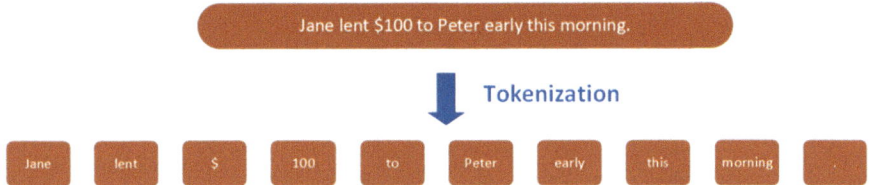

Fig. 11.5 Tokenization example of a sample utterance "*Jane lent $100 to Peter early this morning*"

Tokenization is an NLP process of dividing sentences/utterances from a text, document, or speech into chunks called *tokens*. By using tokenization, vocabulary from a document or corpus can be formed. Tokenization for sentence/utterances *Jane lent $100 to Peter early this morning* is shown in Fig. 11.5.

NLTK provides flexibility to tokenize any string of text using *tokenize()* function as shown below:

[29]	```
	# Create utterance 3 (utt3) and performs tokenization
	utt3 = 'Jane lent $100 to Peter early this morning.'
	wtokens = nltk.word_tokenize(utt3)
	                              wtokens
	``

	['Jane', 'lent', '$', '100', 'to', 'Peter', 'early', 'this', 'morning', '.']

11.9.2   Different Between Tokenize() vs Split()

Python provides *split()* function to split a sentence of text into words as recalled in Sect. 11.1. Let's see how it works with Tokenize() function.

[30]	```
	# Use split() to perform word tokenization
	words = utt3.split()
	words
	``

	['Jane', 'lent', '$100', 'to', 'Peter', 'early', 'this', 'morning.']

 Why are they different?
How is it important in
1. NLP?
2. Meanings?

WORKSHOP

| **Workshop 1.4 Tokenization on The Adventures of Sherlock Holmes with NLTK** |
| 1. Read Adventures_Holmes.txt text file. |
| 2. Save contents into a string object "holmes_doc". |
| 3. Use split() to cut it into list object "holmes". |
| 4. Count total number of words in the document. |
| 5. Tokenize document using NLTK tokenize() function. |
| 6. Count total number of tokens. |
| 7. Compare the two figures. |
| (The file open part is provided to start with.) |

[31]

```
# Workshop 1.4 Solution
with open('Adventures_Holmes.txt', encoding='utf-8') as
f_lit:
    dholmes = f_lit.read()
    # Count number of words in the literature
...
```

NLTK provides a simple way to count total number of tokens in a Text Document using len() in NLTK package.
Try len(tholmes) will notice:

[32]

```
len(tholmes)
```

128367

11.9.3   Count Distinct Tokens

Text analysis is to study distinct words or vocabulary that occurs in a text document.

When text document is tokenized as token objects, Python can group them easily into a set of distinct objects using Set() method.

Set() in Python is to extract distinct objects of any type from a list of objects with repeated instances.

Try the following using *The Adventures of Sherlock Holmes* will notice:

[33]

```
tholmes?
```

Type:	Text
String form:	<Text: The Adventures of Sherlock Holmes by Arthur Conan…>
Length:	128367
File:	d:\anaconda3\envs\py311nlp\lib\site-packages\nltk\text.py

Docstring:
A wrapper around a sequence of simple (string) tokens, which is
intended to support initial exploration of texts (via the
interactive console). Its methods perform a variety of analyses
on the text's contexts (e.g., counting, concordancing, collocation
discovery), and display the results. If you wish to write a
program which makes use of these analyses, then you should bypass
the "Text" class, and use the appropriate analysis function or
class directly instead.
A "Text" is typically initialized from a given document or
corpus. E.g.:
>>> import nltk.corpus
>>> from nltk.text import Text
>>> moby = Text(nltk.corpus.gutenberg.words('melville-moby_dick.txt'))
Init docstring:
Create a Text object.
:param tokens: The source text.
:type tokens: sequence of str

[34]	`set(tholmes)`
	{'madness', 'thirty-nine', 'inches', 'fuss', 'dense', 'exchange', 'swim', 'alive.', 'geese.', 'straw', 'whipcord', 'ill-kempt', 'ungrateful', 'law', 'distorted', 'chemical', 'autumn', 'landscape', 'discontent', 'Atkinson', 'acts', 'snakish', 'start', 'words.', 'brothers', 'handle', 'green-room', 'ruffians', 's—your', 'trip', 'briefly', 'ladies.', 'tragedy', 'Spaulding', 'tailing', 'bearded', 'when', …}

[35]	`len(set(tholmes))`
	10047

This example shows that *The Adventures of Sherlock Holmes* contains 128,366 tokens i.e. words and punctuations, and 10,048 distinct tokens, or types. Try other literatures and see vocabulary can be learnt from these great literatures.

The following example shows how to sort distinct tokens using sorted() function.

[36]	`sorted(set(tholmes)`

['!', '$', '%', '&', '"', '"', "'AS-IS", "'s", '(', ')', '*', ',', '-', '--', '-the-wisp',
'.', '1', '1,100', '1.A', '1.B', '1.C', '1.D', '1.E', '1.E.1', '1.E.2', '1.E.3', '1.E.4',
'1.E.5', '1.E.6', '1.E.7', '1.E.8', '1.E.9', '1.F', '1.F.1', '1.F.2', '1.F.3', '1.F.4',
'1.F.5', '1.F.6', '10', '100', '1000', '10_s_', '10_s_.', '10th', '117', …]

Since books are tokenized in NLTK as a list book object, contents can be accessed by using list indexing method as below:

[37]	```# Access the First 20 tokens``` ```tholmes[1:20]```
	['Adventures', 'of', 'Sherlock', 'Holmes', 'by', 'Arthur', 'Conan', 'Doyle', 'Contents', 'I', '.', 'A', 'Scandal', 'in', 'Bohemia', 'II', '.', 'The', 'Red-Headed']

[38]	```# Access the MIDDLE content``` ```tholmes[100:150]```
	['IN', 'BOHEMIA', 'I', '.', 'To', 'Sherlock', 'Holmes', 'she', 'is', 'always', '_the_', 'woman', '.', 'I', 'have', 'seldom', 'heard', 'him', 'mention', 'her', 'under', 'any', 'other', 'name', '.', 'In', 'his', 'eyes', 'she', 'eclipses', 'and', 'predominates', 'the', 'whole', 'of', 'her', 'sex', '.', 'It', 'was', 'not', 'that', 'he', 'felt', 'any', 'emotion', 'akin', 'to', 'love', 'for']

[39]	```# Aceess from the END``` ```tholmes[-20:]```
	['the', 'main', 'PG', 'search', 'facility', ':', 'www.gutenberg.org' , 'This', 'website', 'includes', 'information', 'about', 'Project', 'Gutenberg-tm', ',', 'including', 'how', 'to', 'make', 'donations', 'to', 'the', 'Project', 'Gutenberg', 'Literary', 'Archive', 'Foundation', ',', 'how', 'to', 'help', 'produce', 'our', 'new', 'eBooks']

11.9.4   Lexical Diversity

11.9.4.1   Token Usage Frequency (Lexical Diversity)

Token usage frequency, also called *Lexical Diversity* is to divide the total number of tokens by total number of token types as shown:

[40]	```len(text1)/len(set(text1))```
	13.502044830977896

[41]	```len(text2)/len(set(text2))```
	20.719449729255086

[42]	`len(text3)/len(set(text3))`
	16.050197203298673

[43]	`len(text4)/len(set(text4))`
	15.251970074812968

Python codes above analyze token usage frequency of four literatures: *Moby Dick, Sense and Sensibility, Book of Genesis, and Inaugural Address Corpus*. It has usage frequency range from 13.5 to 20.7. What are the implications?

11.9.4.2   Word Usage Frequency

There are many commonly used words in English. The following example shows the pattern of word usage frequency for *the* from above literatures.

[44]	`text1.count('the')`
	13721

[45]	`text1.count('the')/len(text1)*100`
	5.260736372733581

[46]	`text2.count('the')/len(text2)*100`
	2.7271571452788606

[47]	`text3.count('the')/len(text3)*100`
	5.386024483960325

[48]	`text4.count('the')/len(text4)*100`
	6.2491416014283745

1. Are there any patterns found from these literatures?
2. Use other words *of*, *a*, *I* to study if there exists other pattern(s).

11.10   Basic Statistical Tools in NLTK

11.10.1   *Frequency Distribution—FreqDist()*

Text analysis is an NTLK tool that can tokenize a string or a book of text document.

Frequency Distribution—*FreqDist()* is an initial built-in method in NLTK to analyze the frequency distribution of every token type in a text document.

Inaugural Address Corpus is used as an example to show how it works.

[49]	`text4`
	<Text: Inaugural Address Corpus>

[50]	`FreqDist?`
	Init signature: FreqDist(samples=**None**)
	Docstring:
	A frequency distribution for the outcomes of an experiment. A frequency distribution records the number of times each outcome of an experiment has occurred. For example, a frequency distribution could be used to record the frequency of each word type in a document. Formally, a frequency distribution can be defined as a function mapping from each sample to the number of times that sample occurred as an outcome.
	...

[51]	`fd4 = FreqDist(text4)`

[52]	`fd4`
	FreqDist({'the': 9555, ',': 7275, 'of': 7169, 'and': 5226, '.': 5011, 'to': 4477, 'in': 2604, 'a': 2229, 'our': 2062, 'that': 1769, ...})

11.10.1.1   FreqDist() as Dictionary Object

It is noted that *FreqDist()* will return *key-value* pairs from *Dictionary* object to reflect the *Key* that store *Token Type* name and the *Value* which are corresponding frequency of occurrence in a text. Since *FreqDist()* returns a *Dictionary* object, *keys()* can be used to return the list of all *Token Types* as shown below.

| [53] | ```
token4 = fd4.keys()
token4
``` |
|---|---|
| | dict_keys(['Fellow', '-', 'Citizens', 'of', 'the', 'Senate', 'and', 'House', 'Representatives', ':', 'Among', 'vicissitudes', 'incident', 'to', 'life', 'no', 'event', 'could', 'have', 'filled', 'me', 'with', 'greater', 'anxieties', 'than', 'that', 'which', 'notification', 'was', 'transmitted', 'by', 'your', 'order', ',', 'received', 'on', '14th', 'day', 'present', 'month', '.', 'On', 'one', 'hand', 'I', 'summoned', 'my', 'Country', 'whose', 'voice', 'can', 'never', 'hear', 'but', 'veneration', 'love', 'from', 'a', 'retreat', 'had', 'chosen', 'fondest', 'predilection', 'in', 'flattering', 'hopes', 'an', 'immutable', 'decision', 'as', 'asylum', 'declining', 'years', '--', 'rendered', 'every', 'more', 'necessary', 'well', 'dear', 'addition', 'habit', 'inclination', 'frequent', 'interruptions', 'health', 'gradual', 'waste', 'committed', 'it', 'time', 'other', 'magnitude', 'difficulty', 'trust', 'country', 'called', 'being', 'sufficient', 'awaken', 'wisest', 'most', 'experienced', 'her', ...]) |

11.10.1.2   Access FreqDist of Any Token Type

Use list item access method to obtain frequency distribution of any token type. FD value of token type for *the* is shown below.

| [54] | ```
fd4['the']
``` |
|---|---|
| | 9555 |

1. What are the five common word types (token types without punctuation) in any text document?
2. Use FreqDist() to verify.

11.10.1.3   Frequency Distribution Plot from NLTK

NLTK is a useful tool to study the top frequency distribution token types for any document using plot() function with FreqDist() method. FreqDist.plot() can also plot the top XX frequently used token types in a text document.

Use fd3 to study *FreqDist.plot()* documentation using *fd3.plot()*.

Plot top 30 frequently used token types from the *Book of Genesis* (Non-Cumulative mode).

11.11   Do the Same Plot with *Cumulative* Mode

[55]	`fd4.plot?`
|      | ```
Signature:
fd4.plot(
    *args,
    title='',
    cumulative=False,
    percents=False,
    show=False,
    **kwargs,
)
Docstring:
Plot samples from the frequency distribution
displaying the most frequent sample first. If an integer
parameter is supplied, stop after this many samples
have been
plotted. For a cumulative plot, specify cumulative=True.
Additional
''**kwargs'' are passed to matplotlib's plot function.
(Requires Matplotlib to be installed.)
...
``` |

[56]	`fd4.plot (30,cumulative=False)`
	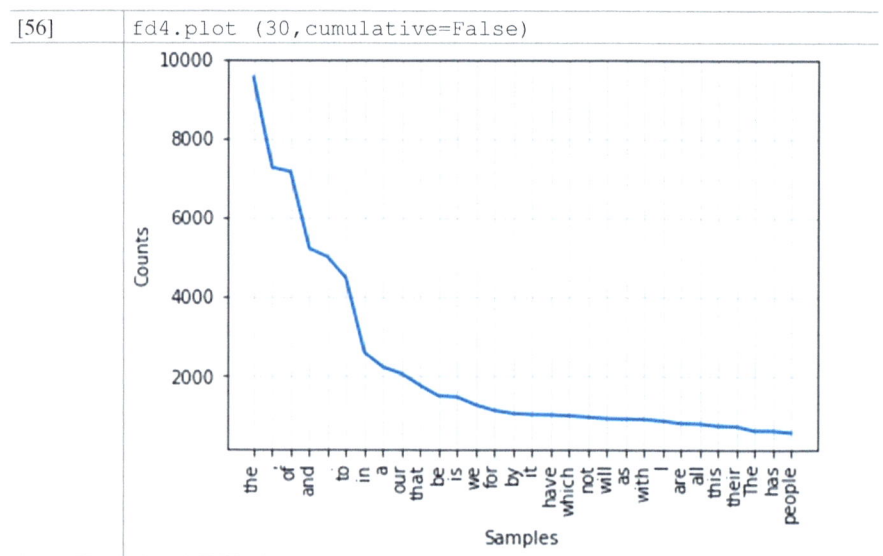
```

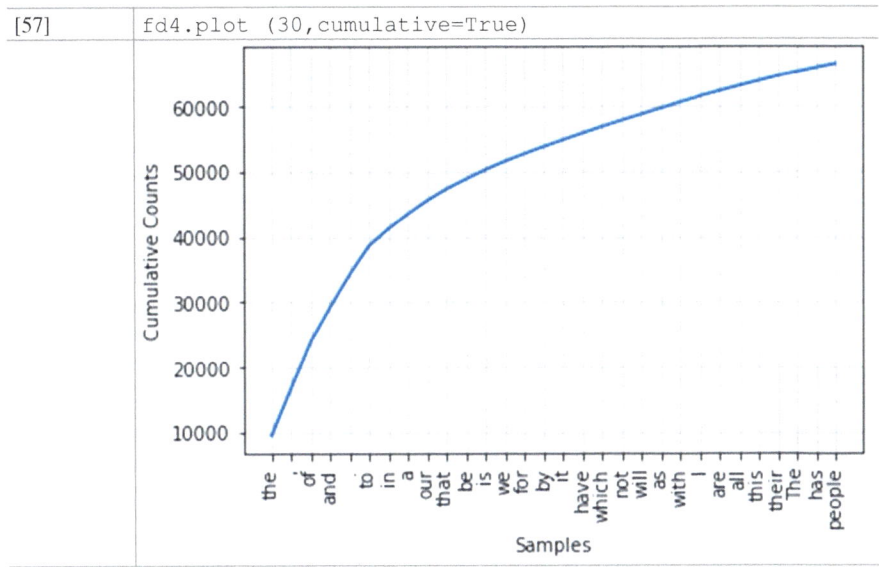

| [57] | `fd4.plot (30,cumulative=True)` |

WORKSHOP

Workshop 1.5 Frequency Distribution Analysis on Classics Literatures
1. What are the top 5 frequently used word types in the *Book of Genesis* (ignore punctuations)?
2. Will it be the same with other great literatures?
3. Verify against *(1) Moby Dick, (2) Sense and Sensibility,* and *(3) Inaugural Address Corpus* to see if they have the same patterns. Why or why not?
4. Why the study of common word types is also important in cryptography?

11.11.1   *Rare Words—Hapax*

Hapaxes are words that occur only once in a body of work whether it is a publication or an entire language.

Ancient texts are full of hapaxes. For instance, in Shakespeare's *Love's Labour's Lost* contains hapax *honorificabilitudinitatibus* which means able to achieve honors.

NLTK provides method hapaxes() under FreqDist object to list out all word types that occurred once in text document.

Try FreqDist() with *The Adventures of Sherlock Holmes* and see how useful it is.

| [58] | `tholmes` |
| | <Text: The Adventures of Sherlock Holmes by Arthur Conan...> |

| [59] | `fd = FreqDist(tholmes)` |

| [60] | ```
hap = fd.hapaxes()
hap[1:50]
``` |
|------|------|
| | ['Adventures', 'Conan', 'Doyle', 'Contents', 'Red-Headed', 'Case', 'Identity', 'Mystery', 'Orange', 'Pips', 'Twisted', 'Lip', 'Blue', 'Carbuncle', 'Speckled', 'Band', 'Engineer', 'Thumb', 'Noble', 'Bachelor', 'SCANDAL', 'BOHEMIA', 'eclipses', 'predominates', 'sex', 'emotions', 'abhorrent', 'balanced', 'softer', 'passions', 'gibe', 'observer—excellent', 'intrusions', 'finely', 'temperament', 'distracting', 'mental', 'Grit', 'sensitive', 'instrument', 'high-power', 'lenses', 'disturbing', 'dubious', 'home-centred', 'establishment', 'absorb', 'loathed', 'alternating'] |

WORKSHOP

Workshop 1.6 Learn Vocabulary using Hapaxes

Hapaxes are helpful to learn vocabulary containing more than 12 characters. The following example uses hapaxes() with Python in-line function to implement [w for w in hap1 if len(w) > 12]:

1. Run Python script and extract vocabulary containing more than 12 characters from *Moby Dick*.
2. Select five meaningful vocabularies with their meanings.
3. Check with *The Adventures of Sherlock Holmes* to learn another five vocabularies.

(Python script to generate vocabulary with over 12 characters is given.)

| [61] | ```
# Workshop 1.6 Solutions
voc12 = [w for w in hap if len(w) > 12]
voc12
``` |
|------|------|
| | ['observer—excellent', 'establishment', 'well-remembered', 'boot-slitting', 'Peculiar—that', 'Eglonitz—here', 'German-speaking', 'glass-factories', 'authoritative', 'double-breasted', 'Cassel-Felstein', 'staff-commander', 'Contralto—hum', 'indiscretion.', 'reproachfully', 'Saxe-Meningen', 'drunken-looking', 'side-whiskered', 'half-and-half', 'moustached—evidently', 'expostulating', 'arrangements.', 'co-operation.', 'unpleasantness', 'self-lighting', 'simple-minded', 'Nonconformist', 'ill—gentlemen', …] |

11.11.2   Collocations

11.11.2.1   What Are Collocations?

A *collocation* is a work grouping for a set of words usually appeared together to convey semantic meanings. The word *collocation* is originated from the Latin word meaning place together and was first introduced by Prof. John R Firth (1890–1960) with his famous quote "You shall know a word by the company it keeps."

There are many collocations cases in English where strong collocations are word pairings always appear together such as *make* and *do,* e.g., You make a cup of coffee, but you do your work.

Collocations are frequently used in business settings when nouns are combined with verbs or adjectives, e.g., setup an appointment, conduct a meeting, set the price, etc.

11.11.2.2   Collocations in NLTK

NLTK also provides a build-in method to handle collocations using NLTK method—*collocations( )*.

The following example is to generate collocations lists from *Moby Dick, Sense and Sensibility, Book of Genesis,* and *Inaugural Address Corpus.*

Let's look at some extracted collocation terms:

[62]	`text1.collocations()`
	Sperm Whale; Moby Dick; White Whale; old man; Captain Ahab; sperm whale; Right Whale; Captain Peleg; New Bedford; Cape Horn; cried Ahab; years ago; lower jaw; never mind; Father Mapple; cried Stubb; chief mate; white whale; ivory leg; one hand

[63]	`text2.collocations()`
	Colonel Brandon; Sir John; Lady Middleton; Miss Dashwood; every thing; thousand pounds; dare say; Miss Steeles; said Elinor; Miss Steele; every body; John Dashwood; great deal; Harley Street; Berkeley Street; Miss Dashwoods; young man; Combe Magna; every day; next morning

[64]	`text3.collocations()`
	said unto; pray thee; thou shalt; thou hast; thy seed; years old; spake unto; thou art; LORD God; every living; God hath; begat sons; 7 years; shalt thou; little ones; living creature; creeping thing; savoury meat; 30 years; every beast

[65]	`text4.collocations()`
	United States; fellow citizens; years ago; 4 years; Federal Government; General Government; American people; Vice President; God bless; Chief Justice; one another; fellow Americans; Old World; Almighty God; Fellow citizens; Chief Magistrate; every citizen; Indian tribes; public debt; foreign nations

References

Albrecht, J., Ramachandran, S. and Winkler, C. (2020) Blueprints for Text Analytics Using Python: Machine Learning-Based Solutions for Common Real World (NLP) Applications. O'Reilly Media.

Antic, Z. (2021) Python Natural Language Processing Cookbook: Over 50 recipes to understand, analyze, and generate text for implementing language processing tasks. Packt Publishing.

Arumugam, R. and Shanmugamani, R. (2018) Hands-On Natural Language Processing with Python: A practical guide to applying deep learning architectures to your NLP applications. Packt Publishing.

Bird, S., Klein, E., and Loper, E. (2009). Natural language processing with python. O'Reilly.

Doyle, A. C. (2019) The Adventures of Sherlock Holmes (AmazonClassics Edition). AmazonClassics.

Gutenberg (2024) Project Gutenberg official site. https://www.gutenberg.org/ Accessed 17 Dec 2024.

Hardeniya, N., Perkins, J. and Chopra, D. (2016) Natural Language Processing: Python and NLTK. Packt Publishing.

Jupyter (2024) Jupyter official site. https://jupyter.org/. Accessed 17 Dec 2024.

Kedia, A. and Rasu, M. (2020) Hands-On Python Natural Language Processing: Explore tools and techniques to analyze and process text with a view to building real-world NLP applications. Packt Publishing.

NLPGitHub (2024) URL: https://github.com/raymondshtlee/NLP/. Accessed 17 Dec 2024.

NLTK (2024) NLTK official site. https://www.nltk.org/. Accessed 17 Dec 2024.

Perkins, J. (2014). Python 3 text processing with NLTK 3 cookbook. Packt Publishing Ltd.

Wintjen, M. and Vlahutin, A. (2020) Practical Data Analysis Using Jupyter Notebook: Learn how to speak the language of data by extracting useful and actionable insights using Python. Packt Publishing.

WordNet (2024) WordNet official site. https://wordnet.princeton.edu/. Accessed 17 Dec 2024.

Chaper 12
Workshop#2: N-Grams Modeling
with Natural Language Toolkit (Hour 3–4)

12.1   Introduction

Workshop 2 consists of two parts:

Part I will introduce N-gram language model using NLTK in Python and N-grams class to generate N-gram statistics on any sentence, text objects, whole document, literature to provide a foundation technique for text analysis, parsing, and semantic analysis in subsequent workshops.

Part II will introduce spaCy, the second important NLP Python implementation tools not only for teaching and learning (like NLTK), but widely used for NLP applications including text summarization, information extraction, and Q&A chatbot. It is a critical mass to integrate with Transformer technology in subsequent workshops.

Please ensure that the following Python packages are installed before starting the workshop:

- Python (demo version 3.11.9)
- tensorflow (demo version 2.17.0)
- NLTK (demo version 3.9.1)
- spacy (demo version 3.4.4)

If these packages are not installed on PC/laptop, use pip install xxx command. The detailed requirements list and Python package version used in this workshop can be found in the requirements.txt file stored in the NLP GitHub repository (NLPGitHub 2024).

© The Author(s), under exclusive license to Springer Nature Singapore Pte
Ltd. 2025
R. Lee, *Natural Language Processing*,
https://doi.org/10.1007/978-981-96-3208-4_12

12.2 What Is N-Gram?

N-gram is an algorithm based on a statistical language model (Bird et al. 2009; Perkins 2014; Arumugam and Shanmugamani 2018), its basic idea is that contents such as phonemes, syllables, letters, words, or base pairs in texts are operated by a sliding window of size N to form a byte fragments sequence of length N (Sidorov 2019).

N can be 1, 2, or another positive integer, although usually large N is not considered because they rarely occur.

Each byte fragment is called a gram, and the frequency of all grams is counted and filtered according to a pre-set threshold to form a list of key grams, which is the text's vector feature space, and each kind of gram in the list is a feature vector dimension.

12.3 Applications of N-Grams in NLP

N-gram models are widely used (Albrecht et al. 2020; Arumugam and Shanmugamani 2018; Hardeniya et al. 2016; Kedia and Rasu 2020) in:

- Speech recognition where phonemes and sequences of phonemes are modeled using an N-gram distribution.
- Parsing on words is modeled so that each N-gram is composed of N words. For language identification, sequences of characters/graphemes (e.g., letters of the alphabet) are modeled for different languages.
- Auto sentences completion.
- Auto spell-check.
- Semantic analysis.

12.4 Generation of N-Grams in NLTK

NLTK (NLTK 2024; Bird et al. 2009; Perkins 2014) offers useful tools in NLP processing.

Ngrams() function in NLTK facilitates N-gram operation.

Python code uses N-grams in NLTK to generate N-grams for any text string. Try it and study how it works.

The following example is the first sentence of *A Scandal in Bohemia* from *The Adventures of Sherlock Holmes* (Doyle 2019): *To Sherlock Holmes she is always The Woman. I have seldom heard him mention her under any other name.* Demonstrating how N-gram generator works in NLTK.

| [1] | ```import nltk
from nltk import ngrams
sentence = input("Enter the sentence: ")
n = int(input("Enter the value of n: "))
n_grams = ngrams(sentence.split(), n)
for grams in n_grams:
    print(grams)``` |
| --- | --- |
| | Enter the sentence: To Sherlock Holmes she is always "The Woman". I have seldom heard him mention her under any other name.
Enter the value of n: 2
('To', 'Sherlock')
('Sherlock', 'Holmes')
('Holmes', 'she')
('she', 'is')
('is', 'always')
('always', '"The')
('"The', 'Woman".')
('Woman".', 'I')
('I', 'have')
('have', 'seldom')
('seldom', 'heard')
('heard', 'him')
('him', 'mention')
('mention', 'her')
('her', 'under')
('under', 'any')
('any', 'other')
('other', 'name.') |

Here are the Bigrams. Let's try Trigrams N = 3.

| [2] | ```import nltk
from nltk import ngrams
sentence = input("Enter the sentence: ")
n = int(input("Enter the value of n: "))
n_grams = ngrams(sentence.split(), n)
for grams in n_grams:
    print(grams)``` |
| --- | --- |

Enter the sentence: To Sherlock Holmes she is always "The Woman". I have seldom heard him mention her under any other name.
Enter the value of n: 3
('To', 'Sherlock', 'Holmes')
('Sherlock', 'Holmes', 'she')
('Holmes', 'she', 'is')
('she', 'is', 'always')
('is', 'always', '"The')
('always', '"The', 'Woman".')
('"The', 'Woman".', 'I')
('Woman".', 'I', 'have')
('I', 'have', 'seldom')
('have', 'seldom', 'heard')
('seldom', 'heard', 'him')
('heard', 'him', 'mention')
('him', 'mention', 'her')
('mention', 'her', 'under')
('her', 'under', 'any')
('under', 'any', 'other')
('any', 'other', 'name.')

How about Quadrigram N = 4? Let's use the same sentence.

[3]
```
import nltk
from nltk import ngrams
sentence = input( "Enter the sentence: " )
n = int(input( "Enter the value of n: " ))
n_grams = ngrams(sentence.split(), n)
for grams in n_grams:
    print(grams)
```

Enter the sentence: To Sherlock Holmes she is always "The Woman". I have seldom heard him mention her under any other name.
Enter the value of n: 4
('To', 'Sherlock', 'Holmes', 'she')
('Sherlock', 'Holmes', 'she', 'is')
('Holmes', 'she', 'is', 'always')
('she', 'is', 'always', '"The')
('is', 'always', '"The', 'Woman".')
('always', '"The', 'Woman".', 'I')
('"The', 'Woman".', 'I', 'have')
('Woman".', 'I', 'have', 'seldom')
('I', 'have', 'seldom', 'heard')
('have', 'seldom', 'heard', 'him')
('seldom', 'heard', 'him', 'mention')
('heard', 'him', 'mention', 'her')
('him', 'mention', 'her', 'under')
('mention', 'her', 'under', 'any')
('her', 'under', 'any', 'other')
('under', 'any', 'other', 'name.')

	NLTK offers an easy solution to generate N-gram of any N-number which are useful in N-gram probability calculations and text analysis
 **WORKSHOP**	**Workshop 2.1 N-Grams on The Adventures of Sherlock Holmes** 1. Read Adventures_Holmes.txt text file. 2. Save contents into a string object "holmes_doc." 3. Extract favorite paragraph from "holmes_doc" into "holmes_para." 4. Use above Python code to generate N-grams for N=3, N=4 and N=5.

12.5   Generation of N-Grams Statistics

Once N-grams are generated, the next step is to calculate the term frequency (TF) of each N-gram from a document to list out top items.

NLTK-based Python codes extend previous example to create N-grams statistics to list out top 10 N-grams.

Let's try first two sentences of *A Scandal in Bohemia* from *The Adventures of Sherlock Holmes*.

[4]	sentence
	'To Sherlock Holmes she is always "The Woman". I have seldom heard him mention her under any other name.'

Import RE package to do some simple text pre-processing:

```
[5]  import re, string
     # get rid of all the XML markup
     sentence = re.sub ('<.*>', ' ', sentence)
     # get rid of punctuation (except periods!)
     punctuationNoPeriod = "[" + re.sub("\.","",string.
     punctuation) + "]"
     sentence = re.sub(punctuationNoPeriod, "", sentence)
     # first get individual words
     tokenized = sentence.split()
     # and get a list of all the bi-grams
     Bigrams = ngrams(tokenized, 2)
```

Review N-grams to see how they work:

[6]	`ngrams?`
	Signature: ngrams(sequence, n, **kwargs)
	Docstring:
	Return the ngrams generated from a sequence of items, as an iterator.
	For example:
	>>> from nltk.util import ngrams
	>>> list(ngrams([1,2,3,4,5], 3))
	[(1, 2, 3), (2, 3, 4), (3, 4, 5)]
	Wrap with list for a list version of this function. Set pad_left
	or pad_right to true in order to get additional ngrams:
	>>> list(ngrams([1,2,3,4,5], 2, pad_right=True))
	[(1, 2), (2, 3), (3, 4), (4, 5), (5, None)]
	>>> list(ngrams([1,2,3,4,5], 2, pad_right=True, right_pad_symbol='</s>'))
	[(1, 2), (2, 3), (3, 4), (4, 5), (5, '</s>')]
	>>> list(ngrams([1,2,3,4,5], 2, pad_left=True, left_pad_symbol='<s>'))
	[('<s>', 1), (1, 2), (2, 3), (3, 4), (4, 5)]
	>>> list(ngrams([1,2,3,4,5], 2, pad_left=True, pad_right=True, left_pad_
	symbol='<s>', right_pad_symbol='</s>'))
	[('<s>', 1), (1, 2), (2, 3), (3, 4), (4, 5), (5, '</s>')]
	…

To generate N-gram statistics, first import "collections" class and invoke Counter() method over Bigrams.

| [7] | ```
import collections
# get the frequency of each bigram in our corpus
BigramFreq = collections.Counter(Bigrams)
# what are the ten most popular ngrams in this corpus?
BigramFreq.most_common(10)
``` |
|---|---|
	[(('To', 'Sherlock'), 1),
	(('Sherlock', 'Holmes'), 1),
	(('Holmes', 'she'), 1),
	(('she', 'is'), 1),
	(('is', 'always'), 1),
	(('always', 'The'), 1),
	(('The', 'Woman'), 1),
	(('Woman', 'I'), 1),
	(('I', 'have'), 1),
	(('have', 'seldom'), 1)]

It is noted that the top 10 bigram frequency are all with count 1.
This is because the sample sentence is short and doesn't contain any bigram(s) with a frequent bigram statistic. To sort out this problem, let's try a longer text. The following example uses the whole first paragraph of *A Scandal in Bohemia* from *The Adventures of Sherlock Holmes* and see whether it has a preferable result.
```

The first paragraph looks like this:

```
[8]  first_para = "To Sherlock Holmes she is always the woman I
     have seldom heard him mention her under any other name In
     his eyes she eclipses and predominates the whole of her sex
     It was not that he felt any emotion akin to love for Irene
     Adler All emotions and that one particularlywere abhorrent
     to his cold precise but admirably balanced mind He was I
     take it the most perfect reasoning and observing machine
     that the world has seen but as a lover he would have placed
     himself in a false position He never spoke of the softer
     passions save with a gibe and a sneer They were admirable
     things for the observer—excellent for drawing the veil from
     men's motives and actions But for the trained reasoner to
     admit such intrusions into his own delicate and finely
     adjusted temperament was to introduce a distracting factor
     which might throw a doubt upon all his mental results Grit
     in a sensitive instrument or a crack in one of his own
     highpower lenses would not be more disturbing than a strong
     emotion in a nature such as his And yet there was but one
     woman to him and that woman was the late Irene Adler of
     dubious and questionable memory."
```

Let's review this first paragraph:

```
[9]  first_para
```

'To Sherlock Holmes she is always the woman I have seldom heard him mention her under any other name In his eyes she eclipses and predominates the whole of her sex It was not that he felt any emotion akin to love for Irene Adler All emotions and that one particularlywere abhorrent to his cold precise but admirably balanced mind He was I take it the most perfect reasoning and observing machine that the world has seen but as a lover he would have placed himself in a false position He never spoke of the softer passions save with a gibe and a sneer They were admirable things for the observer—excellent for drawing the veil from men's motives and actions But for the trained reasoner to admit such intrusions into his own delicate and finely adjusted temperament was to introduce a distracting factor which might throw a doubt upon all his mental results Grit in a sensitive instrument or a crack in one of his own highpower lenses would not be more disturbing than a strong emotion in a nature such as his And yet there was but one woman to him and that woman was the late Irene Adler of dubious and questionable memory.'

Use Python script to remove punctuation marks and tokenize the first_para object:

```python
[10]  import re, string
      # get rid of all the XML markup
      first_para = re.sub ('<.*>', ' ', first_para)
      # get rid of punctuation (except periods!)
      punctuationNoPeriod = "[" + re.sub("\.","",string.
      punctuation) + "]"
      first_para = re.sub(punctuationNoPeriod, "", first_para)
      # first get individual words
      tokenized = first_para.split()
      # and get a list of all the bi-grams
      Bigrams = ngrams(tokenized, 2)
```

Use Counter() method of collections class to calculate bigram statistics of first_para:

| [11] | ```
import collections
# get the frequency of each bigram in our corpus
BigramFreq = collections.Counter(Bigrams)
# what are the ten most popular ngrams in this corpus?
BigramFreq.most_common(10)
``` |
|---|---|
| | [(('in', 'a'), 3),
(('Irene', 'Adler'), 2),
(('and', 'that'), 2),
(('for', 'the'), 2),
(('his', 'own'), 2),
(('To', 'Sherlock'), 1),
(('Sherlock', 'Holmes'), 1),
(('Holmes', 'she'), 1),
(('she', 'is'), 1),
(('is', 'always'), 1)] |

The results are satisfactory. It is noted that bigram *in a* has the most occurrence frequency, i.e., three times while four other bigrams: *Irene Adler*, *and that*, *for the*, *his own* have occurred twice each within the paragraph. Bigram *in a*, *and that* and *for the* are frequently used English phrases which occurred in almost every text document. How about *to Sherlock* and *Irene Adler*? There are two N-gram types frequently used in N-gram language model studied in Chap. 2. One is the frequently used N-gram phrase in English like *in a*, *and that* and *for that* in our case. These bigrams are common phrases in other documents and literature writings. Another is domain-specific N-grams. These types are only frequently used in specific domain, documents, and genre of literatures. Hence, *to Sherlock* and *Irene Adler* are frequently used related to this story only and not in other situations

Workshop 2.2 N-grams Statistics on The Adventures of Sherlock Holmes
1. Read Adventures_Holmes.txt text file.
2. Save contents into a string object "holmes_doc."
3. Generate a representative N-gram statistic using the whole holmes_doc.
4. Generate a top 10 N-grams summary for N=3, N=4 and N=5.
5. Review results and comments on pattern(s) found.

Bigram analysis is required to examine which bigrams are commonly used not only in a single paragraph but for the whole document or literature. Remember in Workshop 1 NLTK has a built-in list of tokenized sample literatures in nltk.book. Let's refer to them first by using the nltk.book import statement.
```

| [12] | ```# Let's load some sample books from the nltk databank
import nltk
from nltk.book import *``` |
| --- | --- |
| | *** Introductory Examples for the NLTK Book ***
Loading text1, ..., text9 and sent1, ..., sent9
Type the name of the text or sentence to view it.
Type: 'texts()' or 'sents()' to list the materials.
text1: Moby Dick by Herman Melville 1851
text2: Sense and Sensibility by Jane Austen 1811
text3: The Book of Genesis
text4: Inaugural Address Corpus
text5: Chat Corpus
text6: Monty Python and the Holy Grail
text7: Wall Street Journal
text8: Personals Corpus
text9: The Man Who Was Thursday by G. K. Chesterton 1908 |

Check with text1 to see what they are:

[13]	```text1```
	<Text: Moby Dick by Herman Melville 1851>

or download using nltk.corpus.gutenberg.words() from Project Gutenberg of copyright clearance classic literature (Gutenberg 2024). Let's use this method to download *Moby Dick* (Melville 2006).

| [14] | ```import nltk.corpus
from nltk.text import Text
moby = Text(nltk.corpus.gutenberg.words('melville-moby_
dick.txt'))``` |
| --- | --- |

[15]	```moby```
	<Text: Moby Dick by Herman Melville 1851>

Review the first 50 elements of *Moby Dick* text object to see whether they are tokenized.

[16]	```moby [1:50]```
	['Moby', 'Dick', 'by', 'Herman', 'Melville', '1851', ']', 'ETYMOLOGY', '.', '(', 'Supplied', 'by', 'a', 'Late', 'Consumptive', 'Usher', 'to', 'a', 'Grammar', 'School', ')', 'The', 'pale', 'Usher', '--', 'threadbare', 'in', 'coat', ',', 'heart', ',', 'body', ',', 'and', 'brain', ';', 'I', 'see', 'him', 'now', '.', 'He', 'was', 'ever', 'dusting', 'his', 'old', 'lexicons', 'and']

Use *collections* class and *ngrams()* method for bigram statistics to identify the top 20 most frequently bigrams occurred for the entire *Moby Dick* literature.

| [17] | ```python
import collections
# and get a list of all the bi-grams
Bigrams = ngrams(moby, 2)
# get the frequency of each bigram in our corpus
BigramFreq = collections.Counter(Bigrams)
# what are the 20 most popular ngrams in this corpus?
BigramFreq.most_common(20)
``` |
|---|---|
| | [((',', 'and'), 2607),
(('of', 'the'), 1847),
(("'", 's'), 1737),
(('in', 'the'), 1120),
((',', 'the'), 908),
((';', 'and'), 853),
(('to', 'the'), 712),
(('.', 'But'), 596),
((',', 'that'), 584),
(('.', '"'), 557),
((',', 'as'), 523),
((',', 'I'), 461),
((',', 'he'), 446),
(('from', 'the'), 428),
((',', 'in'), 402),
(('of', 'his'), 371),
(('the', 'whale'), 369),
(('.', 'The'), 369),
(('and', 'the'), 357),
((';', 'but'), 340)] |

WORKSHOP

Workshop 2.3 N-grams Statistics with removal of unnecessary punctuations

The results are average and unsatisfactory. It is noted that *and*, *of the*, *s* and *in the* are the top 4 bigrams occurred in the entire Moby Dick literature. It is average since these bigrams are common English usage but original bigram statistics in simple sentences required to remove all punctuations by:

1. List out all punctuations required to remove.
2. Revise the above Python script to remove these punctuation symbols from the token list.
3. Generate a top 20 bigram summary for *Moby Dick* without punctuations.
4. Use sample method to generate (cleaned) bigram statistics from *Moby Dick, Adventures of Sherlock Holmes, Sense and Sensibility, Book of Genesis, Inaugural Address Corpus,* and *Wall Street Journal.*
5. Verify results and comments of any pattern(s) found.
6. Try the same analysis for Trigram (N=3) and Quadrigram (N=4) to find any pattern(s).
```

12.6   spaCy in NLP

12.6.1   What Is spaCy?

SpaCy (spaCy 2024) is a free, open-source library for advanced NLP written in Python and Cython programming languages.

The library is published under an MIT license developed by Dr. Matthew Honnibal and Dr. Ines Montani, founders of the software company Explosion.

SpaCy is designed specifically for production use and build NLP applications to process large volumes of text (Altinok 2021; Srinivasa-Desikan 2018; Vasiliev 2020) different from NLTK focused on teaching and learning perspective.

It also provides workflow pipelines for machine learning and deep learning tools that can integrate with common platforms such as PyTorch, MXNet, and TensorFlow with its machine learning library called *Thinc*. spaCy provides recurrent neural models such as convolution neural networks (CNN) by adopting Thinc for NLP implementation such as dependency parsing (DP), named entity recognition (NER), POS tagging and text classification, and other advanced NLP applications such as natural language understanding (NLU) systems, information retrieval (IR), information extraction (IE) systems, and question-and-answer chatbot systems.

A spaCy system architecture is shown in Fig. 12.1, its major features support:

- NLP-based statistical models for over 19 commonly used languages.
- tokenization tools implementation for over 60 international languages.
- NLP pipeline components include NER, POS Tagging, DP, text classification, and chatbot implementation.
- integration with common Python platforms such as TensorFlow, PyTorch, and other high-level frameworks.
- integration with the latest Transformer and BERT technologies.
- user-friendly modular system packaging, evaluation, and deployment tools.

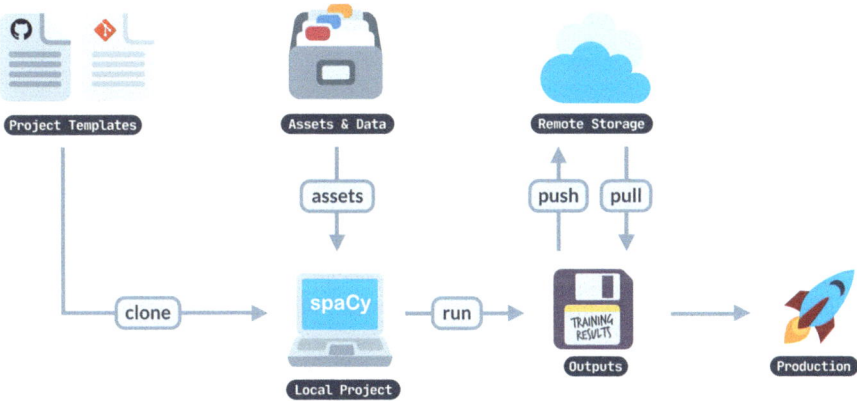

Fig. 12.1 System architecture of spaCy

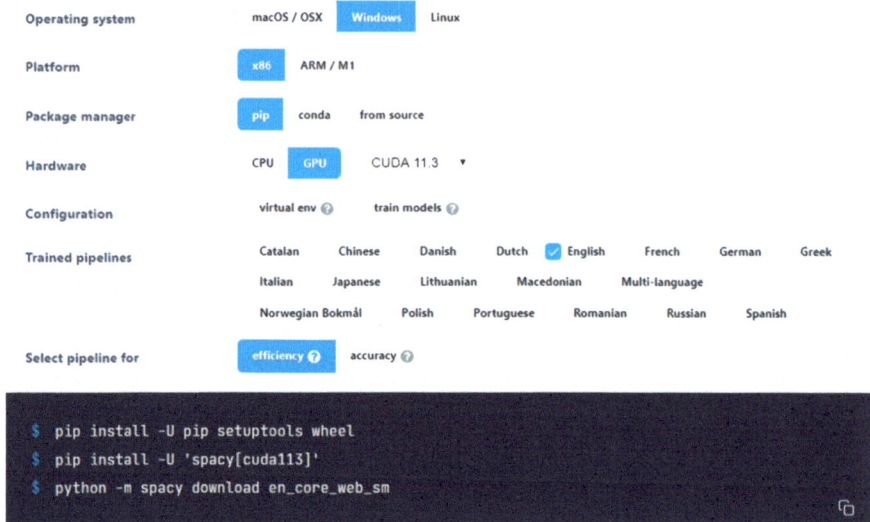

Fig. 12.2 Screenshot of spaCy configuration selection

12.7   How to Install spaCy?

SpaCy can be installed in MacOS/OSX, MS Windows, and Linux platforms (spaCy 2024) as per other Python-based development tools like NLTK.

spaCy.io provides a one-stop-process for users to select their own spaCy (1) language(s) as trained pipelines, (2) optimal target in system efficiency vs. accuracy for NLP applications development based a large dataset and lexical database, and (3) download appropriate APIs and modules to maximize efficiency under CPU and GPU hardware configuration. Figure 12.2 shows a Windows-based PIP download environment using CUDA 11.3 GPU in English as trained pipelines and target for speed efficiency over accuracy.

12.8   Tokenization Using spaCy

Tokenization is an operation in NLP. spaCy provides an easy-to-use scheme to tokenize any text document into sentences like NLTK, and further tokenize sentences into words.

This section uses *Adventures_Holmes.txt* as example to demonstrate tokenization in spaCy.

Step 1: Import spaCy module

Step 2: Load spaCy module "en_core_web_sm"

```
[18]    import spacy
```

Use en_core_web_md-3.2.0 package for English pipeline optimized for CPU in the current platform with components include: tok2vec, tagger, parser, senter, ner, attribute_ruler, lemmatizer.

```
[19]    nlp = spacy.load( "en_core_web_sm" )
```

Step 3: Open and read text file "Adventures_Holmes.txt" into file_handler "fholmes"

Note: Since text file already exists, skip the try-except module to save programming steps

```
[20]    fholmes = open( "Adventures_Holmes.txt", "r",
        encoding="utf-8")
```

Step 4: Read Adventures of Sherlock Holmes

Use read() method to read whole text document as a complex string object "holmes."

```
[21]    holmes = fholmes.read()
        holmes
```
\ufeffThe Adventures of Sherlock Holmes\n\nby Arthur Conan Doyle\n\n\nContents\n\n I. A Scandal in Bohemia\n II. The Red-Headed League\n III. A Case of Identity\n IV. The Boscombe Valley Mystery\n V. The Five Orange Pips\n VI. The Man with the Twisted Lip\n VII. The Adventure of the Blue Carbuncle\n VIII. The Adventure of the Speckled Band\n IX. The Adventure of the Engineer's Thumb\n X. The Adventure of the Noble Bachelor\n XI. The Adventure of the Beryl Coronet\n XII. The Adventure of the Copper Beeches\n\n\n\n\nI. A SCANDAL IN BOHEMIA\n\n\nI.\n\nTo Sherlock Holmes she is always _the_ woman. I have seldom heard him\nmention her under any other name. In his eyes she eclipses and\npredominates the whole of her sex. It was not that he felt any emotion\nakin to love for Irene Adler. All emotions, and that one particularly,\nwere abhorrent to his cold, precise but admirably balanced mind. He\nwas, I take it, the most perfect reasoning and observing machine that\nthe world has seen, but as a lover he would have placed himself in a\nfalse position. …

Step 5: Replace all newline symbols

Replace all newline characters "\n" into space characters.

```
[22]    holmes = holmes.replace( "\n", " " )
```

Step 6: Simple counting

Review total number of characters in *The Adventures of Sherlock Holmes* and examine the result document.

[23]	`len (holmes)`
	580632

[24]	`holmes`
	\ufeffThe Adventures of Sherlock Holmes by Arthur Conan Doyle Contents I. A Scandal in Bohemia II. The Red-Headed League III. A Case of Identity IV. The Boscombe Valley Mystery V. The Five Orange Pips VI. The Man with the Twisted Lip VII. The Adventure of the Blue Carbuncle VIII. The Adventure of the Speckled Band IX. The Adventure of the Engineer's Thumb X. The Adventure of the Noble Bachelor XI. The Adventure of the Beryl Coronet XII. The Adventure of the Copper Beeches I. A SCANDAL IN BOHEMIA I. To Sherlock Holmes she is always _the_ woman. I have seldom heard him mention her under any other name. In his eyes she eclipses and predominates the whole of her sex. It was not that he felt any emotion akin to love for Irene Adler. All emotions, and that one particularly, were abhorrent to his cold, precise but admirably balanced mind. He was, I take it, the most perfect reasoning and observing machine that the world has seen, but as a lover he would have placed himself in a false position. ...

Step 7: Invoke nlp() method in spaCy

SpaCy nlp() method is an important Text Processing Pipeline to initialize nlp object (English in our case) for NLP processing such as tokenization. It will convert any text string object into an NLP object.

Study nlp() docstring to see how it works.

[25]	`nlp?`

	Signature:
	nlp(
	    text: Union[str, spacy.tokens.doc.Doc],
	    *,
	    disable: Iterable[str] = [],
	    component_cfg: Optional[Dict[str, Dict[str, Any]]] = **None**,
	) -> spacy.tokens.doc.Doc
	Type: English
	String form: <spacy.lang.en.English object at 0x000001A70F4CCDD0>
	File: d:\anaconda3\envs\py311nlp\lib\site-packages\spacy\lang\en__init__.py
	Docstring: <no docstring>
	Class docstring:
	A text-processing pipeline. Usually you'll load this once per process, and pass the instance around your application.
	...

```
[26]  holmes_doc = nlp(holmes)
```

```
[27]  holmes_doc
```

The Adventures of Sherlock Holmes by Arthur Conan Doyle Contents I. A Scandal in Bohemia II. The Red-Headed League III. A Case of Identity IV. The Boscombe Valley Mystery V. The Five Orange Pips VI. The Man with the Twisted Lip VII. The Adventure of the Blue Carbuncle VIII. The Adventure of the Speckled Band IX. The Adventure of the Engineer's Thumb X. The Adventure of the Noble Bachelor XI. The Adventure of the Beryl Coronet XII. The Adventure of the Copper Beeches I. A SCANDAL IN BOHEMIA I. To Sherlock Holmes she is always _the_ woman. I have seldom heard him mention her under any other name. In his eyes she eclipses and predominates the whole of her sex. It was not that he felt any emotion akin to love for Irene Adler. All emotions, and that one particularly, were abhorrent to his cold, precise but admirably balanced mind. He was, I take it, the most perfect reasoning and observing machine that the world has seen, but as a lover he would have placed himself in a false position. He never spoke of the softer passions, save with a gibe and a sneer. They were admirable things for the observer— excellent for drawing the veil from men's motives and actions. But for the trained reasoner to admit such intrusions into his own delicate and finely adjusted temperament was to introduce a distracting factor which might throw a doubt upon all his mental results. Grit in a sensitive instrument, or a crack in one of his own high-power lenses, would not be more disturbing than a strong emotion in a nature such as his. And yet there was but one woman to him, and that woman was the late Irene Adler, of dubious and questionable memory.

...

Step 8: Convert text document into sentence object

SpaCy is practical for text document tokenization to convert text document object into (1) sentence objects and (2) tokens.

This example uses *for-in* statement to convert the whole Sherlock Holmes document into holmes_sentences.

```
[28]  holmes_sentences = [sentence.text for sentence in holmes_doc.
      sents]
      holmes_sentences
```

['\ufeffThe Adventures of Sherlock Holmes by Arthur Conan Doyle Contents I. A Scandal in Bohemia II. ',
'The Red-Headed League III. ',
'A Case of Identity IV. ',
'The Boscombe Valley Mystery V. The Five Orange Pips VI. ',
'The Man with the Twisted Lip VII. ',
'The Adventure of the Blue Carbuncle VIII. ',
'The Adventure of the Speckled Band IX. ',
'The Adventure of the Engineer's Thumb X. The Adventure of the Noble Bachelor XI. ',
'The Adventure of the Beryl Coronet XII. ',
'The Adventure of the Copper Beeches I. A SCANDAL IN BOHEMIA I. To Sherlock Holmes',
'she is always _the_ woman.', ...

Examine the structure of spaCy sentences and see what can be found.

[29]	`holmes_sentences?`

Type: list
String form: ['\ufeffThe Adventures of Sherlock Holmes by Arthur Conan Doyle
Contents I. A Scandal <...> oduce our new eBooks, and how to subscribe to our email
newsletter to hear about new eBooks.']
Length: 6625
Docstring:
Built-in mutable sequence.
If no argument is given, the constructor creates a new empty list.
The argument must be an iterable if specified.
Check how many sentences Adventures of Sherlock Holmes contains.

Study the number of sentences contained in *The Adventures of Sherlock Holmes*.

[30]	`len (holmes_sentences)`
	6625

List out sentence numbers 50th–59th to review.

[31]	`holmes_sentences[50:60]`

['"My dear Holmes," said I, "this is too much.',
'You would certainly have been burned, had you lived a few centuries ago.',
'It is true that I had a country walk on Thursday and came home in a dreadful mess, but as
I have changed my clothes I can't imagine how you deduce it.',
'As to Mary Jane, she is incorrigible, and my wife has given her notice, but there, again, I
fail to see how you work it out." ',
'He chuckled to himself and rubbed his long, nervous hands together. ',
'"It is simplicity itself," said he; "my eyes tell me that on the inside of your left shoe, just
where the firelight strikes it, the leather is scored by six almost parallel cuts.',
'Obviously they have been caused by someone who has very carelessly scraped round the
edges of the sole in order to remove crusted mud from it.',
'Hence, you see, my double deduction that you had been out in vile weather, and that you
had a particularly malignant boot-slitting specimen of the London slavey.',
'As to your practice, if a gentleman walks into my rooms smelling of iodoform, with a
black mark of nitrate of silver upon his right forefinger, and a bulge on the right side of his
top-hat to show where he has secreted his stethoscope, I must be dull, indeed, if I do not
pronounce him to be an active member of the medical profession." ',
'I could not help laughing at the ease with which he explained his process of deduction.']

Step 9: Directly tokenize text document

Tokenize text document into word tokens by using "token" object in spaCy instead of text document object extraction into sentence list object. Study how it operates.

[32]   `holmes_words = [token.text for token in holmes_doc]`
      `holmes_words [130:180]`

['To', 'Sherlock', 'Holmes', 'she', 'is', 'always', '_', 'the', '_', 'woman', '.', 'I', 'have', 'seldom', 'heard', 'him', 'mention', 'her', 'under', 'any', 'other', 'name', '.', 'In', 'his', 'eyes', 'she', 'eclipses', 'and', 'predominates', 'the', 'whole', 'of', 'her', 'sex', '.', 'It', 'was', 'not', 'that', 'he', 'felt', 'any', 'emotion', 'akin', 'to', 'love', 'for', 'Irene', 'Adler']

[33]   `holmes_words?`

Type: list
String form: ['\ufeffThe', 'Adventures', 'of', 'Sherlock', 'Holmes', ' ', 'by', 'Arthur', 'Conan', 'Doyle', '<...> ubscribe', 'to', 'our', 'email', 'newsletter', 'to', 'hear', 'about', 'new', 'eBooks', '.', ' ']
Length: 133749
Docstring:
Built-in mutable sequence.
If no argument is given, the constructor creates a new empty list.
The argument must be an iterable if specified.

[34]   `len (holmes_words)`

133749

[35]   `nltk_homles_tokens = nltk.word_tokenize(holmes)`

[36]   `nltk_homles_tokens [104:153]`

['To', 'Sherlock', 'Holmes', 'she', 'is', 'always', '_the_', 'woman', '.', 'I', 'have', 'seldom', 'heard', 'him', 'mention', 'her', 'under', 'any', 'other', 'name', '.', 'In', 'his', 'eyes', 'she', 'eclipses', 'and', 'predominates', 'the', 'whole', 'of', 'her', 'sex', '.', 'It', 'was', 'not', 'that', 'he', 'felt', 'any', 'emotion', 'akin', 'to', 'love', 'for', 'Irene', 'Adler', '.']

According to the extracted tokens, they seem to be identical
1. Are they 100% identical?
2. What is/are the difference(s)?
3. Which one is better?

WORKSHOP

Workshop 2.4 SpaCy or NLTK - Which one is Faster?
In many applications, especially in AI and NLP applications, *speed* (i.e., efficiency) is one of the most important considerations because:
1. Many AI and NLP applications involve a huge data/database/databank for system training with a huge population size, e.g., Lexical database of English and Chinese. So, whether an NLP engine/application is fast enough in every NLP operation such as tokenization, tagging, POS tagging, and parsing is an important factor.
2. In many AI-based related NLP applications such as Deep Learning for real-time information extraction, it involves tedious network training and learning process, how efficient of every NLP operation is a critical process to decide whether NLP application can be used in real-world scenario.
This workshop studies how efficient NLTK vs. spaCy in terms of text document Tokenization.
To achieve this, integrate Python codes of NTLK/spaCy document tokenization with Timer object-time.
1. Implement tokenization codes in NTLK and spaCy to time tokenization time by using a time object, the following codes can be used as reference.
2. Examine time taken for Tokenization process of "Adventures_Holmes.txt" using NLTK vs. spaCy methods.
3. Which one is faster? or are they similar? Why?
4. How about Document→Text efficiency? Compare NLTK vs. spaCy on Doc→Text efficiency.
Hint: Like spaCy, NLTK can also implement Document→Text by two simple codes:
nltk_tokenizer = nltk.data.load("tokenizers/punkt/english.pickle")
nltk_sentences = tokenizer.tokenize(holmes) # holmes is the text document string object

[37]
```
# Sample code for Efficiency Performance of the NLP Engine
import nltk # or spacy
import time
start = time.time()
#
# YOUR NTLK or spaCy Tokenization codes
#
print( "Time taken: %s s" % (time.time() - start))
```
Time taken: 0.0 s

References

Albrecht, J., Ramachandran, S. and Winkler, C. (2020) Blueprints for Text Analytics Using Python: Machine Learning-Based Solutions for Common Real World (NLP) Applications. O'Reilly Media.

Altinok, D. (2021) Mastering spaCy: An end-to-end practical guide to implementing NLP applications using the Python ecosystem. Packt Publishing.

Arumugam, R., & Shanmugamani, R. (2018). Hands-on natural language processing with python. Packt Publishing.

Bird, S., Klein, E., and Loper, E. (2009). Natural language processing with python. O'Reilly.

Doyle, A. C. (2019) The Adventures of Sherlock Holmes (AmazonClassics Edition). AmazonClassics.

Gutenberg (2024) Project Gutenberg official site. https://www.gutenberg.org/ Accessed 17 Dec 2024.

Hardeniya, N., Perkins, J. and Chopra, D. (2016) Natural Language Processing: Python and NLTK. Packt Publishing.

Melville, H. (2006) Moby Dick. Hard Press.

Kedia, A. and Rasu, M. (2020) Hands-On Python Natural Language Processing: Explore tools and techniques to analyze and process text with a view to building real-world NLP applications. Packt Publishing.

NLPGitHub (2024) URL: https://github.com/raymondshtlee/NLP/. Accessed 17 Dec 2024.

NLTK (2024) NLTK official site. https://www.nltk.org/. Accessed 17 Dec 2024.

Perkins, J. (2014). Python 3 text processing with NLTK 3 cookbook. Packt Publishing Ltd.

SpaCy (2024) spaCy official site. https://spacy.io/. Accessed 17 Dec 2024.

Sidorov, G. (2019) Syntactic n-grams in Computational Linguistics (SpringerBriefs in Computer Science). Springer.

Srinivasa-Desikan, B. (2018). Natural language processing and computational linguistics: A practical guide to text analysis with python, gensim, SpaCy, and keras. Packt Publishing, Limited.

Vasiliev, Y. (2020) Natural Language Processing with Python and spaCy: A Practical Introduction. No Starch Press.

Chapter 13
Workshop#3: Part-of-Speech Tagging Using Natural Language Toolkit (Hour 5–6)

13.1 Introduction

In Chap. 3, we studied basic concepts and theories related to Part-of-Speech (POS) and various POS tagging techniques. This workshop will explore how to implement POS tagging using NLTK by starting from a simple recap on tokenization techniques and two fundamental processes in word-level progressing: stemming and stop-word removal. There are two types of stemming techniques: Porter Stemmer and Snowball Stemmer that can be integrated with WordCloud commonly used in data visualization followed by the main theme of this workshop, and introduce PENN Treebank Tagset to create your own POS tagger.

Please ensure that the following Python packages are installed before starting the workshop:

- Python (demo version 3.11.9)
- NLTK (demo version 3.9.1)
- matplotlib (demo version 3.9.2)
- WordCloud (demo version 1.9.3)
- svgling (demo version 0.5.0)
- svgwrite (demo version 1.4.3)
- scikit-learn (demo version 1.5.1)
- spacy (demo version 3.4.4)

If these packages are not installed on PC/laptop, use pip install xxx command. The detailed requirements list and Python package version used in this workshop can be found in the requirements.txt file stored in the NLP GitHub repository (NLPGitHub 2024).

R. Lee, *Natural Language Processing*, https://doi.org/10.1007/978-981-96-3208-4_13

13.2   A Revisit on Tokenization with NLTK

Text sentences are divided into subunits first and map into vectors in most NLP tasks. These vectors are fed into a model to encode where output is sent to a downstream task for results. NLTK (2024) provides methods to divide text into subunits as tokenizers. Twitter sample corpus is extracted from NLTK to perform tokenization (Hardeniya et al. 2016; Kedia and Rasu 2020; Perkins 2014) in procedures below (Albrechit et al. 2020; Antic 2021, Bird et al. 2009):

1. Import NLTK package
2. Import twitter sample data
3. List out fields
4. Get Twitter string list
5. List out first 15 Twitters
6. Tokenize the twitter

Let's start with the import of NLTK package and download twitter samples provided by NLTK platform.

```
[1]     # Import NLTK
        import nltk
        # Download twitter_samples
        # nltk.download('twitter_samples')
```

Import twitter samples dataset as *twtr* and check file id using *fileids()* method:

```
[2]     # Import twitter samples from NTLK corpus (twtr)
        from nltk.corpus import twitter_samples as twtr
```

```
[3]     # Display Field IDs
        twtr.fileids()
        ['negative_tweets.json','positive_tweets.json', 'tweets.20150430-23406.json']
```

Review first 5 twitter messages:

```
[4]     # Assign sample twitters (stwtr)
        stwtr = twtr.strings('tweets.20150430-223406.json')
```

```
[5]     # Display the first 5 sample twitters
        stwtr[:5]
```

['RT @KirkKus: Indirect cost of the UK being in the EU is estimated to be costing
Britain £170 billion per year! #BetterOffOut #UKIP',
'VIDEO: Sturgeon on post-election deals http://t.co/BTJwrpbmOY',
'RT @LabourEoin: The economy was growing 3 times faster on the day David
Cameron became Prime Minister than it is today.. #BBCqt http://t.co…',
'RT @GregLauder: the UKIP east lothian candidate looks about 16 and still has an
msn addy http://t.co/7eIU0c5Fm1',
"RT @thesundaypeople: UKIP's housing spokesman rakes in £800k in housing
benefit from migrants. http://t.co/GVwb9Rcb4w http://t.co/c1AZxcLh…"]

Import word_tokenize method from NLTK, name as *w_tok* to perform tokenization on 5th twitter message:

| [6] | ```# Import NLTK word tokenizer
from nltk.tokenize import word_tokenize as w_tok``` |

| [7] | ```# tokenize stwtr[4]
w_tok(stwtr[4])``` |
| | ['RT', '@', 'thesundaypeople', ':', 'UKIP', "'s", 'housing', 'spokesman', 'rakes', 'in', '£800k', 'in', 'housing', 'benefit', 'from', 'migrants', '.', 'http', ':', '//t.co/ GVwb9Rcb4w', 'http', ':', '//t.co/c1AZxcLh…'] |

NLTK offers tokenization for punctuation and spaces *wordpunct_tokenize*. Let's
use the 5th twitter message to see how it works.

| [8] | ```from nltk.tokenize import wordpunct_tokenize as wp_tok
wp_tok(stwtr[4])``` |
| | ['RT', '@', 'thesundaypeople', ':', 'UKIP', "'", 's', 'housing', 'spokesman', 'rakes', 'in', '£', '800k', 'in', 'housing', 'benefit', 'from', 'migrants', '.', 'http', '://', 't', '.', 'co', '/', 'GVwb9Rcb4w', 'http', '://', 't', '.', 'co', '/', 'c1AZxcLh', '…'] |

It can also tokenize words between hyphens and other punctuations. Further,
NLTK's regular expression (RegEx) tokenizer can build custom tokenizers:

[9]	``` # Import the RegEx tokenizer from nltk import regexp_tokenize as rx_tok rx_pattern1 = '\w+' rx_tok(stwtr[4],rx_pattern1) ```
	['RT', 'thesundaypeople', 'UKIP', 's', 'housing', 'spokesman', 'rakes', 'in', '800k', 'in', 'housing', 'benefit', 'from', 'migrants', 'http', 't', 'co', 'GVwb9Rcb4w', 'http', 't', 'co', 'c1AZxcLh']

 A simple regular expression filtered out words with alphanumeric characters only, but not punctuations in previous code. Another regular expression can detect and filter out both words containing alphanumeric characters and punctuation marks in the following code:

[10]	```
	# Create Rx pattern2 and perform the RX tokenize again
	rx_pattern2 = '\w+|[!,\-,]'
	rx_tok(stwtr[4],rx_pattern2)
	``

	['RT', 'thesundaypeople', 'UKIP', 's', 'housing', 'spokesman', 'rakes', 'in', '800k', 'in', 'housing', 'benefit', 'from', 'migrants', 'http', 't', 'co', 'GVwb9Rcb4w', 'http', 't', 'co', 'c1AZxcLh']

13.3   Stemming Using NLTK

After tokenization has sentences divided into words, stemming is a procedure to unify words and extract the root, base form of each word, e.g., stemming of word *compute* is shown in Fig. 13.1.

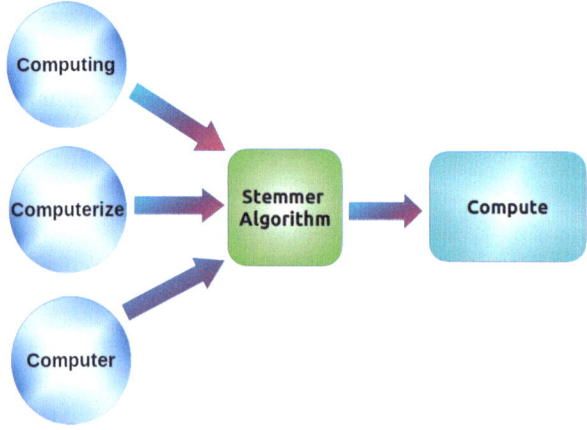

Fig. 13.1 Stemming of Compute

13.3.1 What Is Stemming?

Stemming usually removes prefixes or suffixes such as *-er, -ion, -ization* from words to extract the base or root form of a word, e.g., computers, computation, and computerization. Although these words are spelled differently but share identical concepts related to compute, so compute is the stem of these words.

13.3.2 Why Stemming?

It is needless to extract every single word in a document but only the concept or notion they represent such as information extraction and topic summarization in NLP applications. It can save computational capacity and preserve the overall meaning of the passage. The stemming technique is to extract the overall meaning or words' base form instead of distinct words.

Let's look at how to perform stemming on text data.

13.3.3 How to Perform Stemming?

NLTK provides a practical solution to implement stemming without sophisticated programming. Let's try two commonly used methods (1) Porter Stemmer and (2) Snowball Stemmer in NLP.

13.3.3.1 Porter Stemmer

Porter Stemmer is the earliest stemming technique used in 1980s, its key procedure is to remove words common endings and parse into generic forms. This method is simple and used in many NLP applications effectively.

Import Porter Stemmer from NLTK library:

| [11] | ```
# Import PorterStemmer as p_stem
from nltk.stem.porter import PorterStemmer as p_stem
``` |

Try to stem words like *computer*.

| [12] | ```
p_stem().stem("computer")
``` |
| | 'comput' |

PorterStemmer simply removes suffix *-er* when processing *computer* to acquire *compute* which is incorrect. Hence this stemmer is basic.

Next, try to stem *dogs* to see what happens.

[13]	`p_stem().stem("dogs")`
	'dog'

 For the above code, *dogs* are converted from plural to singular, remove suffix *-s* and convert to *dog*

Let's try more, say *traditional*.

[14]	`p_stem().stem("traditional")`
	'tradit'

 Stemmer may output an invalid word when dealing with special words, e.g., tradit is acquired if suffix *-ional* is removed. *Tradit* is not a word in English, it is a root form.

Let's work on words in plural form. There are 26 words extracted from a to z in plural form to perform PorterStemming:

| [15] | ```
# Define some plural words
w_plu = ['apes','bags','computers','dogs','egos','fresc
oes','generous','hats','igloos','jungles', 'kites','lea
rners','mice','natives','openings','photos','queries','
rats','scenes', 'trees','utensils','veins','wells','xyl
ophones','yoyos','zens']
``` |
|------|------|

| [16] | ```
from nltk.stem.porter import PorterStemmer as p_stem
w_sgl = [p_stem().stem(wplu) for wplu in w_plu]
print(' '.join(w_sgl))
``` |
|------|------|
| | ape bag comput dog ego fresco gener hat igloo jungl kite learner mice nativ open photo queri rat scene tree utensil vein well xylophon yoyo zen |

Porter stemming will remove suffixes *-s* or *-es* to extract root form, that may result in single form such as *apes*, *bags*, *dogs*, etc. but in some cases, it will generate non-English words such as *gener*, *jungl* and *queri*.

WORKSHOP

Workshop 3.1 Try to stem a paragraph from The Adventures of Sherlock Holmes
1. Read Adventures_Holmes.txt *text file from The Adventures of Sherlock Holmes* (Doyle 2019; Gutenberg 2024).
2. Save contents into a string object "holmes_doc."
3. Extract a paragraph and tokenize it.
4. Use porter stemming and output a list of stemmed words.

13.3.3.2   Snowball Stemmer

Snowball Stemmer provides improvement in stemming results as compared with Porter Stemmer and provides multi-language stemming solution. One can check languages using languages() method. Import from NLTK package to invoke Snowball Stemmer:

```
[17]   # Import Snowball Stemmer as s_stem
       from nltk.stem.snowball import SnowballStemmer as s_stem
```

Review what languages Snowball Stemmer can support:

```
[18]   # Display the s_stem language set
       print(s_stem.languages)
```
('arabic', 'danish', 'dutch', 'english', 'finnish', 'french', 'german', 'hungarian', 'italian', 'norwegian', 'porter', 'portuguese', 'romanian', 'russian', 'spanish', 'swedish')

Snowball Stemmer provides a variety of solutions in commonly used languages from Arabic to Swedish.

| [19] | ```
# Import Snowball Stemmer as s_stem and assign to
English language
from nltk.stem.snowball import SnowballStemmer as s_stem
s_stem_ENG = s_stem(language="english")
``` |

Use same list of plural words (w_plu) to check how it works in Snowball Stemmer for comparison:

| [20] | ```
# Display the list of plural words
w_plu
``` |
|------|------|
| | ['apes', 'bags', 'computers', 'dogs', 'egos', 'frescoes', 'generous', 'hats', 'igloos', 'jungles', 'kites', 'learners', 'mice', 'natives', 'openings', 'photos', 'queries', 'rats', 'scenes', 'trees', 'utensils', 'veins', 'wells', 'xylophones', 'yoyos', 'zens'] |

| [21] | ```
# Apply Snowball Stemmer onto the plural words
sgls = [s_stem_ENG.stem(wplu) for wplu in w_plu]
print(' '.join(sgls))
``` |
|------|------|
| | Ape bag comput dog ego fresco generous hat igloo jungl kite learner mice nativ open photo queri rat scene tree utensil vein well xylophon yoyo zen |

Try to compare with previous stemmer. What are the differences?

1. Snowball Stemmer achieved similar results as porter Stemmer in most cases except in *generously* where snowball stemmer came up with a meaningful root form *generous* instead of *gener* in porter stemmer.
2. Try some plural words to compare performance between porter Stemmer vs snowball stemmer.

13.4   Stop-Words Removal with NLTK

13.4.1   *What Are Stop-Words?*

There are input words and utterances to filter out impractical stop-words in NLP preprocessing such as *a*, *is*, *the*, *of*, etc.

NLTK already provides a built-in stop-words package for this function. Let's see how it works.
```

13.4.2 NLTK Stop-Words List

Import stop-words module and call stopwords.words() method to list out all stop-words in English.

| [22] | ```# Import NLTK stop-words as wstops
from nltk.corpus import stopwords as wstops
print(wstops.words('english'))``` |
| --- | --- |
| | ['i', 'me', 'my', 'myself', 'we', 'our', 'ours', 'ourselves', 'you', "you're", "you've", "you'll", "you'd", 'your', 'yours', 'yourself', 'yourselves', 'he', 'him', 'his', 'himself', 'she', "she's", 'her', 'hers', 'herself', 'it', "it's", 'its', 'itself', 'they', 'them', 'their', 'theirs', 'themselves', 'what', 'which', 'who', 'whom', 'this', 'that', "that'll", 'these', 'those', 'am', 'is', 'are', 'was', 'were', 'be', 'been', 'being', 'have', 'has', 'had', 'having', 'do', 'does', 'did', 'doing', 'a', 'an', 'the', 'and', 'but', 'if', 'or', 'because', 'as', 'until', 'while', 'of', 'at', 'by', 'for', 'with', 'about', 'against', 'between', 'into', 'through', 'during', 'before', 'after', 'above', 'below', 'to', 'from', 'up', 'down', 'in', 'out', 'on', 'off', 'over', 'under', 'again', 'further', 'then', 'once', 'here', 'there', 'when', 'where', 'why', 'how', 'all', 'any', 'both', 'each', 'few', 'more', 'most', 'other', 'some', 'such', 'no', 'nor', 'not', 'only', 'own', 'same', 'so', 'than', 'too', 'very', 's', 't', 'can', 'will', 'just', 'don', "don't", 'should', "should've", 'now', 'd', 'll', 'm', 'o', 're', 've', 'y', 'ain', 'aren', "aren't", 'couldn', "couldn't", 'didn', "didn't", 'doesn', "doesn't", 'hadn', "hadn't", 'hasn', "hasn't", 'haven', "haven't", 'isn', "isn't", 'ma', 'mightn', "mightn't", 'mustn', "mustn't", 'needn', "needn't", 'shan', "shan't", 'shouldn', "shouldn't", 'wasn', "wasn't", 'weren', "weren't", 'won', "won't", 'wouldn', "wouldn't"] |

1. Stop-words corpus size is not large.
2. All stop-words are commonly used in many documents. They affect storage and system efficiency in NLP applications if they are not removed.
3. This stop-word corpus is incomplete and subjective. There may be words considered as stop-words not included in this databank.

Use *stopwords.fileids()* function to review how many languages library of stop-words NLTK contains.

| [23] | ```# Import NLTK stop-words as wstops and display the
FILE_IDs
from nltk.corpus import stopwords as wstops
print(wstops.fileids())``` |
| --- | --- |
| | ['arabic', 'azerbaijani', 'basque', 'bengali', 'catalan', 'chinese', 'danish', 'dutch', 'english', 'finnish', 'french', 'german', 'greek', 'hebrew', 'hinglish', 'hungarian', 'indonesian', 'italian', 'kazakh', 'nepali', 'norwegian', 'portuguese', 'romanian', 'russian', 'slovene', 'spanish', 'swedish', 'tajik', 'turkish'] |

13.4.3   Try Some Texts

The above list shows all stop-words. Let's use a simple utterance:

[24]	``` # Import NLTK stop-words as wstops from nltk.corpus import stopwords as wstops wstops_ENG = wstops.words('english') utterance = "Try to test for the stop word remove function to see how it works." utterance_clean =[w for w in utterance.split() if w not in wstops_ENG] ```

Review results:

[25]	``` # Display the cleaned utterance utterance_clean ```
	['Try', 'test', 'stop', 'word', 'remove', 'function', 'see', 'works.']

1. All commonly used stop-words such as *to, for, the, it,* are removed as shown in the example.
2. It has little effect on the overall meaning of the utterance.
3. It requires the same computational time and effort.

The following example uses *Hamlet* from *The Complete Works of Shakespeare* to demonstrate how stop-words are removed from text processing in NLP.

[26]	``` # Import the Gutenberg library from NLTK from nltk.corpus import gutenberg as gub hamlet = gub.words('shakespeare-hamlet.txt') hamlet_clean = [w for w in hamlet if w not in wstops_ENG] ```

[27]	``` len(hamlet_clean)*100.0/len(hamlet) ```
	69.26124197002142

This classic literature contains deactivated words. Nevertheless, these stop-words are unmeaningful in many NLP tasks that may affect results, so most of them are removed during preprocessing

13.4.4   Create your Own Stop-Words

Stop-word corpus can extract a list of strings that can add any stop-words with simple *append( )* function, but it is advisable to create a new stop-word library object name to begin.

Step 1: Create own stop-word library list.

Step 2: Check object type and just see it has a simple list

[28]	`My_sws = wstops.words('english')`

[29]	`My_sws?`
	Type: list
	String form: ['i', 'me', 'my', 'myself', 'we', 'our', 'ours', 'ourselves', 'you', "you're", "you've", "you'll" <...> houldn', "shouldn't", 'wasn', "wasn't", 'weren', "weren't", 'won', "won't", 'wouldn', "wouldn't"]
	Length: 179
	Docstring:
	Built-in mutable sequence.
	If no argument is given, the constructor creates a new empty list.
	The argument must be an iterable if specified.

Step 3: Study stop-word list

Step 4: Add new stop-word "sampleSW" using append()

[30]	`My_sws`
	['i', 'me', 'my', 'myself', 'we', 'our', 'ours', 'ourselves', 'you', "you're", "you've", "you'll", "you'd", 'your', 'yours', 'yourself', 'yourselves', 'he', 'him', 'his', 'himself', 'she', "she's", 'her', 'hers', 'herself', 'it', "it's", 'its', 'itself', 'they', 'them', 'their', 'theirs', 'themselves', 'what', 'which', 'who', 'whom', ...

[31]	`My_sws.append('sampleSW')`
	`My_sws[160:]`
	['ma', 'mightn', "mightn't", 'mustn', "mustn't", 'needn', "needn't", 'shan', "shan't", 'shouldn', "shouldn't", 'wasn', "wasn't", 'weren', "weren't", 'won', "won't", 'wouldn', "wouldn't", 'sampleSW']

Try this to see how it works.

[32]	```python
	# Import word_tokenize as w_tok
	from nltk.tokenize import word_tokenize as w_tok
	# Create the sample utterance
	utterance = "This is a sample utterance which consits
	of eg as stop word sampleSW."
	# Tokenize the utterance
	utt_toks = w_tok(utterance)
	# Stop word removal
	utt_nosw = [w for w in utt_toks if not w in My_sws]
	# Display utterance without My stopwords
	print(utt_nosw)
	``

	['This', 'sample', 'utterance', 'consits', 'eg', 'stop', 'word', '.']

WORKSHOP

Workshop 3.2 Stop-word Filtering on The Adventures of Sherlock Holmes
Use stop-word filtering technique for *The Adventures of Sherlock Holmes*:
1. Read Adventures_Holmes.txt text file.
2. Save contents into a string object "holmes_doc."
3. Use stop-word technique just learnt to tokenize holmes_doc.
4. Generate a list of word tokens with stop-words removed.
5. Check any 3 possible stop-words to add into own stop-word list.
6. Regenerate a new token list with additional stop-word removed.

13.5   Text Analysis with NLTK

When text data has been processed and tokenized, basic analysis is required to calculate words or tokens, their distribution and usage frequency in NLP tasks. This allows understanding of main contents and topics accuracy in the document. Import a sample web text (Firefox.txt) from NLTK library.

[33]	```python
	# Import webtext as wbtxt
	from nltk.corpus import webtext as wbtxt
	# Create sample webtext
	wbtxt_s = wbtxt.sents('firefox.txt')
	wbtxt_w = wbtxt.words('firefox.txt')
	# Display total nos of webtext sentences in firefox.txt
	len(wbtxt_s)
	``

	1144

Review the number of words as well.

[34]	```python
	# Display total nos of webtext words in firefox.txt
	len(wbtxt_w)
	``

	102457

FireFox.Txt contains sample texts extracted from the Firefox discussion forum to serve as a useful dataset for basic text-level analysis in NLP.

It can also obtain vocabulary size by passing through a set as shown in the following code:

| [35] | ```
# Define vocabulary object (vocab)
vocab = set(wbtxt_w)
# Display the size of Vocab
len(vocab)
``` |
|------|------|
| | 8296 |

nltk.FreqDist( ) function is used to generate words frequency distribution occurred in the whole text as shown:

| [36] | ```
# Define Frequency Distribution object
fdist = nltk.FreqDist(wbtxt_w)
``` |
|------|------|

| [37] | ```
sorted(fdist, key=fdist.__getitem__,reverse=True)
[0:30]
``` |
|------|------|
| | ['.', 'in', 'to', '', 'the', '"', 'not', '-', 'when', 'on', 'a', 'is', 't', 'and', 'of', '(', 'page', 'for', 'with', ')', 'window', 'Firefox', 'does', 'from', 'open', ':', 'menu', 'should', 'bar', 'tab'] |

The above code generates the top 30 frequently used words and punctuations in the whole text. *In*, *to* and *the* are top 3 on the list like other literatures as Firefox. Txt text is the collection of users' discussion messages and contents about Firefox browser like conversations.

To exclude stop-words such as *the*, and *not*, use the following code to see f words frequency distribution longer than 3.

| [38] | ```
# Import Matplotlib pyplot object
import matplotlib.pyplot as pyplt
pyplt.figure(figsize=(20, 8))
lwords = dict([(k,v) for k,v in fdist.items() if len(k)>3])
fdist = nltk.FreqDist(lwords)
fdist.plot(50,cumulative=False)
``` |
|------|------|

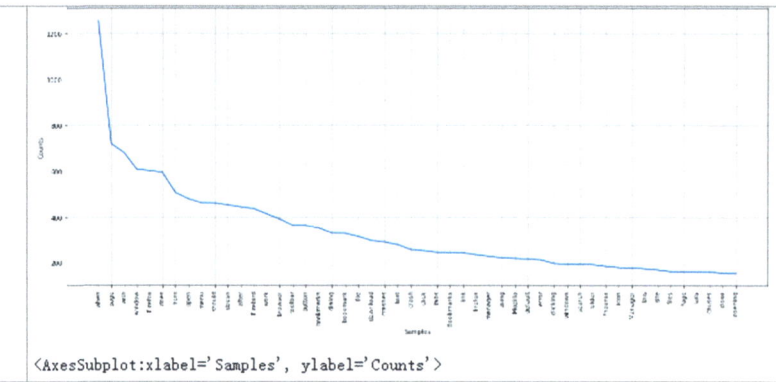

```
<AxesSubplot:xlabel='Samples', ylabel='Counts'>
```

Exclude stop-words such as *the*, *and*, *is*, *and* create a tuple dictionary to record words frequency. Visualize and transform them into an NLTK frequency distribution graph based on this dictionary as shown above

WORKSHOP

Workshop 3.3 Text Analysis on The Adventures of Sherlock Holmes
1. Read Adventures_Holmes.txt text file.
2. Save contents into a string object "holmes_doc."
3. Use stop-word technique from tokenize holmes_doc.
4. Generate a word tokens list with stop-words removed.
5. Use the technique learnt to plot the first 30 frequently occurred words from this literature.
6. Identify any special pattern related to word distribution. If not, try the first 50 ranking words.

13.6   Integration with WordCloud

13.6.1   What Is WordCloud?

Wordcloud, also known as tag cloud, is a data visualization method commonly used in many web statistics and data analysis scenarios. It is a graphical representation of all words and keywords in sizes and colors. A word has the largest and bold in word cloud means it occurs frequently in the text (dataset), as illustrated in Fig. 13.2.

To generate frequency distribution of all words that occur in a text document, the most natural way is to generate statistics in a WordCloud.

Python provides a built-in WordCloud package "WordCloud."

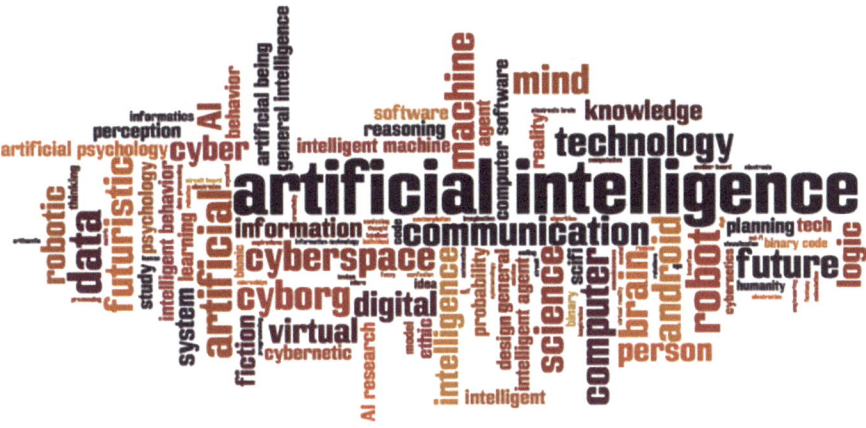

Fig. 13.2 A sample WordCloud (Tuchong 2024)

It can obtain an intuitive visualization of words used in the text from the frequency distribution.

Install WordCloud package first using the *pip install* command:

```
pip install WordCloud
```

Once WordCloud package is installed, import WordCloud package using import command and invoke the frequency generator with *generate_from frequencies() method*:

```
[39]    # Import WordCloud as wCloud
        from wordcloud import WordCloud as wCloud
```

```
[40]    wcld = wCloud().generate_from_frequencies(fdist)
```

```
[41]    Import matplotlib.pyplot as pyplt
        pyplt.figure(figsize=(20, 8))
        pyplt.imshow(wcld, interpolation='bilinear')
        pyplt.axis("off")
        pyplt.show()
```

Workshop 3.4 WordCloud for The Adventures of Sherlock Holmes
1. Read Adventures_Holmes.txt text file.
2. Save contents into a string object "holmes_doc."
3. Use stop-word technique from tokenize holmes_doc.
4. Generate word tokens list with stop-words removed.
5. Extract the top 100 frequent words that occurred from this literature.
6. Generate WordCloud for this literature.

WORKSHOP

13.7   POS Tagging with NLTK

The earlier part of this workshop had studied several NLP preprocessing tasks: tokenization, stemming, stop-word removal, word distribution in text corpus, and data visualization using WordCloud. This section will explore POS tagging in NLTK.

13.7.1   What Is POS Tagging?

POS refers to the process of classifying words in a sentence/utterance into specific syntactic or grammatical functions.

There are nine major POS in English: Nouns, Pronouns, Adjectives, Verbs, Prepositions, Adverbs, Determiners, Conjunctions and Interjections. POS tagging is to assign POS tags into each word token in the sentence/utterance.

NTLK supports commonly used tagset such as PENN Treebank (Treebank 2024) and Brown Corpus to create own tags used for specific NLP applications.

Table 13.1 Table of Universal POS Tagset in English

Tag	Meaning	English Examples
ADJ	adjective	*new, good, high, special, big, local*
ADP	adposition	*on, of, at, with, by, into, under*
ADV	adverb	*really, already, still, early, now*
CONJ	conjunction	*and, or, but, if, while, although*
DET	determiner, *article*	*the, a, some, most, every, no, which*
NOUN	noun	*year, home, costs, time, Africa*
NUM	numeral	*twenty-four, fourth, 1991, 14:24*
PRT	particle	*at, on, out, over, per, that, up, with*
PRON	pronoun	*he, their, her, its, my, I, us*
VERB	verb	*is, say, told, given, playing, would*
.	punctuation marks	*. , ; !*
x	other	*ersatz, esprit, dunno, qr8, university*

13.7.2   Universal POS Tagset

A tagset consists of 12 universal POS categories and is constructed to facilitate future requirements for unsupervised induction of syntactic structure. When combined with original treebank data, this universal tagset and mapping produce a dataset consisting of common POS in 22 languages (Albrechit et al. 2020; Antic 2021, Bird et al. 2009).

Table 13.1 shows a table of universal POS tagset in English.

13.7.3   PENN Treebank Tagset (English & Chinese)

English PENN Treebank Tagset is used with English corpora developed by Prof. Helmut Schmid in TC project at the Institute for Computational Linguistics of the University of Stuttgart (TreeBank 2024). Table 13.2 shows an original 45 used PENN Treebank Tagset.

A recent version of this English POS Tagset can be found at Sketchengine.eu (Sketchengine 2024a), and Chinese POS Tagset (Sketchengine 2024b).

NLTK provides direct mapping from tagged corpus such as Brown Corpus (NLTK 2024) to universal tags for implementation, e.g., tags VBD (for past tense verb) and VB (for base form verb) map to VERB only in universal tag set.

[42]	```# Import Brown Corpus as bwn``` ```from nltk.corpus import brown as bwn```

Table 13.2 Original 45 used PENN Treebank Tagset

No	POS Tag	Description	Example	No	POS Tag	Description	Example
1	CC	coordinating conjunction	and, but, or	24	SYM	Symbol	$ / [= *
2	CD	cardinal number	1, third	25	TO	infinitive 'to'	to
3	DT	determiner	a, the	26	UH	interjection	haha, oops
4	EX	existential there	there is	27	VB	verb - base form	drink
5	FW	foreign word	les	28	VBD	verb - past tense	drank
6	IN	preposition, sub-conj	in, of, by, like	29	VBG	verb - gerund	drinking
7	JJ	adjective	big, wide, green	30	VBN	verb - past participle	drunk
8	JJR	adjective, comparative	bigger, wider, greener	31	VBP	verb - non-3sg pres	drink
9	JJS	adjective, superlative	biggest, wildest, greenest	32	VBZ	verb - 3sg pres	drinks
10	LS	list marker	1), One, i	33	WDT	wh-determiner	which, that
11	MD	modal	can, could, shall, will	34	WP	wh-pronoun	who, what
12	NN	noun, singular or mass	table, shop	35	WP$	possessive wh-pronoun	whose, those
13	NNS	noun plural	tables, shops	36	WRB	wh-abverb	where, when, how
14	NNP	proper noun, singular	Samsung	37	#	#	#
15	NNPS	proper noun, plural	Vikings	38	$	$	$
16	PDT	predeterminer	all/both the students	39	"	Left quotation	' "
17	POS	possessive ending	friend's	40	``	right quotation	' "
18	PP	personal pronoun	I, he, it, you	41	(	Opening brackets	({
19	PPZ	possessive pronoun	my, his, your, one's	42	)	Closing brackets	) }
20	RB	adverb	however, quickly, here	43	,	Comma	,
21	RBR	adverb, comparative	better, quicker	44	:	Sent-final punc	. ! ?
22	RBS	adverb, superlative	best, quickest	45	:	Mid-sentence punc	: ; ... -
23	RP	particle	of, up (e.g. give up)				

[43]	`bwn.tagged_words()[0:30]`
	[('The', 'AT'), ('Fulton', 'NP-TL'), ('County', 'NN-TL'), ('Grand', 'JJ-TL'), ('Jury', 'NN-TL'), ('said', 'VBD'), ('Friday', 'NR'), ('an', 'AT'), ('investigation', 'NN'), ('of', 'IN'), ("Atlanta's", 'NP$'), ('recent', 'JJ'), ('primary', 'NN'), ('election', 'NN'), ('produced', 'VBD'), ('\`\`', '\`\`'), ('no', 'AT'), ('evidence', 'NN'), ("''", "''"), ('that', 'CS'), ('any', 'DTI'), ('irregularities', 'NNS'), ('took', 'VBD'), ('place', 'NN'), ('.', '.'), ('The', 'AT'), ('jury', 'NN'), ('further', 'RBR'), ('said', 'VBD'), ('in', 'IN')]

Fulton is tagged as NP-TL in example code above, a *proper noun (NP)* appears in a title (TL) context in Brown corpus that mapped to *noun* in universal tag set. These sub-categories are to be considered instead of generalized universal tags in NLP application

13.7.4 Applications of POS Tagging

POS tagging is commonly used in many NLP applications ranging from IE and NER to sentiment analysis and question-&-answering systems.

Try the following and see how it works:

| [44] | ```# Import word_tokenize and pos_tag as w_tok and p_tag
from nltk.tokenize import word_tokenize as w_tok
from nltk import pos_tag as p_tag
# Create and tokenizer two sample utterances utt1 and utt2
utt1 = w_tok("Give me a call")
utt2 = w_tok("Call me later")``` |
|------|---|

Review these two utterances' POS tags:

[45]	`p_tag(utt1, tagset='universal' )`
	[('Give', 'VERB'), ('me', 'PRON'), ('a', 'DET'), ('call', 'NOUN')]

[46]	`p_tag(utt2, tagset='universal' )`
	[('Call', 'VERB'), ('me', 'PRON'), ('later', 'ADV')]

1. The word *call* is a noun in text 1 and a verb in text 2.
2. POS tagging is used to identify a person, a place, or a location, based on the tags in NER.
3. NLTK also provides a classifier to identify such entities in text as shown in the following code:

[47]	`utt_untag = w_tok("My dad was born in South America")` `utt_untag`
	['My', 'dad', 'was', 'born', 'in', 'South', 'America']

[48]	`utt_tagged = p_tag(utt_untag)` `utt_tagged`
	[('My', 'PRP$'), ('dad', 'NN'), ('was', 'VBD'), ('born', 'VBN'), ('in', 'IN'), ('South', 'NNP'), ('America', 'NNP')]

| [49] | ```
# Import svgling package
import svgling
# Import NLTK.ne_chunk as chunk
from nltk import ne_chunk as chunk
# Display POS Tags chunk
chunk(utt_tagged)
``` |
| --- | --- |

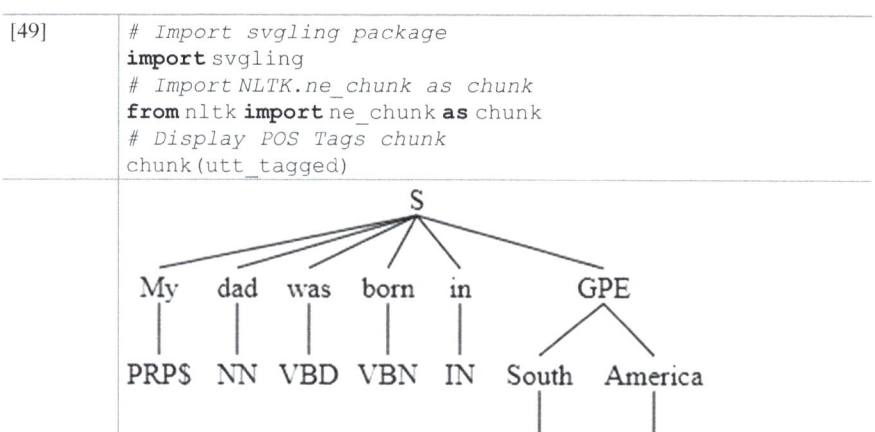
```

NLTK chunk() function is applied to NER to identify the chunker South America as a geopolitical entity (GPE) in this example. So far, there are examples using NLTK's built-in taggers. The next section will look at how to develop own POS tagger

[50]
```
# Try another example
utt_tok = w_tok("Can you please buy me Haagen-Dazs
Icecream? It's $30.8.")
print("Tokens are: ", utt_tok)
```

Tokens are: ['Can', 'you', 'please', 'buy', 'me', 'Haagen-Dazs', 'Icecream', '?', 'It', "'s", '$', '30.8', '.']

[51]
```
utt_tagged = p_tag(utt_tok)
chunk(utt_tagged)
```

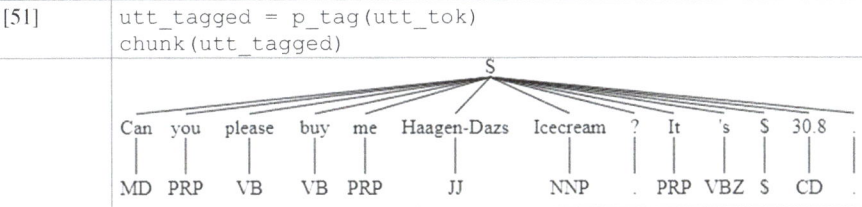

1. The system treats '$', '30.8', and '.' as separate tokens in this example. It is crucial because contractions have their own semantic meanings and own POS leading to the ensuing part of NLTK library POS tagger.
2. POS tagger in NLTK library outputs specific tags for certain words.
3. However, it makes a mistake in this example. Where is it?
4. Compare POS Tagging for the following sentence to identify problem. Explain.

[52]
```
# Try one more example
utt_tok = w_tok("Can you please buy me New-Zealand
Icecream? It's $30.8.")
print("Tokens are: ", utt_tok)
utt_tagged = nltk.pos_tag(utt_tok)
chunk(utt_tagged)
```

Tokens are: ['Can', 'you', 'please', 'buy', 'me', 'New-Zealand', 'Icecream', '?', 'It', "'s", '$', '30.8', '.']

```
                                    S
      Can   you   please   buy   me   New-Zealand   Icecream   ?   It   's   $   30.8   .
       |     |      |       |     |        |            |       |    |   |    |    |    |
      MD    PRP    VB      VB    PRP      NNP          NNP      .   PRP VBZ  $   CD    .
```

Workshop 3.5 POS Tagging on *The Adventures of Sherlock Holmes*
1. Read Adventures_Holmes.txt text file.
2. Save contents into a string object "holmes_doc."
3. Extract three typical sentences from three stories of this literature.
4. Use POS Tagging to these sentences.
5. Use ne_chunk function to display POS tagging tree for these three sentences.
6. Compare POS Tags among these example sentences and examine on how they work.

13.8   Create Own POS Tagger with NLTK

This section will create own POS tagger using NLTK's tagged set corpora and sklearn Random Forest machine learning model.

The following example demonstrates a classification task to predict POS tag for a word in a sentence using NLTK treebank dataset for POS tagging, and extract word prefixes, suffixes, previous and neighboring words as features for system training.

Import all necessary Python packages as below:

[53]	
| | ```
# Import all necessary Python packages
import nltk
import numpy as np
from nltk import word_tokenize as w_tok
import matplotlib.pyplot as pyplt
%matplotlib inline
from sklearn.feature_extraction import DictVectorizer as DVect
from sklearn.model_selection import train_test_split as tt_split
from sklearn.ensemble import RandomForestClassifier as RFClassifier
from sklearn.metrics import accuracy_score as a_score
from sklearn.metrics import confusion_matrix as c_matrix
``` |
```

[54]
```
# Define the ufeatures() class
def ufeatures(utt, idx):
    ftdist = {}
    ftdist['word'] = utt[idx]
    ftdist['dist_from_first'] = idx - 0
    ftdist['dist_from_last'] = len(utt) - idx
    ftdist['capitalized'] = utt[idx][0].upper() ==
 utt[idx][0]
    ftdist['prefix1'] = utt[idx][0]
    ftdist['prefix2'] = utt[idx][:2]
    ftdist['prefix3'] = utt[idx][:3]
    ftdist['suffix1'] = utt[idx][-1]
    ftdist['suffix2'] = utt[idx][-2:]
    ftdist['suffix3'] = utt[idx][-3:]
    ftdist['prev_word'] = '' if idx==0 else utt[idx-1]
    ftdist['next_word'] = '' if idx==(len(utt)-1) else
 utt[idx+1]
    ftdist['numeric'] = utt[idx].isdigit()
    return ftdist
```

[55]
```
# Define the Retreive Untagged Utterance (RUutterance)
class
def RUutterance(utt_tagged):
    [utt,t] = zip(*utt_tagged)
    return list(utt)
```

Function *ufeatures()* converts input text into a dict object of features, whereas each utterance is passed with corresponding index of current token word from which features are extracted. Let's use treebank tagged utterances with universal tags to label and train data:

[56]
```
utt_tagged = nltk.corpus.treebank.tagged_
sents(tagset='universal')
```

[57]
```
utt_tagged
```
[[('Pierre', 'NOUN'), ('Vinken', 'NOUN'), (',', '.'), ('61', 'NUM'), ('years', 'NOUN'), ('old', 'ADJ'), (',', '.'), ('will', 'VERB'), ('join', 'VERB'), ('the', 'DET'), ('board', 'NOUN'), ('as', 'ADP'), ('a', 'DET'), ('nonexecutive', 'ADJ'), ('director', 'NOUN'), ('Nov.', 'NOUN'), ('29', 'NUM'), ('.', '.')], [('Mr.', 'NOUN'), ('Vinken', 'NOUN'), ('is', 'VERB'), ('chairman', 'NOUN'), ('of', 'ADP'), ('Elsevier', 'NOUN'), ('N.V.', 'NOUN'), (',', '.'), ('the', 'DET'), ('Dutch', 'NOUN'), ('publishing', 'VERB'), ('group', 'NOUN'), ('.', '.')], ...]

1. In this example, universal tags are used for simplicity.
2. Of course, one can also use fine-grained treebank POS tags for implementation.
3. Once do so, can now extract the features for each tagged utterance in corpus with training labels.
Use the following code to extract the features:

[58]
```
# Define Extract Feature class (exfeatures)
def exfeatures(utt_tag):
    utt, tag = [], []
    for ut in utt_tag:
        for idx in range(len(ut)):
            utt.append(ufeatures(RUutterance(ut), idx))
            tag.append(ut[idx][1])
    return utt, tag
```

[59]
```
X,y = exfeatures(utt_tagged)
```

This example uses DVect to convert feature-value dictionary into training vectors.

If the number of possible values for suffix3 feature is 40, there will be 40 features in output. Use following code to DVect:

[60]
```
# Define sample size
nsize = 10000
# Invoke Dict Vectorizer
dvect = DVect(sparse=False)
Xtran = dvect.fit_transform(X[0:nsize])
ysap = y[0:nsize]
```

This example has a sample size of 10,000 utterances which 80% of the dataset is used for training and the other 20% is used for testing. Random forecast (RF) classifier is used as POS tagger model as shown:

[61]
```
Xtrain,Xtest,ytrain,ytest = tt_split(Xtran, ysap,
test_size=0.2,
random_state=123)
```

[62]
```
rfclassifier = RFClassifier(n_jobs=4)
rfclassifier.fit(Xtrain,ytrain)
```

```
▾        RandomForestClassifier      ● ●

RandomForestClassifier(n_jobs=4)
```

After system training, can perform POS Tagger validation by using some sample utterances. But before passing to *ptag_predict()* method, extract features are required by *ufeatures()* method as shown:

[63]
```
# Define the POS Tags Predictor class (ptag_predict)
def ptag_predict(utt):
    utt_tagged = []
    fts = [ufeatures(utt, idx) for idx in range(len(utt))]
    fts = dvect.transform(fts)
    tgs = rfclassifier.predict(fts)
    return zip(utt, tgs)
```

Convert utterance into corresponding features with *ufeatures()* method. The features dictionary extracted from this method is vectorized using previously trained *dvect*:

[64]
```
# Test with a sample utterance (utt3)
utt3 = "It is an example for POS tagger"
for utt_tagged in ptag_predict(utt3.split()):
                print(utt_tagged)
```
('It', 'PRON')
('is', 'VERB')
('an', 'DET')
('example', 'NOUN')
('for', 'ADP')
('POS', 'NOUN')
('tagger', 'NOUN')

Use a sample utterance "utt3" and invoke ptag_predict() method to output tags for each word token inside utt3 and review for accuracy afterward.

[65]	`predict = rfclassifier.predict(Xtest)`

[66]	`a_score(ytest,predict)`
	0.9365

The overall a_score has approximately 93.6% accuracy rate and satisfactory. Next, let's look at confusion matrix (c-mat) to check how well can POS tagger perform

[67]	`c_mat = c_matrix(ytest,predict)`

[68]	```pyplt.figure(figsize=(10,10))```

```
pyplt.figure(figsize=(10,10))
pyplt.xticks(np.arange(len(rfclassifier.
classes_)),rfclassifier.classes_)
pyplt.yticks(np.arange(len(rfclassifier.
classes_)),rfclassifier.classes_)
pyplt.imshow(c_mat, cmap=pyplt.cm.Blues)
pyplt.colorbar()
```

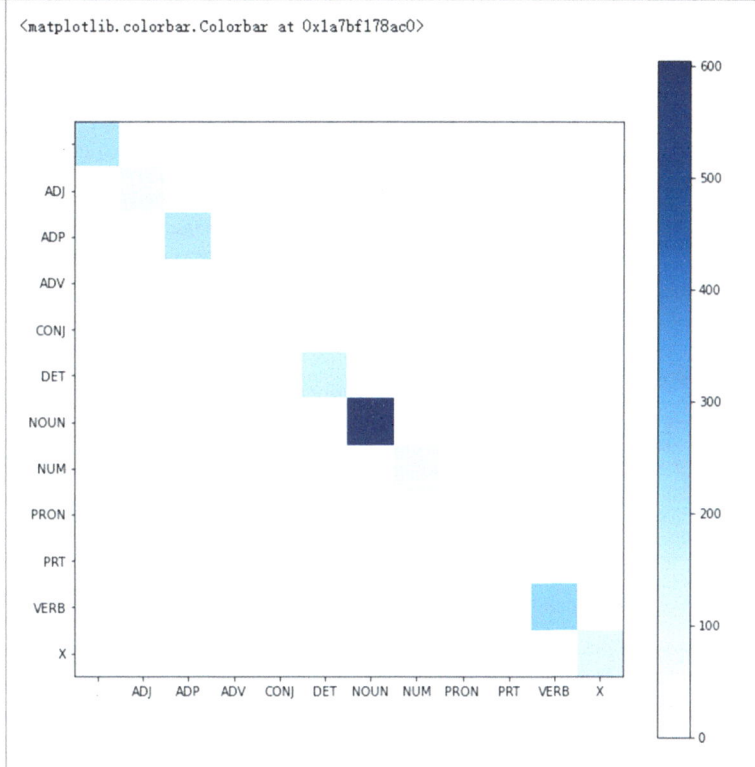

```
<matplotlib.colorbar.Colorbar at 0x1a7bf178ac0>
```

Use classes from RF classifier as x and y labels to create a c-mat (confusion matrix). These labels are POS tags used for system training. The plot that follows shows a pictorial representation of the confusion matrix

Use classes from Random Forest classifier as x and y labels in the code for plotting confusion matrix.

It looks like the tagger performs relatively well for nouns, verbs, and determiners in sentences reflected in dark regions of the plot. Let's look at some top features of the model from the following code:

[69]
```
flist = zip(dvect.get_feature_names_out(),
rfclassifier.feature_importances_)
sfeatures = sorted(flist,key=lambda x: x[1],
reverse=True)
print(sfeatures[0:20])
```

[('prefix1=*', 0.018027666774656427), ('capitalized', 0.014560536843271387), ('dist_from_last', 0.013067122358738224), ('prefix2=th', 0.011508009921371423), ('suffix2=he', 0.010995137232578216), ('prefix2=,', 0.010843292840402313), ('suffix2=ed', 0.01065544048464163), ('prefix1=.', 0.010442335119192925), ('suffix1=d', 0.010042966512875777), ('dist_from_first', 0.010020085126011984), ('word=the', 0.009518771614554129), ('numeric', 0.008902146191801517), ('prefix1=t', 0.008316132993197207), ('suffix1=s', 0.008264535100812235), ('word=,', 0.00802534859280316), ('suffix3=,', 0.007825258495661055), ('prefix3=the', 0.007328831604325926), ('prefix2=.', 0.007323771203933977), ('suffix3=the', 0.006992438126037274), ('prefix1=,', 0.006974723571575362)]

1. The RF feature importance is stored in the python feature_importances list. Some of the suffix features have higher importance scores than others.
2. For instances, words ending with *-ed* are usually verbs in past tense which make sense in many situations, and punctuations like commas may affect POS tagging performance in some situations.

Workshop 3.6 Revisit POS Tagging on The Adventures of Sherlock Holmes with Additional Tagger
1. Read Adventures_Holmes.txt text file.
2. Save contents into a string object "holmes_doc."
3. Extract three typical sentences from three stories of this literature.
4. Use method learnt to create own POS taggers. What are the new POS tags to add or use?
5. Try new POS taggers for these three typical sentences and compare results with previous workshop.

References

Albrechit, J., Ramachandran, S. and Winkler, C. (2020) Blueprints for Text Analytics Using Python: Machine Learning-Based Solutions for Common Real World (NLP) Applications. O'Reilly Media.

Antic, Z. (2021) Python Natural Language Processing Cookbook: Over 50 recipes to understand, analyze, and generate text for implementing language processing tasks. Packt Publishing.

Bird, S., Klein, E., and Loper, E. (2009). Natural language processing with python. O'Reilly.

Doyle, A. C. (2019) The Adventures of Sherlock Holmes (AmazonClassics Edition). AmazonClassics.

Gutenberg (2024) Project Gutenberg official site. https://www.gutenberg.org/ Accessed 17 Dec 2024.

Hardeniya, N., Perkins, J. and Chopra, D. (2016) Natural Language Processing: Python and NLTK. Packt Publishing.

Kedia, A. and Rasu, M. (2020) Hands-On Python Natural Language Processing: Explore tools and techniques to analyze and process text with a view to building real-world NLP applications. Packt Publishing.

NLTK (2024) NLTK official site. https://www.nltk.org/. Accessed 17 Dec 2024.

NLPGitHub (2024) URL: https://github.com/raymondshtlee/NLP/. Accessed 17 Dec 2024.

Perkins, J. (2014). Python 3 text processing with NLTK 3 cookbook. Packt Publishing Ltd.

Sketchengine (2024a) Recent version of English POS Tagset by Sketchengine. https://www.sketchengine.eu/english-treetagger-pipeline-2/. Accessed 17 Dec 2024.

Sketchengine (2024b) Recent version of Chinese POS Tagset by Sketchengine. https://www.sketchengine.eu/chinese-penn-treebank-part-of-speech-tagset/. Accessed 17 Dec 2024.

Treebank (2024) Penn TreeBank Release 2 official site. https://catalog.ldc.upenn.edu/docs/LDC95T7/treebank2.index.html. Accessed 17 Dec 2024.

Tuchong (2024) A sample WordCloud. URL: https://stock.tuchong.com/image/detail?imageId=920077068255363088. Accessed 28 Dec 2024

Chapter 14
Workshop#4 Semantic Analysis and Word Vectors Using spaCy (Hour 7–8)

14.1 Introduction

In Chaps. 5 and 6, we studied the basic concepts and theories related to meaning representation and semantic analysis. This workshop will explore how to use spaCy technology to perform semantic analysis starting from a revisit on word vectors concept, implement and pretrain them followed by the study of similarity method and other advanced semantic analysis.

Please ensure that the following Python packages are installed before starting the workshop:

- Python (demo version 3.11.9)
- NLTK (demo version 3.9.1)
- matplotlib (demo version 3.9.2)
- scikit-learn (demo version 1.5.1)
- spacy (demo version 3.4.4)

If these packages are not installed on PC/laptop, use pip install xxx command. The detailed requirements list and Python package version used in this workshop can be found in the requirements.txt file stored in the NLP GitHub repository (NLPGitHub 2024).

14.2 What Are Word Vectors?

Word vectors (Albrecht et al. 2020; Bird et al. 2009; Hardeniya et al. 2016; Kedia and Rasu 2020; NLTK 2024) are practical tools in NLP.

A word vector is a dense representation of a word. Word vectors are important for semantic similarity applications like similarity calculations between words,

© The Author(s), under exclusive license to Springer Nature Singapore Pte Ltd. 2025
R. Lee, *Natural Language Processing*,
https://doi.org/10.1007/978-981-96-3208-4_14

phrases, sentences, and documents, e.g., they provide information about synonymity, semantic analogies at word level.

Word vectors are produced by algorithms to reflect similar words that appear in similar contexts. This paradigm captures target word meaning by collecting information from surrounding words is called distributional semantics.

They are accompanied by associative semantic similarity methods including word vector computations such as distance, analogy calculations, and visualization to solve NLP problems.

This workshop will cover the following topics (Altinok 2021; Arumugam and Shanmugamani 2018; Perkins 2014; spaCy 2024; Srinivasa-Desikan 2018; Vasilev 2020):

- Understanding word vectors
- Using spaCy's pretrained vectors
- Advanced semantic similarity methods

14.3   Understanding Word Vectors

Word vectors, or word2vec are important quantity units in statistical methods to represent text in statistical NLP algorithms. There are several ways of text vectorization to provide words semantic representation.

14.3.1   Example: A Simple Word Vector

Let's look at a basic way to assign words vectors:

- Assign an index value to each word in vocabulary and encode this value into a sparse vector.
- Consider *tennis* as vocabulary and assign an index to each word according to vocabulary order as in Table 14.1.

Table 14.1 A basic word vector example consists of nine words

1	a
2	go
3	I
4	tennis
5	play
6	outside
7	hot
8	swim
9	rest

Table 14.2 Word Vectors corresponding index value consists of nine words

word									
a	1	0	0	0	0	0	0	0	0
go	0	1	0	0	0	0	0	0	0
I	0	0	1	0	0	0	0	0	0
tennis	0	0	0	1	0	0	0	0	0
play	0	0	0	0	1	0	0	0	0
outside	0	0	0	0	0	1	0	0	0
hot	0	0	0	0	0	0	1	0	0
swim	0	0	0	0	0	0	0	1	0
today	0	0	0	0	0	0	0	0	1

Table 14.3 Word Vector matrix for *I play tennis today*

word									
I	0	0	1	0	0	0	0	0	0
play	0	0	0	0	1	0	0	0	0
tennis	0	0	0	1	0	0	0	0	0
today	0	0	0	0	0	0	0	0	1

Vocabulary word vector will be 0, except for word corresponding index value position as in Table 14.2.

Since each row corresponds to one word, a sentence represents a matrix, e.g., *I play tennis today* is represented by a matrix as in Table 14.3.

Vectors length is equal to word numbers in vocabulary as shown above. Each dimension is apportioned to one word explicitly. When applying this encoding vectorization to text, each word is replaced by its vector, and the sentence is transformed into a (N, V) matrix, where N is words number in sentence and V is vocabulary size.

This text representation is easy to compute, debug, and interpret. It looks good so far but there are potential problems:

• Vectors are sparse. Each vector contains many 0 s but has one 1. If words have similar meanings and can group to share dimensions, this vector will deplete space. Also, numerical algorithms don't accept high dimension and sparse vectors in general.

- A sizeable vocabulary is comparable to high dimensions vectors are impractical for memory storage and computation.
- Similar words do not assign with similar vectors resulting unmeaningful vectors, e.g., *cheese, topping, salami,* and *pizza* have related meanings but have unrelated vectors. These vectors depend on corresponding word's index and assign randomly in vocabulary, indicating that one-hot encoded vectors are incapable to capture semantic relationships and against word vectors' purpose to answer preceding list concerns.

14.4   A Taste of Word Vectors

A word vector is a fixed-size, dense, and real-valued vector. It is a learnt representation of text where semantic similar words correspond to similar vectors and a solution to preceding problems.

```
the 0.418 0.24968 -0.41242 0.1217 0.34527 -0.044457 -0.49688 -
0.17862 -0.00066023 -0.6566 0.27843 -0.14767 -0.55677 0.14658 -
0.0095095 0.011658 0.10204 -0.12792 -0.8443 -0.12181 -0.016801 -
0.33279 -0.1552 -0.23131 -0.19181 -1.8823 -0.76746 0.099051 -
0.42125 -0.19526 4.0071 -0.18594 -0.52287 -0.31681 0.00059213
0.0074449 0.17778 -0.15897 0.012041 -0.054223 -0.29871 -0.15749 -
0.34758 -0.045637 -0.44251 0.18785 0.0027849 -0.18411 -0.11514 -
0.78581
```

 This is a 50-dimensional vector for word *the,* these dimensions have floating points
1. What do dimensions represent?
2. These individual dimensions don't have inherent meanings typically but instead they represent vector space locations, and the distance between these vectors indicates the similarity of corresponding words' meanings.
3. Hence, a word's meaning is distributed across dimensions.
4. This type of word's meaning representation is called distributional semantics.

Use word vector visualizer for TensorFlow from (TensorFlow 2024) Google offers word vectors for 10,000 words. Each vector is 200-dimensional and projected into three dimensions for visualization. Let's look at the representation of *tennis* as in Fig. 14.4.

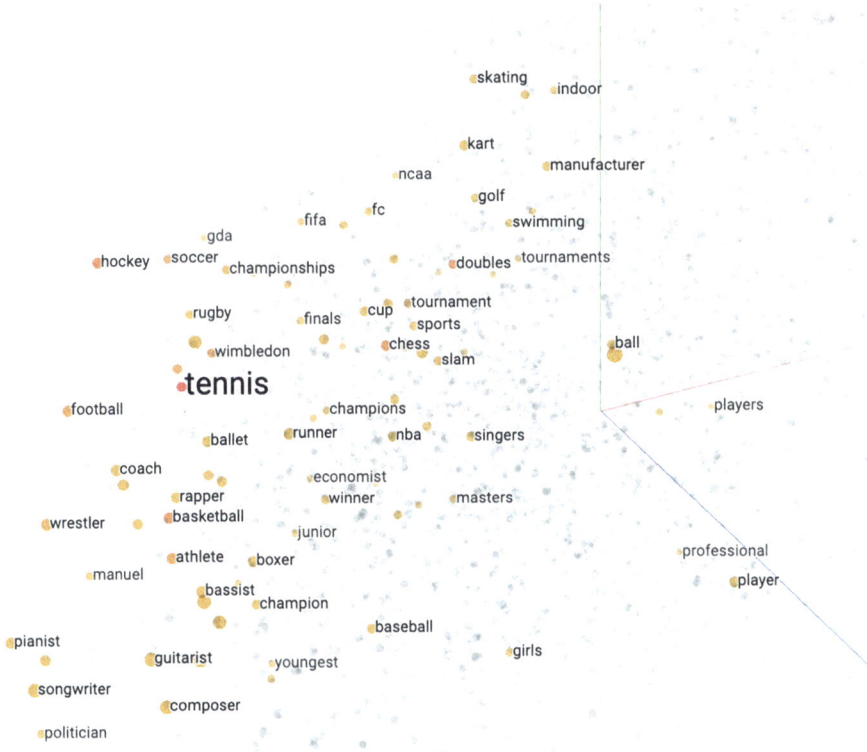

Fig. 14.4 Vector representation of *tennis* and semantic similar words

 *tennis* is semantically grouped with other sports i.e. hockey, basketball, chess etc. Words in proximity are calculated by their cosine distances as shown in Fig. 14.5

Word vectors are trained on a large corpus such as Wikipedia which included to learn proper nouns representations, e.g., *Alice* is a proper noun represented by vector as in Fig. 14.6.

Fig. 14.5 *tennis* proximity words in three-dimensional space

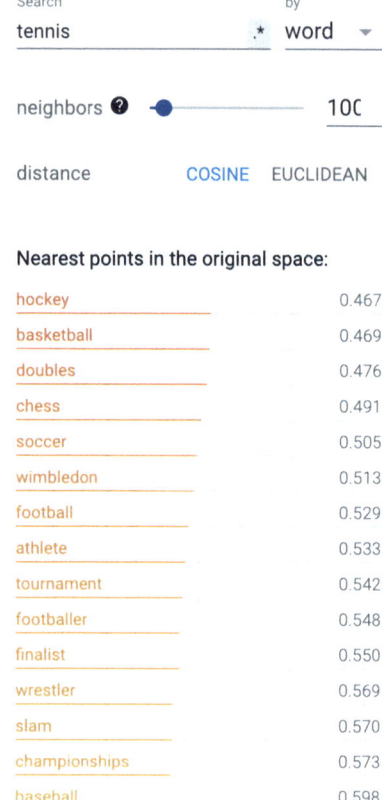

It showed that all vocabulary input words are in lower cases to avoid multiple representations of the same word. *Alice* and *Bob* are person names to be listed. In addition, *lewis* and *carroll* have relevance to *Alice* because *of the famous literature Alice's Adventures in Wonderland* written by *Lewis Carroll*. Further, it also showed syntactic category of all neighboring words are nouns but not verbs.

Word vectors can capture synonyms, antonyms, and semantic categories such as animals, places, plants, names, and abstract concepts.

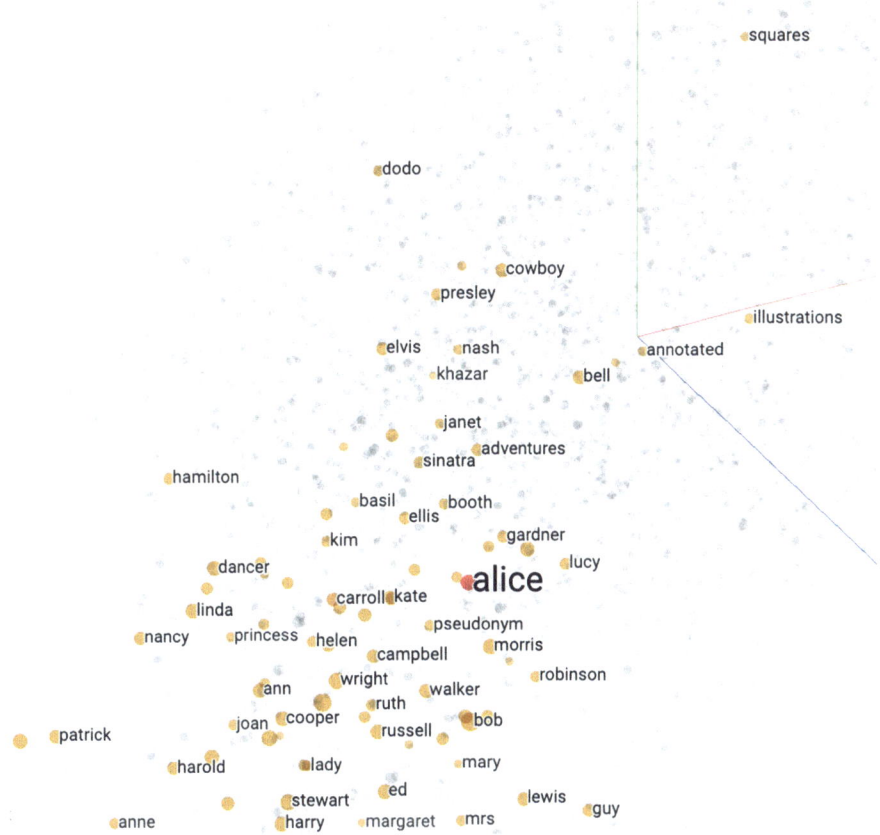

Fig. 14.6 Vector representation of *alice*

14.5   Analogies and Vector Operations

Word vectors capture semantics, support vector addition, subtraction, and analogies. A word analogy is a semantic relationship between a pair of words. There are many relationship types such as synonymity, anonymity, and whole-part relation. Some example pairs are (*King—man*, *Queen—woman*), (*airplane—air*, *ship - sea*), (*fish—sea*, *bird - air*), (*branch—tree*, *arm—human*), (*forward—backward*, *absent—present*), etc.

For example, gender mapping represents *Queen* and *King* as *Queen—Woman* + *Man* = *King*. If *woman* is subtracted by *Queen* and add *Man* instead to obtain *King*. Then, this analogy interprets queen is attributed to king as woman is attributed to man. Embeddings can generate analogies such as gender, tense, and capital city as shown in Fig. 14.7.

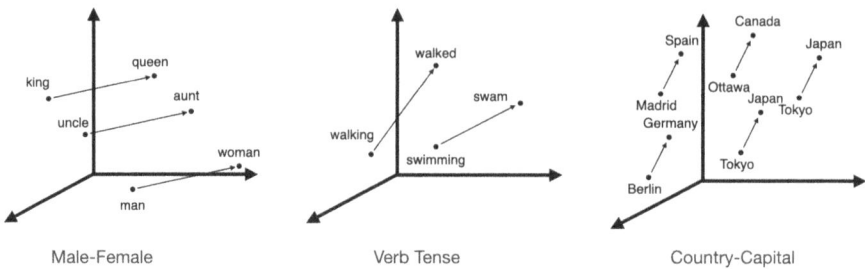

Fig. 14.7 Analogies created by word vectors

14.6   How to Create Word Vectors?

There are many ways to produce and pretrained word vectors:

1. word2vec is a name of statistical algorithm created by Google to produce word vectors. Word vectors are trained with a neural network architecture to process windows of words and predict each word vector depending on surrounding words. These pretrained word vectors can be downloaded from Synthetic (2024).
2. Glove vectors are invented by Stanford NLP group. This method depends on singular value decomposition used in word co-occurrences matrix. The pretrained vectors are available at nlp.stanford.edu (Stanford 2024).
3. fastText (FastText 2024) was created by Facebook Research like word2vec. word2vec predicts words based on their surrounding context, while fastText predicts subwords, i.e., character N-grams. For example, the word *chair* generates the following subwords:

```
ch, ha, ai, ir, cha, hai, air
```

14.7   spaCy Pretrained Word Vectors

Word vectors are part of many spaCy language models. For instance, en_core_web_md model ships with 300-dimensional vectors for 20,000 words, while en_core_web_lg model ships with 300-dimensional vectors with a 685,000 words vocabulary.

Typically, small models (names end with sm) do not include any word vectors but context-sensitive tensors. Semantic similarity calculations can perform but results will not be as accurate as word vector computations.

A word's vector is via token.vector method. Let's look at this method using code query word vector for *banana*:

[1]
```
# Import spaCy and load the en_core_web_md model
import spacy
nlp = spacy.load("en_core_web_md")
# Create a sample utterance (utt1)
utt1 = nlp("I ate a banana.")
```

[2]
```
import en_core_web_md
nlp = en_core_web_md.load()
```

Use the following script to show Word Vector for *banana*:

[3]
```
utt1[3].vector
```
array([0.20778, −2.4151, 0.36605, 2.0139, −0.23752, −3.1952,
        −0.2952, 1.2272, −3.4129, −0.54969, 0.32634, −1.0813,
        0.55626, 1.5195, 0.97797, −3.1816, −0.37207, −0.86093,
        2.1509, −4.0845, 0.035405, 3.5702, −0.79413, −1.7025,
        −1.6371, −3.198, −1.9387, 0.91166, 0.85409, 1.8039,
        −1.103, −2.5274, 1.6365, −0.82082, 1.0278, −1.705,
        1.5511, −0.95633, −1.4702, −1.865, −0.19324, −0.49123,
        2.2361, 2.2119, 3.6654, 1.7943, −0.20601, 1.5483,
        −1.3964, −0.50819, 2.1288, −2.332, 1.3539, −2.1917,

        ...
        −1.354, 2.6261, 1.9156, −1.5651, 1.8315, −1.4257,
        −1.6861, −0.51953, 1.7635, −0.50722, 1.388, −1.1012],
        dtype=float32)

In this example, *token.vector* returns a NumPy ndarray.
Use the following command to call NumPy methods for result.

[4]
```
type(utt1[3].vector)
```
numpy.ndarray

[5]	`utt1[3].vector.shape`
	(300,)

Query Python type of word vector in this code segment. Then, invoke shape() method of NumPy array on the vector.

Doc and Span object also have vectors. A sentence vector or a span is the average of words' vectors. Run the following code and see results:

| [6] | ```
# Create second utterance (utt2)
utt2 = nlp("I like a banana,")
utt2.vector
utt2[1:3].vector
``` |
| --- | --- |
| | array([−5.84815, 3.9533, −4.2019, 1.851645,
        4.2339, −3.74201, 2.1273, 6.0418997,
        2.7598, 0.40665, 11.029249, 2.792575,
        −5.2807, −0.47160006, 2.38658, 2.2019,
        4.65584, 0.33210003, 0.76987505, 0.72405005,
        1.9154, 2.24705, −0.748515, −1.29685,
        1.0118049, −5.3013496, −5.97755, −1.618835,
        −0.23785007, −2.2115, −0.61186, −3.56615,
        ….
        −1.32008, −4.63445, −2.8069, 1.747215,
        7.172359, −2.6399, 1.54486, −1.320575,
        −5.26095, 5.7922, −5.7227497, −0.20825005,
        0.47510207, 2.4512, −1.01646, 4.55843,
        1.4716, 4.96085, −4.954, 1.50534],
        dtype=float32) |

Only words in model's vocabulary have vectors, words that are not in vocabulary are called out-of-vocabulary (OOV) words. token.is_oov and token.has_vector are two methods to query whether a token is in the model's vocabulary and has a word vector:

| [7] | ```
# Create the utterance 3
utt3 = nlp("You went there afskfsd.")
``` |
| --- | --- |

[8]	```
	for token in utt3:
	    print("Token is: ",token, "OOV: ", token.is_oov,
	    "Token has vector:", token.has_vector)
	``

	Token is: You OOV: False Token has vector: True Token is: went OOV: False Token has vector: True Token is: there OOV: False Token has vector: True Token is: afskfsd OOV: True Token has vector: False Token is: OOV: False Token has vector: True

This is basically how to use spaCy's pretrained word vectors. Next, discover how to invoke spaCy's semantic similarity method on Doc, Span, and Token objects.

14.8 Similarity Method in Semantic Analysis

Every container type object has a similarity method to calculate the semantic similarity of other container objects by comparing word vectors in spaCy. Semantic similarity between two container objects is different container types. For instance, a Token object to a Doc object and a Doc object to a Span object.

The following example computes two Span objects similarity:

[9]	```
	# Create utt4 and utt5 and measure the similarity
	utt4 = nlp("I visited England.")
	utt5 = nlp("I went to London.")
	utt4[1:3].similarity(utt5[1:4])
	``

	0.45464012026786804

Compare two Token objects, London and England:

[10]	`utt4[2]`
	England

[11]	`utt4[2].similarity(utt5[3])`
	0.6339874267578125

The sentence's similarity is computed by calling similarity() on Doc objects:

[12]	utt4.similarity(utt5)
	0.8206949942253569

1. The preceding code segment calculates semantic similarity between two sentences *I visited England* and *I went to London*.
2. Similarity score is high enough to consider both sentences are similar (similarity degree ranges from 0 to 1, 0 represents unrelated and 1 represents identical).

similarity() method returns 1 compare an object to itself unsurprisingly:

[13]	utt4.similarity(utt4)
	1.0

Judge the distance with numbers is complex but review vectors on paper can understand how vocabulary word groups are formed.

Code snippet below visualizes a vocabulary of two graphical semantic classes. The first word class is for animals and the second class is for food.

| [14] | ```
import matplotlib.pyplot as plt
from sklearn.decomposition import PCA
import numpy as np
import spacy
nlp = spacy.load( "en_core_web_md" )
vocab = nlp( "cat dog tiger elephant bird monkey lion
cheetah burger pizza food cheese wine salad noodles
macaroni fruit vegetable" )
words = [word.text for word in vocab]
``` |
|---|---|

Create Word Vector vecs:

| [15] | ```
vecs = np.vstack([word.vector for word in vocab if word.
has_vector])
``` |
|---|---|

Use principal component analysis (PCA) similarity analysis and plot similarity results with plt class.

[16]
```
pca = PCA(n_components=2)
vecs_transformed = pca.fit_transform(vecs)
plt.figure(figsize=(20,15))
plt.
scatter(vecs_transformed[:,0], vecs_transformed[:,1])
for word, coord in zip(words, vecs_transformed):
    x,y = coord
    plt.text(x,y,word, size=15)
plt.show()
```

1. Import matplotlib library to create a graph.
2. Next two imports are for vectors calculation.
3. Import spaCy and create a nlp object.
4. Create a Doc object from vocabulary.
5. Stack word vectors vertically by calling np.vstack.
6. Project vectors into a two-dimensional space for visualization since they are 300-dimensional. Extract two principal components via PCA for projection.
7. Create a scatter plot for rest of the code to deal with matplotlib function calls.

It shows that spaCy word vectors can visualize two semantic classes are grouped. The distance between animals is reduced and uniformly distributed, while food class formed groups within the group.

Workshop 4.1 Word Vector Analysis on The Adventures of Sherlock Holmes
In this workshop, we have just learnt how to use spaCy to produce word vector to compare the similarity of two text objects/document. Try to use *The Adventures of Sherlock Holmes* (Doyle 2019; Gutenberg 2024) to select two "presentative" texts from this detective story:
1. Read Adventures_Holmes.txt text file.
2. Save contents into a string object "holmes_doc."
3. Plot Semantic Graphs for these two texts.
4. Perform Similarity text for these two documents. See what can be found.

14.9   Advanced Semantic Similarity Methods with spaCy

It has learnt that spaCy's similarity method can calculate semantic similarity to obtain scores but there are advanced semantic similarity methods to calculate words, phrases, and sentences similarity.

14.9.1   Understanding Semantic Similarity

It is necessary to identify examples characteristics when collecting data or text data (any sort of data), i.e., calculate two text similarity scores. Semantic similarity is a metric to define the distance between texts based on semantics texts.

Metrics in mathematics are basically distance functions. Each metric produces a topology on the vector space. Word vectors are vectors that can be used to calculate the distance between them as a similarity score.

There are two commonly used distance functions (1) Euclidian distance and (2) cosine distance.

14.9.2   Euclidean Distance

Euclidian distance counts on vector magnitude and disregards orientation. If a vector is drawn from an origin, let's call it a *dog* vector to another point, call a *cat* vector and subtract one vector from and other, the distance represents the magnitude of vectors is shown in Fig. 14.8.

If two more semantically similar words (*canine, terrier*) to *dog* and make it a text of three words, i.e., *dog canine terrier*. Obviously, the *dog* vector will now grow in magnitude, possibly in the same direction. This time, the distance will be much bigger due to geometry, although the semantics of first piece of text (now *dog canine terrier*) remain the same.

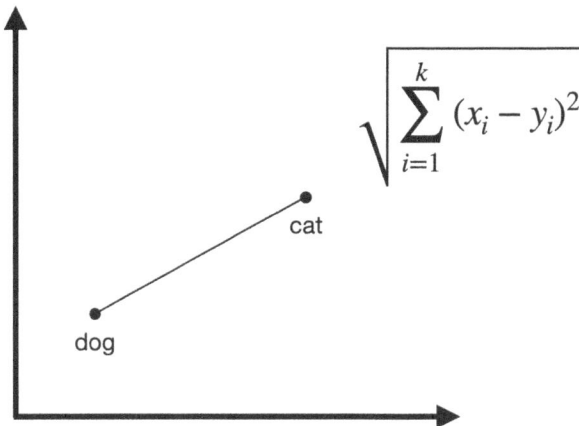

Fig. 14.8 Euclidian distance between two vectors: *dog* and *cat*

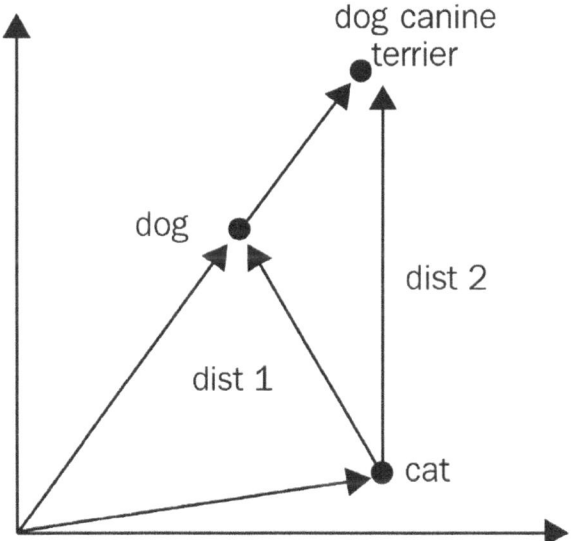

Fig. 14.9 Distance between *dog* and *cat*, as well as the distance between *dog canine terrier* and *cat*

This is the main drawback of using Euclidian distance for semantic similarity as the orientation of two vectors in the space is not considered. Figure 14.9 illustrates the distance between *dog* and *cat*, and the distance between *dog canine terrier* and *cat*.

How can we fix this problem? There's another way of calculating similarity called cosine similarity to address this problem.

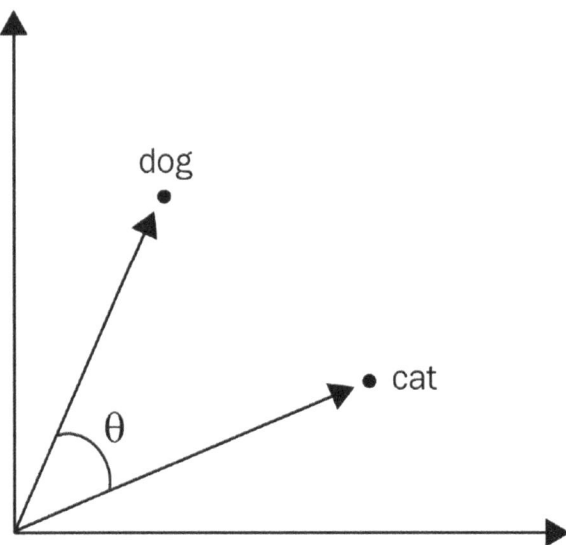

Fig. 14.10 The angle between *dog* and *cat* vectors. Here, the semantic similarity is calculated by cos(θ)

14.9.3   Cosine Distance and Cosine Similarity

Contrary to Euclidian distance, cosine distance is more concerned with the orientation of two vectors in the space. The cosine similarity of two vectors is basically the cosine angle created by these two vectors. Figure 14.10 shows the angle between

I purchased a science fiction book last week.

I loved everything related to this fragrance: Light, floral and feminine …

I purchased a bottle of wine.

dog and *cat* vectors.

The maximum similarity score that's allowed by cosine similarity is 1. This is obtained when the angle between two vectors is 0 degree (hence, the vectors coincide). The similarity between two vectors is 0 when the angle between them is 90 degrees.

Cosine similarity provides scalability when vectors grow in magnitude. If one of the input vectors is expanded as in Fig. 14.10, the angle between them remains the same and so is the cosine similarity score.

Note that here is to calculate semantic similarity score and not distance. The highest possible value is 1 when vectors coincide, while the lowest score is 0 when two vectors are perpendicular. The cosine distance is $1-\cos(\theta)$ which is a distance function.

spaCy uses cosine similarity to calculate semantic similarity. Hence, calling the similarity method helps to perform cosine similarity calculations.

So far, we've learnt to calculate similarity scores, but still haven't discovered words meaning. Obviously, not all words in a sentence have the same impact on the semantics of sentence. The similarity method will only calculate the semantic similarity score, but the right keywords are required for calculation results comparison.

Consider the following text snippet:

> Blue whales are the biggest mammals in the world. They're observed in California coast during spring.

If interested in finding the biggest mammals on the planet, the phrases *biggest mammals* and *in the world* will be keywords. Comparing these phrases with the search phrases *largest mammals* and *on the planet* should give a high similarity score. But if is interested in finding out about places in the world, *California* will be a keyword. *California* is semantically like word *geography* and more suitably, the entity type is a geographical noun.

Since we have learnt how to calculate similarity score, the next section will learn about where to look for the meaning. It will cover a case study on text categorization before improving task results via key phrase extraction with similarity score calculations.

A dog

My dog

My beautiful dog

A beautiful dog

A beautiful and happy dog

My happy and cute dog

14.9.4   Categorizing Text with Semantic Similarity

Determining two sentences' semantic similarity can categorize texts into predefined categories or spot only the relevant texts. This case study will filter users' comments in an e-commerce website related to the word *perfume*. Suppose to evaluate the following comments:

Here, it is noted that only the second sentence is related. This is because it contains the word *fragrance* and adjectives describing scents. To understand which sentences are related, can try several comparison strategies.

To start, compare *perfume* to each sentence. Recall that spaCy generates a word vector for a sentence by averaging the word vector of its tokens. The following code snippet compares preceding sentences to *perfume* search key:

[17]	```python
	utt6 = nlp("I purchased a science fiction book last
	week. I loved everything related to this fragrance:
	light, floral and feminine... I purchased a bottle of
	wine. ")
	key = nlp("perfume")
	for utt in utt6.sents:
	    print(utt.similarity(key))
	``

	0.2950337433100861 0.4292321445243577 0.4216416633742172

The following steps are performed:

Create a Doc object with three preceding sentences. For each sentence, calculate similarity score with *perfume* and print the score by invoking similarity() method on the sentence. The degree of similarity between *perfume* and the first sentence is minute, indicating that this sentence is irrelevant to the search key. The second sentence looks relevant which means that semantic similarity is correctly identified.

How about the third sentence? The script identified that the third sentence is relevant somehow, most probably because it includes the word *bottle,* and perfumes are sold in bottles. The word *bottle* appears in similar contexts with the word *perfume*. For this reason, the similarity score of this sentence and search key is not small enough; also, the scores of second and third sentences are not distant enough to make the second sentence significant.

In practice, long texts such as web documents can be dealt with but averaging over them diminishes the importance of keywords.

Let's look at how to identify key phrases in a sentence to improve performance.

14.9.5 *Extracting Key Phrases*

Semantic categorization is effective to extract important words phrases and compare them to the search key. Instead of comparing the key to different parts of speech, we can compare the key to noun phrases. Noun phrases are subjects, direct objects, and indirect objects of sentences that convey high percentages of sentences semantics.

For example, in sentence *Blue whales live in California*, focuses will likely be on *blue whales, whales, California,* or *whales in California*.

Similarly, in the preceding sentence about *perfume*, the focus is to pick out *fragrance as* the noun. Different semantic tasks may need other context words such as verbs to decide what the sentence is about, but for semantic similarity, noun phrases convey significant weights.

What is a noun phrase? A noun phrase (NP) is a group of words that consist of a noun and its modifiers. Modifiers are usually pronouns, adjectives, and determiners. The following phrases are noun phrases:

spaCy extracts noun phases by parsing the output of the dependency parser. It can identify noun phrases of a sentence by using doc.noun_chunks method:

[18]	`utt7 = nlp( "My beautiful and cute dog jumped over the fence" )`

[19]	`utt7.noun_chunks`
	\<generator at 0x1be166595a0\>

[20]	`list(utt7.noun_chunks)`
	[My beautiful and cute dog, the fence]

Let's modify the preceding code snippet. Instead of comparing the search key perfume to the entire sentence, this time will only compare it with sentence's noun chunks:

| [21] | ```
for utt in utt7.sents:
    nchunks = [nchunk.text for nchunk in utt.noun_chunks]
    nchunk_utt = nlp(" ".join(nchunks))
    print(nchunk_utt.similarity(key))
``` |
|---|---|
| | 0.28984869889342174 |
```

The following is performed for the preceding code:
1. Iterate over sentences.
2. Extract noun chunks and store them in a python list for each sentence.
3. Join noun chunks in the list into a python string and convert it into a doc object.
4. Compare this doc object of noun chunks to search key *perfume* to determine semantic similarity scores.

If these scores are compared with previous scores, it is noted that the first sentence remains irrelevant, so its score decreased marginally but the second sentence's score increased significantly. Also, the second and third sentences scores are distant from each other to reflect that second sentence is the most related sentence.

14.9.6   Extracting and Comparing Named Entities

In some cases, it can focus on extracting proper nouns instead of every noun. Hence, it is required to extract named entities. Let's compare the following paragraphs:

```
"Google Search, often referred as Google, is the most popular search engine nowadays. It
answers a huge volume of queries every day."
"Microsoft Bing is another popular search engine. Microsoft is known by its star product
Microsoft Windows, a popular operating system sold over the world."
"The Dead Sea is the lowest lake in the world, located in the Jordan Valley of Israel. It
is also the saltiest lake in the world."
```

The codes should be able to recognize that first two paragraphs are about large technology companies and their products whereas the third paragraph is about a geographic location.

Comparing all noun phrases in these sentences may not be helpful because many of them such as volume are irrelevant to categorization. The topics of these paragraphs are determined by phrases within them, that is, *Google Search, Google, Microsoft Bing, Microsoft, Windows, Dead Sea, Jordan Valley,* and *Israel.* spaCy can identify these entities:

```
[22]    utt8 = nlp( "Google Search, often referred as Google,
        is the most popular search engine nowadays. It answers
        a huge volume of queries every day." )
        utt9 = nlp( "Microsoft Bing is another popular search
        engine. Microsoft is known by its star product
        Microsoft Windows, a popular operating system sold over
        the world." )
        utt10 = nlp( "The Dead Sea is the lowest lake in the
        world, located in the Jordan Valley of Israel. It is
        also the saltiest lake in the world." )
```

[23]	`utt8.ents`
	(Google Search, Google, every day)

[24]	`utt9.ents`
	(Microsoft Bing, Microsoft, Microsoft Windows)

[25]	`utt10.ents`
	(The Dead Sea, the Jordan Valley, Israel)

Since words are extracted for comparison, let's calculate similarity scores:

| [26] | ```
ents1 = [ent.text for ent in utt8.ents]
ents2 = [ent.text for ent in utt9.ents]
ents3 = [ent.text for ent in utt10.ents]
ents1 = nlp(" ".join(ents1))
ents2 = nlp(" ".join(ents2))
ents3 = nlp(" ".join(ents3))
``` |
|---|---|

[27]	`ents1.similarity(ents2)`
	0.5618316156902609

[28]	`ents1.similarity(ents3)`
	0.12924407611214866

[29]	`ents2.similarity(ents3)`
	0.11911278371814159

These figures revealed that the highest level of similarity exists between first and second paragraphs, which are both about large tech companies. The third paragraph is unlike other paragraphs. How can this calculation be obtained by using word vectors only? It is probably because words *Google* and *Microsoft* often appear together in news and other social media text corpora, hence producing similar word vectors

This is the conclusion of advanced semantic similarity methods section with different ways to combine word vectors with linguistic features such as key phrases and named entities.
```

WORKSHOP

Workshop 4.2 Further Semantic Analysis on The Adventures of Sherlock Holmes
It has learnt to further improve Semantic Analysis results on document similarity comparison by extracting (1) key phrases; (2) and comparing names entities. Try to use these techniques on *The Adventures of Sherlock Holmes*:
1. Extract three "representative texts" from this novel.
2. Perform key phrases extraction to improve the similarity rate as compared with Workshop 4.1 results.
3. Extract and compare name entities to identify significant name entities from this literature to further improve semantic analysis performance.
4. Remember to plot semantic diagram to show how these entities and keywords are related.
5. Discuss and explain what can be found.

References

Albrecht, J., Ramachandran, S. and Winkler, C. (2020) Blueprints for Text Analytics Using Python: Machine Learning-Based Solutions for Common Real World (NLP) Applications. O'Reilly Media.

Altinok, D. (2021) Mastering spaCy: An end-to-end practical guide to implementing NLP applications using the Python ecosystem. Packt Publishing.

Arumugam, R., & Shanmugamani, R. (2018). Hands-on natural language processing with python. Packt Publishing.

Bird, S., Klein, E., and Loper, E. (2009). Natural language processing with python. O'Reilly.

Doyle, A. C. (2019) The Adventures of Sherlock Holmes (AmazonClassics Edition). AmazonClassics.

FastText (2024) FastText official site. https://fasttext.cc/. Accessed 17 Dec 2024.

Gutenberg (2024) Project Gutenberg official site. https://www.gutenberg.org/. Accessed 17 Dec 2024.

Hardeniya, N., Perkins, J. and Chopra, D. (2016) Natural Language Processing: Python and NLTK. Packt Publishing.

Kedia, A. and Rasu, M. (2020) Hands-On Python Natural Language Processing: Explore tools and techniques to analyze and process text with a view to building real-world NLP applications. Packt Publishing.

NLTK (2024) NLTK official site. https://www.nltk.org/. Accessed 17 Dec 2024.

NLPGitHub (2024) URL: https://github.com/raymondshtlee/NLP/. Accessed 17 Dec 2024.

Perkins, J. (2014). Python 3 text processing with NLTK 3 cookbook. Packt Publishing Ltd.

SpaCy (2024) spaCy official site. https://spacy.io/. Accessed 17 Dec 2024.

Srinivasa-Desikan, B. (2018). Natural language processing and computational linguistics: A practical guide to text analysis with python, gensim, SpaCy, and keras. Packt Publishing, Limited.

Stanford (2024) NLP.stanford.edu Glove official site. https://nlp.stanford.edu/projects/glove/. Accessed 17 Dec 2024.

Synthetic (2024) Synthetic Intelligent Network site on Word2Vec Model. https://developer.syn.co.in/tutorial/bot/oscova/pretrained-vectors.html#word2vec-and-glove-models. Accessed 17 Dec 2024.

TensorFlow (2024) TensorFlow official site. https://projector.tensorflow.org/. Accessed 17 Dec 2024.

Vasiliev, Y. (2020) Natural Language Processing with Python and spaCy: A Practical Introduction. No Starch Press.

Chapter 15
Workshop#5: Sentiment Analysis and Text Classification (Hour 9–10)

15.1 Introduction

NLTK and spaCy are two major NLP Python implementation tools for basic text processing, N-gram modeling, POS tagging, and semantic analysis introduced in the last four workshops. Workshop 5 will explore how to position these NLP implementation techniques into two important NLP applications: text classification and sentiment analysis. TensorFlow and Kera are two vital components to implement LSTM, a commonly used RNN on machine learning, especially in NLP applications.

This workshop will:

1. Study text classification concepts in NLP and how spaCy NLP pipeline works on text classifier training.
2. Use movie reviews as a problem domain to demonstrate how to implement sentiment analysis with spaCy.
3. Introduce Artificial Neural Networks (ANN) concepts, TensorFlow, and Kera technologies.
4. Introduce sequential modeling scheme with LSTM technology using movie reviews domain as example to integrate these technologies for text classification and movie sentiment analysis.

15.2 Text Classification with spaCy and LSTM Technology

Text classification is a vital component in sentiment analysis application.

TextCategorizer is a spaCy's text classifier component applied in dataset for sentiment analysis to perform text classification with two vital Python frameworks (1) TensorFlow Keras API and (2) spaCy technology.

© The Author(s), under exclusive license to Springer Nature Singapore Pte
Ltd. 2025
R. Lee, *Natural Language Processing*,
https://doi.org/10.1007/978-981-96-3208-4_15

Neural networks basics, sequential data modeling with LSTM technology to process text for machine learning tasks with Keras's text preprocessing module and implement a neural network with tf.keras.

This workshop will cover the following topics:

- Basic concept and knowledge of text classification
- Model training of spaCy text classifier
- Sentiment Analysis with spaCy
- Sequential modeling with LSTM Technology

15.3   Technical Requirements

Please ensure that the following Python packages are installed before starting the workshop:

- Python (demo version 3.11.9)
- spacy (demo version 3.4.4)
- keras (demo version 3.5.0)
- tensorflow (demo version 2.17.0)
- numPy (demo version 1.26.4)
- pandas (demo version 2.2.2)
- matplotlib (demo version 3.9.2)

If these packages are not installed on PC/laptop, use pip install xxx command. The detailed requirements list and Python package version used in this workshop can be found in the requirements.txt file stored in the NLP GitHub repository (NLPGitHub 2024).

15.4   Text Classification in a Nutshell

15.4.1   What Is Text Classification?

Text Classification (Albrecht et al. 2020; Bird et al. 2009; George 2022; Sarkar 2019; Siahaan and Sianipar 2022; Srinivasa-Desikan 2018) is the task of assigning a set of predefined labels to text.

They are classified by manual tagging, but machine learning techniques are applied progressively to train classification system with known examples, or train samples to classify unseen cases. It is a fundamental task of NLP (Perkins 2014, Sarkar 2019) using various machine learning method such as LSTM technology (Arumugam and Shanmugamani 2018; Géron 2019; Kedia and Rasu 2020).

Text classification types are (Agarwal 2020; George 2022; Pozzi et al. 2016):

Fig. 15.1 Example of top detection for customer complaint in customer service automation system (CSAS)

- Language detection is the first step of many NLP systems, i.e., machine translation.
- Topic generation and detection are the process of summarization, or classification of a batch of sentences, paragraphs, or texts into certain topic of interest (TOI) or topic titles, e.g., customers' email request refund or complaints about products or services.
- Sentiment analysis to classify or analyze users' responses, comments, and messages on a particular topic attribute to positive, neutral, or negative sentiments. It is an essential task in e-commerce and social media platforms.

Text classifiers can emphasize overall text sentiments, text language detection, and words levels, i.e., verbs. A text classifier of a customer service automation system is shown in Fig. 15.1.

15.4.2 Text Classification as AI Applications

Text classification is considered as supervised-learning (SL) task in AI which means that the classifier can predict the class label of a text based on sample input text-class label pairs. It must require sufficient input (text)-output (classified labels) pairs databank for network training, testing, and validation. Hence, a labeled dataset is a list of text-label pairs required to train a text classifier. An example dataset of five training sentences with sentiment labels is shown in Table 15.1.

When a classifier encounters a new text not in the training text, it predicts a class label of this unseen text based on examples during the training phase to induce a text classifier output is always a class label.

Text classification can also be divided into (1) binary, (2) multi-class, and (3) multi-label categories:

1. Binary text classification refers to categorizing text into two classes.
2. Multi-class text classification refers to categorizing texts with more than two classes. Each class is mutually exclusive where one text is associated with a single class, e.g., rating customer reviews are represented by a 1–5 stars category single class label.

Table 15.1 Sample input texts and their corresponding output class labels

This TV has brought me so much joy.	Pos
This is the best soccer game I have ever seen.	Pos
This dress is so ordinary not worth this expensive selling price.	Neg
Mom makes the best dinners.	Pos
Shut up, you can't talk to me like that.	Neg

3. Multi-label text classification system is to generalize its multi-class counterpart assigned to each example text, e.g., *toxic, severe toxic, insult, threat, obscenity* levels of negative sentiment. What are Labels in Text Classification?

Labels are class names for output. A class label can be categorical (string) or numerical (a number).

Text classification has the following class labels:

• Sentiment analysis has positive and negative class labels abbreviated by pos and neg where 0 represents negative sentiment and 1 represents positive sentiment. Binary class labels are popular as well.
• The identical numeric representation applies to binary classification problems, i.e., use 0–1 for class labels.
• Class labeled with a meaningful name for multi-class and multi-label problems, e.g., movie genre classifier has labels *action, scifi, weekend, Sunday movie*, etc. Numbers are labels for a five-class classification problem, i.e., 1–5.

15.5   Text Classifier with spaCy NLP Pipeline

TextCategorizer (*tCategorizer*) is spaCy's text classifier component (Altinok 2021; SpaCy 2024; Vasiliev 2020). It required class labels and examples in NLP pipeline to perform training procedure as shown in Fig. 15.2.

TextCategorizer provides user-friendly and end-to-end approaches to train classifier so that it does not need to deal with neural network architecture directly.

15.5.1   *TextCategorizer Class*

Import spaCy and load nlp component from "en_core_web_md":

```
[1]    # Load and import spacy package
       import spacy
       # Load the en_core_web_md module
       nlp = spacy.load( "en_core_web_md" )
```

This TV has brought me so much joy.	Pos
This is the best soccer game I have ever seen.	Pos
This dress is so ordinary not worth this expensive selling price.	Neg
Mom makes the best dinners.	Pos
Shut up, you can't talk to me like that.	Neg

Fig. 15.2 *TextCategorizer* in the spaCy NLP pipeline

Import *TextCategorizer* from spaCy pipeline components:

```
[2]        # Import the Single Text Categorizer Model
           from spacy.pipeline.textcat import
           DEFAULT_SINGLE_TEXTCAT_MODEL
```

TextCategorizer consists of (1) single-label and (2) multi-label classifiers.

A multi-label classifier can predict more than a single class. A single-label classifier predicts an individual class for each example and classes are mutually exclusive.

The preceding import line imports single-label classifier, and the following code imports multi-label classifier:

```
[3]        # Import the Multiple Text Categorizer Model
           from spacy.pipeline.textcat_multilabel import
           DEFAULT_MULTI_TEXTCAT_MODEL
```

There are two parameters (1) a threshold value and (2) a model name (either Single or Multi depends on classification task) required for a *TextCategorizer* component configuration.

TextCategorizer generates a probability for each class and a class is assigned to text if the probability of this class is higher than the threshold value.

A traditional threshold value for text classification is 0.5, however, if prediction is required for a higher confidence, it can adjust threshold to 0.6–0.8.

A single-label *TextCategorizer* (*tCategorizer*) component is added to nlp pipeline as follows:

```
[4]        # Import the Single Text Categorizer Model
           # Define the model parameters: threshold and model
           from spacy.pipeline.textcat import
           DEFAULT_SINGLE_TEXTCAT_MODEL
           config = {
               "threshold": 0.5,
               "model": DEFAULT_SINGLE_TEXTCAT_MODEL
           }
```

[5]
```
# Define the Text Categorizer object (tCategorizer)
tCategorizer = nlp.add_pipe("textcat", config=config)
```

Let's look at Text Categorizer object (tCategorizer):

[6]
```
tCategorizer
```
<spacy.pipeline.textcat.TextCategorizer at 0x278850c0830>

Add a multilabel component to nlp pipeline:

[7]
```
# Import the Multiple Text Categorizer Model
# Define the model parameters: threshold and model
from spacy.pipeline.textcat_multilabel import
DEFAULT_MULTI_TEXTCAT_MODEL
config = {
    "threshold": 0.5,
    "model": DEFAULT_MULTI_TEXTCAT_MODEL
}
```

[8]
```
tCategorizer = nlp.add_pipe( "textcat_multilabel",
config=config)
```

[9]
```
tCategorizer
```
<spacy.pipeline.textcat_multilabel.MultiLabel_TextCategorizer at 0x278850c26f0>

Add a *TextCategorizer* pipeline component to nlp pipeline object at the last line of each preceding code blocks. The newly created *TextCategorizer* component is captured by textcat variable and set for training

15.5.2   *Formatting Training Data for the TextCategorizer*

Let's prepare a customer sentiment dataset for binary text classification.

The label (category) will be called sentiment to obtain two possible values, 0 and 1 corresponding to negative and positive sentiments.

There are six examples from IMDB with three each of positive and negative as below:

```
[10]    movie_comment1 = [
            ("This movie is perfect and worth watching. ",
        {"cats": {"Positive Sentiment": 1}}),
            ("This movie is great, the performance of Al Pacino
          is brilliant.", {"cats": {"Positive Sentiment": 1}}),
            ("A very good and funny movie. It should be the best
          this year!", {"cats": {"Positive Sentiment": 1}}),
            ("This movie is so bad that I really want to leave
          after the first hour watching.", {"cats": {"Positive
          Sentiment": 0}}),
            ("Even free I won't see this movie again. Totally
          failure!",
        {"cats": {"Positive Sentiment": 0}}),
            ("I think it is the worst movie I saw so far this
          year.",
        {"cats": {"Positive Sentiment": 0}})
        ]]
```

Check on any movie comment1 element:

```
[11]    movie_comment1 [1]
        ('This movie is great, the performance of Al Pacino is brilliant.',
        {'cats': {'Positive Sentiment': 1}})
```

• Each training example (movie_coment1) is a tuple object consists of a text and a nested dictionary.
• The dictionary contains a class category in a format recognized by spaCy.
• The *cats* field means categories.
• Include class category sentiment and its value. The value should always be a floating-point number.

The code will introduce a class category selected for *TextCategorizer* component.

```
[12]    import random
        import spacy
        from spacy.training import Example
        from spacy.pipeline.textcat import
        DEFAULT_SINGLE_TEXTCAT_MODEL
```

• Import a built-in library random to shuffle dataset.
• Import spaCy as usual, then import *example* to prepare training samples in spaCy format.
• Import *TextCategorizer* model in the final statement.

Initialize pipeline and *TextCategorizer* component.

When a new *TextCategorizer* component *tCategorizer* is created, use calling *add_label* method to introduce category sentiment to *TextCategorizer* component with examples.

The following code adds label to *TextCategorizer* component and initializes *TextCategorizer* model's weights with training samples:

```
[13]    import random
        import spacy
        from spacy.training import Example
        from spacy.pipeline.textcat import
        DEFAULT_SINGLE_TEXTCAT_MODEL
        # Load the spaCy NLP model
        nlp = spacy.load('en_core_web_md')
        # Set the threshold and model
        config = {
            "threshold": 0.5,
            "model": DEFAULT_SINGLE_TEXTCAT_MODEL
        }
        # Define TextCategorizer object (tCategorizer)
        tCategorizer = nlp.add_pipe("textcat", config=config)
```

Let's look at pipe_names:

```
[14]    nlp.pipe_names
        ['tok2vec',
        'tagger',
        'parser',
        'attribute_ruler',
        'lemmatizer',
        'ner',
        'textcat']
```

When a new *TextCategorizer* component textcat is created, use calling *add_label* method to introduce label sentiment to the *TextCategorizer* component and initialize this component with examples.

The following code adds a label to *TextCategorizer* component and initializes *TextCategorizer* model's weights with training samples (*movie_comment_exp*):

```
[15]    # Create the two sentiment categories
        tCategorizer.add_label("Positive Sentiment")
        tCategorizer.add_label("Negative Sentiment")
        # Create the movie comment samples
        movie_comment_exp = [Example.from_dict(nlp.make_
        doc(comments), category) for comments,category in
        movie_comment1]
        tCategorizer.initialize(lambda: movie_comment_exp,
        nlp=nlp)
```

Let's look at movie_comment_exp:

[16]	`movie_comment_exp`
	[{'doc_annotation': {'cats': {'Positive Sentiment': 1}, 'entities': ['O', 'O', 'O', 'O', 'O', 'O', 'O', 'O'], 'spans': {}, 'links': {}}, 'token_annotation': {'ORTH': ['This', 'movie', 'is', 'perfect', 'and', 'worth', 'watching', '.'], 'SPACY': [True, True, True, True, True, True, False, True], 'TAG': ['', '', '', '', '', '', '', ''], 'LEMMA': ['', '', '', '', '', '', '', ''], 'POS': ['', '', '', '', '', '', '', ''], 'MORPH': ['', '', '', '', '', '', '', ''], 'HEAD': [0, 1, 2, 3, 4, 5, 6, 7], 'DEP': ['', '', '', '', '', '', '', ''], 'SENT_START': [1, 0, 0, 0, 0, 0, 0, 0]}}, {'doc_annotation': {'cats': {'Positive Sentiment': 1}, 'entities': ['O', 'O', 'O', 'O', 'O', 'O', 'O', 'O', 'O', 'O', 'O', 'O', 'O'], 'spans': {}, 'links': {}}, 'token_annotation': {'ORTH': ['This', 'movie', 'is', 'great', ',', 'the', 'performance', 'of', 'Al', 'Pacino', 'is', 'brilliant', '.'], 'SPACY': [True, True, True, False, True, True, True, True, True, True, True, False, False], 'TAG': ['', '', '', '', '', '', '', '', '', '', '', '', ''], 'LEMMA': ['', '', '', '', '', '', '', '', '', '', '', '', ''], 'POS': ['', '', '', '', '', '', '', '', '', '', '', '', ''], 'MORPH': ['', '', '', '', '', '', '', '', '', '', '', '', ''], 'HEAD': [0, 1, 2, 3, 4, 5, 6, 7, 8, 9, 10, 11, 12], 'DEP': ['', '', '', '', '', '', '', '', '', '', '', '', ''], 'SENT_START': [1, 0, 0, 0, 0, 0, 0, 0, 0, 0, 0, 0, 0]}}, {'doc_annotation': {'cats': {'Positive Sentiment': 1}, 'entities': ['O', 'O', 'O', 'O', 'O', 'O', 'O', 'O', 'O', 'O'], 'spans': {}, 'links': {}}, 'token_annotation': {'ORTH': ['A', 'very', 'good', 'and', 'funny', 'movie', '.', 'It', 'should', …

15.5.3 System Training

Training loop is all set to be defined.

First, disable other pipe components to allow only textcat to be trained.

Second, create an optimizer object by calling *resume_training* to keep the weights of existing statistical models.

Examine each epoch training example one by one and update the weights of textcat. Examine data for 20 epochs.

Try the whole program with training loop:

[17]	`movie_comment1`
	[('This movie is perfect and worth watching. ', {'cats': {'Positive Sentiment': 1}}), ('This movie is great, the performance of Al Pacino is brilliant.', {'cats': {'Positive Sentiment': 1}}), ('A very good and funny movie. It should be the best this year!', {'cats': {'Positive Sentiment': 1}}), ('This movie is so bad that I really want to leave after the first hour watching.', {'cats': {'Positive Sentiment': 0}}), ("Even free I won't see this movie again. Totally failure!", {'cats': {'Positive Sentiment': 0}}), ('I think it is the worst movie I saw so far this year.', {'cats': {'Positive Sentiment': 0}})]

[18]

```python
# Full implementation of the Movie Sentiment Analysis
System
import random
import spacy
from spacy.training import Example
from spacy.pipeline.textcat import
DEFAULT_SINGLE_TEXTCAT_MODEL
# Load the spaCy NLP model
nlp = spacy.load('en_core_web_md')
# Set the threshold and model
config = {
    "threshold": 0.5,
    "model": DEFAULT_SINGLE_TEXTCAT_MODEL
}
# Create the TextCategorizer object (tCategorizer)
tCategorizer = nlp.add_pipe("textcat", config=config)
# Add the two movie sentiment categories
tCategorizer.add_label("Positive Sentiment")
tCategorizer.add_label("Negative Sentiment")
# Create the movie sample comments
movie_comment_exp = [Example.from_dict(nlp.make_
doc(comments), category) for comments,category in
movie_comment1]
tCategorizer.initialize(lambda: movie_comment_exp,
nlp=nlp)
# Set the training epochs and loss values
epochs=20
losses = {}
# Main program loop
with nlp.select_pipes(enable="textcat"):
    optimizer = nlp.resume_training()
    for i in range(epochs):
        random.shuffle(movie_comment1)
        for comments, category in movie_comment1:
            mdoc = nlp.make_doc(comments)
            exp = Example.from_dict(mdoc, category)
            nlp.update([exp], sgd=optimizer,
    losses=losses)
        print("Epoch #",i, "Losses: ",losses)
```

```
Epoch # 0 Losses: {'textcat': 1.4457833915948868}
Epoch # 1 Losses: {'textcat': 2.6075984984636307}
Epoch # 2 Losses: {'textcat': 3.502597125247121}
Epoch # 3 Losses: {'textcat': 4.506776508176699}
Epoch # 4 Losses: {'textcat': 5.337087038293248}
Epoch # 5 Losses: {'textcat': 6.35534223607101}
Epoch # 6 Losses: {'textcat': 7.171283801217214}
Epoch # 7 Losses: {'textcat': 8.18715655813412}
Epoch # 8 Losses: {'textcat': 8.992988994691586}
Epoch # 9 Losses: {'textcat': 10.04922488262389}
Epoch # 10 Losses: {'textcat': 10.843044139094673}
Epoch # 11 Losses: {'textcat': 11.831995705193918}
Epoch # 12 Losses: {'textcat': 12.701028650988377}
Epoch # 13 Losses: {'textcat': 13.476843157594317}
Epoch # 14 Losses: {'textcat': 14.474409362490363}
Epoch # 15 Losses: {'textcat': 15.255843693623234}
Epoch # 16 Losses: {'textcat': 16.088656657451878}
Epoch # 17 Losses: {'textcat': 16.996663642346313}
Epoch # 18 Losses: {'textcat': 17.76832411575554}
Epoch # 19 Losses: {'textcat': 18.6191813947854}
```

15.5.4   System Testing

Let's test a new text categorizer component, doc.cats property holds the class labels:

| [19] | ```# Test 1: This movie sucks
test1 = nlp("This movie sucks and the worst I ever saw.")
test1.cats``` |
|---|---|
| | {'Positive Sentiment': 0.8156227469444275, 'Negative Sentiment': 0.18437722325325012} |

| [20] | ```# Test 2: I'll watch it again, how amazing.
test2 = nlp("This movie really very great!")
test2.cats``` |
|---|---|
| | {'Positive Sentiment': 0.8716222047805786, 'Negative Sentiment': 0.1283777952194214} |

 The small dataset trained spaCy text classifier successfully for a binary text classification problem to perform correct sentiment analysis. Now, let's perform multi-label classification

15.5.5 *Training TextCategorizer for Multi-Label Classification*

Multi-label classification means the classifier can predict more than single-label for an example text. Naturally, the classes are not mutually exclusive.

Provide training samples with at least more than two categories to train a multiple labeled classifier.

Construct a small training set to train spaCy's *TextCategorizer* for multi-label classification. This time will form a set of movie reviews, where the multi-category is:

- ACTION
- SCIFI
- WEEKEND

Here is a small sample dataset (*movie_comment2*):

| [21] | ```
movie_comment2 = [
    ("This movie is great for weekend watching.",
{"cats": {"WEEKEND": True}}),
    ("This a 100% action movie, I enjoy it.",
{"cats": {"ACTION": True}}),
    ("Avatar is the best Scifi movie I ever seen!",
{"cats": {"SCIFI": True}}),
    ("Such a good Scifi movie to watch during the
 weekend!",
{"cats": {"WEEKEND": True, "SCIFI": True}}),
    ("Matrix a great Scifi movie with a lot of action.
 Pure action, great!", {"cats": {"SCIFI": True,
 "ACTION": True}})
]
``` |
|---|---|

Check dataset first:

| [22] | ```
movie_comment2
``` |
|---|---|
| | [('This movie is great for weekend watching.', {'cats': {'WEEKEND': True}}), ('This a 100% action movie, I enjoy it.', {'cats': {'ACTION': True}}), ('Avatar is the best Scifi movie I ever seen!', {'cats': {'SCIFI': True}}), ('Such a good Scifi movie to watch during the weekend!', {'cats': {'WEEKEND': True, 'SCIFI': True}}), ('Matrix a great Scifi movie with a lot of action. Pure action, great!', {'cats': {'SCIFI': True, 'ACTION': True}})] |

| [23] | ```
movie_comment2[1]
``` |
|---|---|
| | ('This a 100% action movie, I enjoy it.', {'cats': {'ACTION': True}}) |
```

Provide examples with a single-label, such as first example (the first sentence of *movie_comment2*, the second line of preceding code block), and examples with more than single-label, such as fourth example of *movie_comment2*.

Import after the training set is formed.

```
[24]    import random
        import spacy
        from spacy.training import Example
        from spacy.pipeline.textcat_multilabel import
        DEFAULT_MULTI_TEXTCAT_MODEL
        # Load spaCy NLP model
        nlp = spacy.load( 'en_core_web_md' )
```

Note that the last line has different code than the previous section. Import multi-label model instead of single-label model.

Next, add multi-label classifier component to nlp pipeline.

Also note that pipeline component name is textcat_multilabel as compared with previous section's textcat:

```
[25]    # Set the threshold and model
        config = {
            "threshold": 0.5,
            "model": DEFAULT_MULTI_TEXTCAT_MODEL
        }
        # Create the TextCategorizer object (tCategorizer)
        tCategorizer = nlp.add_pipe( "textcat_multilabel",
        config=config)
```

Add categories to *TextCategorizer* component and initialize model like previous text classifier section.

Add three labels instead of one:

```
[26]    # Create the categorizer object with 3 categories
        categories = ["SCIFI", "ACTION", "WEEKEND"]
        # Using For Loop to add the 3 categories
        for category in categories:
            tCategorizer.add_label(category)
        # Create the movie comment sample for training
        movie_comment_exp = [Example.from_dict(nlp.make_
        doc(comments), category) for comments,category in
        movie_comment2]
        # Initializer the tCategorizer
        tCategorizer.initialize(lambda: movie_comment_exp,
        nlp=nlp)
```

Training loop is all set to be defined.

Code functions are like previous section's code, the only difference is component name *textcat_multilabel* in the first line:

[27]
```
# Set the training epochs and loss values
epochs=20
losses = {}
# Main Loop of the program
with nlp.select_pipes(enable="textcat_multilabel"):
    optimizer = nlp.resume_training()
    for i in range(epochs):
        random.shuffle(movie_comment2)
        for comments, category in movie_comment2:
            mdoc = nlp.make_doc(comments)
            exp = Example.from_dict(mdoc, category)
            nlp.update([exp], sgd=optimizer,
  losses=losses)
        print(losses)
```

{'textcat_multilabel': 0.6435152161866426}
{'textcat_multilabel': 0.6444572633295138}
{'textcat_multilabel': 0.644508911859063}
{'textcat_multilabel': 0.6445386177332972}
{'textcat_multilabel': 0.6445531247200567}
{'textcat_multilabel': 0.6445633990291366}
{'textcat_multilabel': 0.6445710929335524}
{'textcat_multilabel': 0.6445779604009196}
{'textcat_multilabel': 0.6445835611851201}
{'textcat_multilabel': 0.644588678256226}
{'textcat_multilabel': 0.6445934303342438}
{'textcat_multilabel': 0.6445977886109024}
{'textcat_multilabel': 0.6446017611581358}
{'textcat_multilabel': 0.644605454090792}
{'textcat_multilabel': 0.6446089753232687}
{'textcat_multilabel': 0.6446123319626418}
{'textcat_multilabel': 0.6446155674397027}
{'textcat_multilabel': 0.6446186350310477}
{'textcat_multilabel': 0.6446215542191567}
{'textcat_multilabel': 0.644624413417668}

The output should look like the output of the previous section but use multiple categories for system training. Let's test the new multi-label classifier:

[28]
```
test3 = nlp("Definitely in my weekend scifi movie night
list")
test3.cats
```

{'SCIFI': 0.9721231460571289,
'ACTION': 0.6180852055549622,
'WEEKEND': 0.9213110208511353}

[29]
```
test4 = nlp("Go to watch action scifi movie this
weekend.")
test4.cats
```

{'SCIFI': 0.9883644580841064,
'ACTION': 0.9022844433784485,
'WEEKEND': 0.9972347617149353}

Although sample size is small, the multiple text categorizer can classify two IMDB user comments correctly into three categories: SCIFI, ACTION, and WEEKEND. Note that over thousands of IMDB, user comments are required to perform satisfactory sentiment analysis in real situations

This section has learnt how to train a spaCy's *TextCategorizer* component for binary and multi-label text classifications.

Now, *TextCategorizer* will be trained on a real-world dataset for a sentiment analysis using IMDB user comments dataset.

Workshop 5.1 Movie comments from IMDB.com
Movie comments is a significant classification in social media. This workshop constructs a simple movie comment classification with millions of user comments from IMDB.com, the world biggest movie social media platform
1. Try to collect 900 comments with 300 *Good*, 300 *Average,* and 300 *Bad* comments to train the system. Make sure they make sense or the system won't function.
2. Construct a Multi-label Classification System to create three movie comments: *Good, average,* or *bad.*
3. Train system with at least 100 epochs.
4. Use 10 examples to test and see whether it works.

15.6   Sentiment Analysis with spaCy

15.6.1   *IMDB Large Movie Review Dataset*

This section will work on a real-world dataset using IMDB (2024) Large Movie Reviews Dataset from Kaggle (2024).

The original *imdb_sup.csv* dataset has 50,000 rows. They need to down-size and select the first 5000 records into datafile imdb_5000.csv to speed up training. This movie reviews dataset consists of movie reviews, reviews sizes, IMDB Ratings (1–10), and Sentiment Ratings (0 or 1).

The dataset can be downloaded from workshop directory namely: imdb_sup.csv (complete dataset) or imdb_5000.csv (5000 records).

15.6.2 Explore the Dataset

Let's have some understanding from dataset prior to sentiment analysis.

1. First, import to read and visualize dataset:

[30]	`import pandas as pd`
	`import matplotlib.pyplot as plt`
	`%matplotlib inline`

2. Read *imdb_5000*.csv datafile into a pandas DataFrame (*mcommentDF*) and output the shape of DataFrame:

[31]	`mcommentDF=pd.read_csv( 'imdb_5000.csv' )`

[32]	`mcommentDF.shape`
	(5000, 3)

Note: This IMDB movie reviews dataset contains 5000 records, each record has three fields' attributes: *Review, Rating,* and *Sentiment*

3. Examine rows and columns of dataset by printing the first few rows using *head( )* method:

[33]	`mcommentDF.head()`			
		Review	**Rating**	**Sentiment**
	0	**Possible Spoilers**	1	0
	1	Read the book, forget the movie!	2	0
	2	**Possible Spoilers Ahead**	2	0
	3	What a script, what a story, what a mess!	2	0
	4	I hope this group of film-makers never re-unites.	1	0

4. Use *Review* and *Sentiment* columns only in this workshop. Hence, drop other columns that won't use, and call *dropna( )* method to drop the rows with missing values:

[34]	`mcommentDF_clean = mcommentDF[[ 'Review', 'Sentiment'` `]].dropna()`

[35]
```
mcommentDF_clean.head()
```

	Review	Sentiment
0	**Possible Spoilers**	0
1	Read the book, forget the movie!	0
2	**Possible Spoilers Ahead**	0
3	What a script, what a story, what a mess!	0
4	I hope this group of film-makers never re-unites.	0

5. Let's look at how review scores are distributed:

[36]
```
axplot=mcommentDF.Rating.value_counts().plot
(kind='bar', colormap='Paired')
plt.show()
```

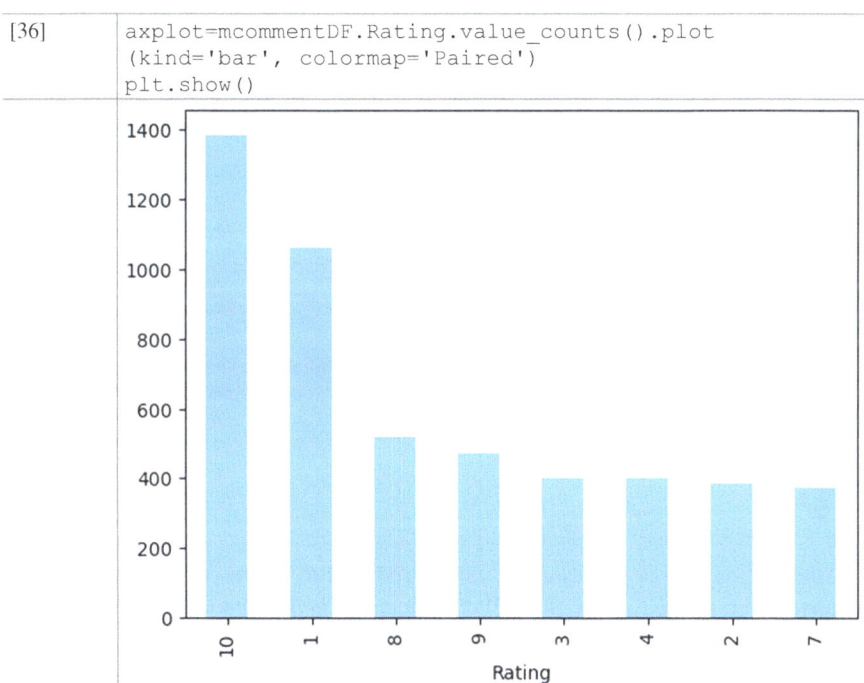

1. Users prefer to give a high rating, i.e., 8 or above, and 10 is the highest as shown.
2. It is better to select a sample set with an even distribution to balance sample data rating.
3. Check system performance first. If it is not as good as predicted, can use fine-tune sampling method to improve system performance.

Here use the sentiments already labeled.

6. Plot ratings distribution:

| [37] | `axplot=mcommentDF.Sentiment.value_counts().plot (kind=`
`'bar', colormap='Paired')`
`plt.show()` |

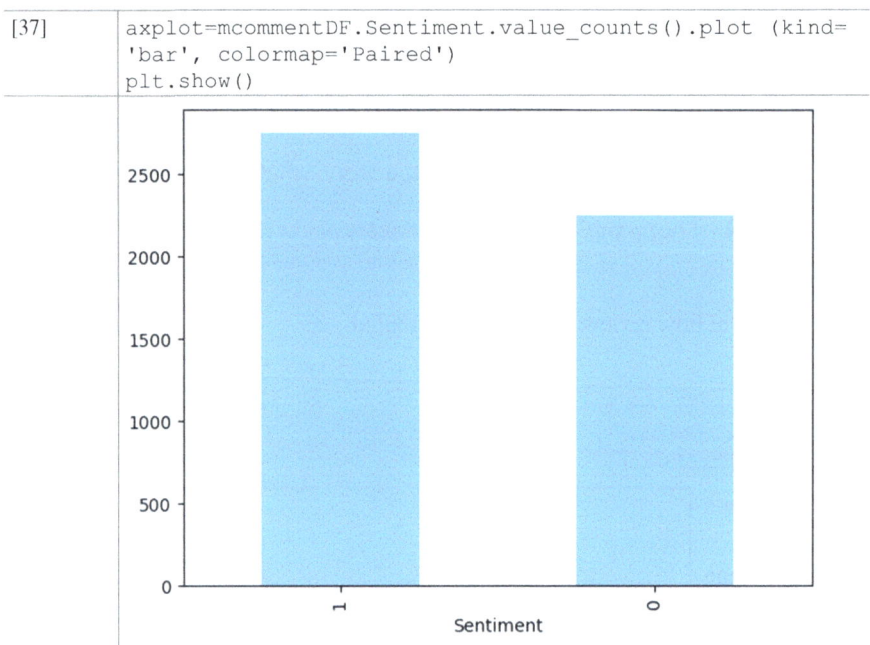

Note that rating distribution has better results than the previous one, it has higher number of positive reviews, but negative reviews is also significant as shown

After the dataset is processed, it can be reduced to a two-column dataset with negative and positive ratings. So, call *mcommentDF.head( )* again and the following result is obtained:

| [38] | `mcommentDF.head()` |

	Review	Rating	Sentiment
0	**Possible Spoilers**	1	0
1	Read the book, forget the movie!	2	0
2	**Possible Spoilers Ahead**	2	0
3	What a script, what a story, what a mess!	2	0
4	I hope this group of film-makers never re-unites.	1	0

Complete dataset exploration and display review scores with class categories distribution. The dataset is ready to be processed. Drop unused columns and convert review scores to binary class labels. Let's begin with the training procedure

15.6.3 Training the TextClassifier

Use a multi-label classifier to train a binary text classifier this time.

1. Import spaCy classes as follows:

```
[39]    import spacy
        import random
        from spacy.training import Example
        from spacy.pipeline.textcat_multilabel import
        DEFAULT_MULTI_TEXTCAT_MODEL
```

2. Create a pipeline object nlp, define classifier configuration, and add *TextCategorizer* component to nlp with the following configuration:

```
[40]    # Load the spaCy NLP model
        nlp = spacy.load( "en_core_web_md" )
        # Set the threshold and model
        config = {
            "threshold": 0.5,
            "model": DEFAULT_MULTI_TEXTCAT_MODEL
        }
        # Create the TextCategorizer object (tCategorizer)
        tCategorizer = nlp.add_pipe("textcat_multilabel",
        config=config)
```

3. When TextCategorizer object is available, create movie comment sample object as a list and load all user comments and categories into it.

```
[41]     # Create the IMDB movie comment sample object
         movie_comment_exp = []
         # Load all the IMDB user comments and categories
         for idx, rw in mcommentDF.iterrows():
             comments = rw["Review"]
             rating = rw["Sentiment"]
             category = {"POS": True, "NEG": False} if rating == 1
             else
         {"NEG": True, "POS": False}
             movie_comment_exp.append(Example.from_dict(nlp.
         make_doc(comments), {"cats": category}))
```

4. Let's check *movie_comment_exp*:

```
[42]     movie_comment_exp[0]
```
{'doc_annotation': {'cats': {'NEG': True, 'POS': False}, 'entities': ['O', 'O', 'O', 'O', 'O', 'O'], 'spans': { }, 'links': { }}, 'token_annotation': {'ORTH': ['*', '*', 'Possible', 'Spoilers', '*', '*'], 'SPACY': [False, False, True, False, False, False], 'TAG': ['', '', '', '', '', ''], 'LEMMA': ['', '', '', '', '', ''], 'POS': ['', '', '', '', '', ''], 'MORPH': ['', '', '', '', '', ''], 'HEAD': [0, 1, 2, 3, 4, 5], 'DEP': ['', '', '', '', '', ''], 'SENT_START': [1, 0, 0, 0, 0, 0]}}

5. Use POS and NEG labels for positive and negative sentiment respectively. Introduce these labels to the new component and initialize it with examples.

```
[43]     # Add the two sentiment categories into tCategorizer
         tCategorizer.add_label("POS")
         tCategorizer.add_label("NEG")
         tCategorizer.initialize(lambda: movie_comment_exp,
         nlp=nlp)
```

```
[44]     tCategorizer
```
<spacy.pipeline.textcat_multilabel.MultiLabel_TextCategorizer at 0x2791f978d70>

6. Define training loop by examining the training set for two epochs but can examine further if necessary. The following code snippet will train the new text categorizer component:

```
[45]        # Set the training epochs to 2 to save time
            epochs = 2
            # Main program loop
            with nlp.select_pipes(enable="textcat_multilabel"):
                optimizer = nlp.resume_training()
                for i in range(epochs):
                    random.shuffle(movie_comment_exp)
                    for exp in movie_comment_exp:
                        nlp.update([exp], sgd=optimizer)
```

7. Test how text classifier component works for two example sentences:

```
[46]        test5 = nlp("This is the best movie that I have ever
            watched")
```

```
[47]        test5.cats
```
{'POS': 0.9747582674026489, 'NEG': 0.017647745087742805}

```
[48]        test6 = nlp("This movie is so bad")
```

```
[49]        test6.cats
```
{'POS': 0.11307813227176666, 'NEG': 0.8834090828895569}

Note both NEG and POS labels appeared in prediction results because it used a multi-label classifier. The results are satisfactory, but it can improve if the numbers for training epochs are increased. The first sentence has a high positive probability output, and the second sentence has predicted as negative with a high probability

SpaCy's text classifier component training is completed.

The next section will explore Kera, a popular deep leaning library, and how to write Keras code for text classification with another machine learning library—TensorFlow's Keras API.

Fig. 15.3 System
architecture of ANN

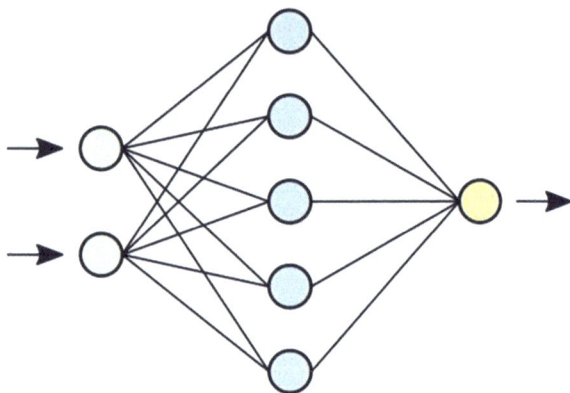

15.7   Artificial Neural Network in a Nutshell

This workshop section will learn how to incorporate spaCy technology with ANN
technology using TensorFlow and its Keras package (Géron 2019; Kedia and Rasu
2020; TensorFlow 2024).

A typical ANN has:

1. Input layer consists of input neurons, or nodes
2. Hidden layer consists of hidden neurons, or nodes
3. Output layer consists of output neurons, or nodes

ANN will learn knowledge by its network weights update through network training with sufficient sample inputs and target outputs pairs. The network can predict
or match unseen inputs to corresponding output result after it had sufficient training
to a predefined accuracy. A typical ANN architecture is shown in Fig. 15.3.

15.8   An Overview of TensorFlow and Keras

TensorFlow (Géron 2019; TensorFlow 2024) is a popular Python tool widely used
for machine learning. It has huge community support and great documentation
available at TensorFlow official site (TensorFlow 2024), while Keras (2024) is a
Python-based deep learning tool that can be integrated with Python platforms such
as TensorFlow, Theano, and CNTK.

TensorFlow 1 was disagreeable to symbolic graph computations and other low-
level computations, but TensorFlow 2 initiated great changes in machine learning
methods allowing developers to use Keras'with TensorFlow's low-level methods.
Keras is popular in R&D because it supports rapid prototyping and user-friendly
API to neural network architectures (Kedia and Rasu 2020; Srinivasa-Desikan 2018).

Neural networks are commonly used for computer vision and NLP tasks including object detection, image classification, scene understanding, text classification,
POS tagging, text summarization, and natural language generation.

TensorFlow 2 will be used to study the details of a neural network architecture for text classification with tf.keras implementation throughout this section.

15.9   Sequential Modeling with LSTM Technology

LSTM is one of the significant recurrent networks used in various machine learning applications such as NLP applications nowadays (Ekman 2021; Korstanje 2021).

RNNs are special neural networks that can process sequential data in steps.

All inputs and outputs are independent but not for text data in neural networks. Every word's presence depends on neighboring words, e.g., a word is predicted by considering all preceding predicted words and stored the past sequence token of words within an LTSM cell in a machine translation task. An LSTM is showed in Fig. 15.4.

An LSTM cell is moderately complex than an RNN cell, but computation logic is identical. A diagram of an LSTM cell is shown in Fig. 15.5. Note that input and output steps are identical to RNN counterparts:

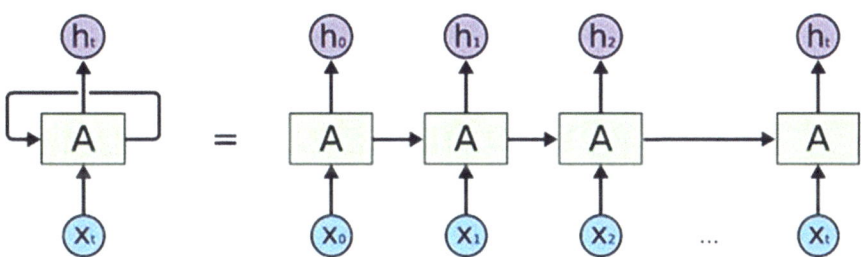

Fig. 15.4 RNN with LSTM technology

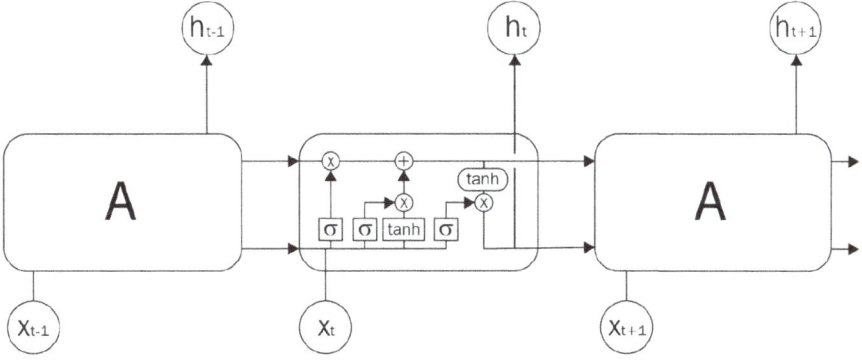

Fig. 15.5 Architecture of LSTM cell

Keras has extensive support for RNN variations GRU, LSTM and simple API for training RNNs. RNN variations are crucial for NLP tasks as language data's nature is sequential, i.e., text is a sequence of words, speech is a sequence of sounds, and so on.

Since the type of statistical model has been identified in the design, it can transform a sequence of words into a word IDs sequence and build vocabulary with Keras preprocessing module simultaneously.

15.10   Keras Tokenizer in NLP

Text is a sequence of words or characters data. A sentence can be fed by a tokens sequence. Hence, tokens are to be vectorized first by the following steps:

1. Tokenize each utterance and turn these utterances into a sequence of tokens.
2. Build a vocabulary from set of tokens presented in Step 1. These are tokens to be recognized by neural network design.
3. Create a vocabulary and assign ID to each token.
4. Map token vectors with corresponding token-IDs.

Let's look at a short example of a corpus for three sentences:

[50]	```testD = ["I am going to buy a gift for Christmas tomorrow morning.",```
	```"Yesterday my mom cooked a wonderful meal.",```
	```"Jack promised he would remember to turn off the lights."]```

[51]	```testD```
	['I am going to buy a gift for Christmas tomorrow morning.', 'Yesterday my mom cooked a wonderful meal.', 'Jack promised he would remember to turn off the lights.']

Let's tokenize words into utterances:

[52]	```import spacy```
	```# Load the NLP model```
	```nlp = spacy.load("en_core_web_md")```
	```# Create the utterances object```
	```utterances = [[token.text for token in nlp(utterance)] for utterance in testD]```
	```for utterance in utterances:```
	``` utterance```

All tokens of Doc object generated by calling nlp(sentence) are iterated in the preceding code. Note that punctuation marks have not been filtered as this filtering depends on the task, e.g., punctuation marks such as '!', correlate to the result in sentiment analysis, they are preserved in this example

Build vocabulary and token sequences into token-ID sequences using *Tokenizer* as shown:

| [53] | ```
# Import Tokenizer
from tensorflow.keras.preprocessing.text import Tokenizer
# Create tokenizer object (ktoken)
ktoken = Tokenizer(lower=True)
``` |

| [54] | ```
ktoken.fit_on_texts(testD)
ktoken
``` |
| | <keras.src.legacy.preprocessing.text.Tokenizer at 0x278fbd58c90> |

| [55] | ```
ktoken.word_index
``` |
| | {'to': 1, 'i': 2, 'am': 3, 'going': 4, 'buy': 5, 'some': 6, 'gift': 7, 'for': 8, 'christmas': 9, 'tomorrow': 10, 'morning': 11, 'yesterday': 12, 'my': 13, 'mom': 14, 'cooked': 15, 'a': 16, 'wonderful': 17, 'meal': 18, 'john': 19, 'promised': 20, 'he': 21, 'would': 22, 'remember': 23, 'turn': 24, 'off': 25, 'the': 26, 'lights': 27} |

The following are performed in the above codes:
1. Import tokenizer from Keras text preprocessing module.
2. Create a tokenizer object (*ktoken*) with parameter lower = true, which means tokenizer should lower all words for vocabulary formation.
3. Call *ktoken.fit_on_texts* on data to form vocabulary. *fit_on_text* work on a tokens sequence; input should always be a list of tokens.
4. Examine vocabulary by printing *ktoken.word_index*. *Word_index* is a dictionary where keys are vocabulary tokens and values are token-IDs.

Call *ktoken.texts_to_sequences()* method to retrieve a token-ID.

Notice that the input to this method should always be a list, even if a single token is fed.

Feed one-word input as a list (notice list brackets) in the following code segment:

| [56] | ```
ktoken.texts_to_sequences(["Christmas"])
``` |
| | [[9]] |

[57]	`ktoken.texts_to_sequences(["cooked", "meal"])`
	[[15], [18]]

 1. Note token-IDs start from 1 and not 0. 0 is a reserved value, which means a padding value with specific meaning.
2. Keras cannot process utterances of different lengths, hence need to pad all utterances.
3. Pad each sentence of dataset to a maximum length by adding padding utterances either at the start or end of utterances.
4. Keras inserts 0 for the padding which means it's a padding value without a token.

Let's understand how padding works with a simple example.

[58]	```
	# Import the pad_sequences package
	from tensorflow.keras.preprocessing.sequence import
	pad_sequences
	# Create the utterance sequences
	seq_utterance = [[7], [8,1], [9,11,12,14]]
	# Define Maximum Length (MLEN)
	MLEN=4
	# Pad the utterance sequences.
	pad_sequences(seq_utterance, MLEN, padding="post")```

	array([[7, 0, 0, 0], 　　[8, 1, 0, 0], 　　[9, 11, 12, 14]])

[59]	`pad_sequences(seq_utterance, MLEN, padding="pre")`
	array([[0, 0, 0, 7], 　　[0, 0, 8, 1], 　　[9, 11, 12, 14]])

Call pad_sequences on this sequences list and every sequence is padded with zeros so that its length reaches MAX_LEN = 4 which is the length of the longest sequence. Then pad sequences from the right or left with *post* and *pre* options. Sentences with post option are padded in the preceding code, hence the sentences are padded from the right.

When these sequences are organized, the complete text preprocessing steps are as follows:

[60]	```
	# Import the Tokenizer and pad sequences package
	from tensorflow.keras.preprocessing.text import Tokenizer
	from tensorflow.keras.preprocessing.sequence import
	pad_sequences
	# Create the token object
	ktoken = Tokenizer(lower=True)
	ktoken.fit_on_texts(testD)
	# Create the sequence utterance object
	sutterance = ktoken.texts_to_sequences(testD)
	MLEN=7
	# Pad the utterance sequences
	pseq_utterance = pad_sequences(sutterance, MLEN,
	padding="post")
	pseq_utterance
	``

	array([[5, 6, 7, 8, 9, 10, 11], [12, 13, 14, 15, 16, 17, 18], [22, 23, 1, 24, 25, 26, 27]])

Transform utterances into a token-IDs sequence for tokens vectorization so that utterances will be ready to feed into neural network

15.10.1 Embedding Words

Tokens can be transformed into token vectors. Embedding tokens into vectors occurred via a lookup embedding table. Each row holds a token vector indexed by token-IDs, hence the flow of obtaining a token vector is as follows:

1. token->token-ID: A token-ID is assigned with each token with Keras'Tokenizer in previous section. Tokenizer holds all vocabularies and maps each vocabulary token to an ID.
2. token-ID->token vector: A token-ID is an integer that can be used as an index to embed table's rows. Each token-ID corresponds to one row and when a token vector is required, first obtain its token-ID and lookup in the embedding table rows with this token-ID.

A sample of embedding words into token vectors is shown in Fig. 15.6.
Remember when a list of utterances began in the previous section:

1. Each utterance is divided into tokens and built a vocabulary with Keras'Tokenizer.
2. The Tokenizer object held a token index with a token->token-ID mapping.
3. When a token-ID is obtained, lookup to embedding table rows with this token-ID to acquire a token vector.

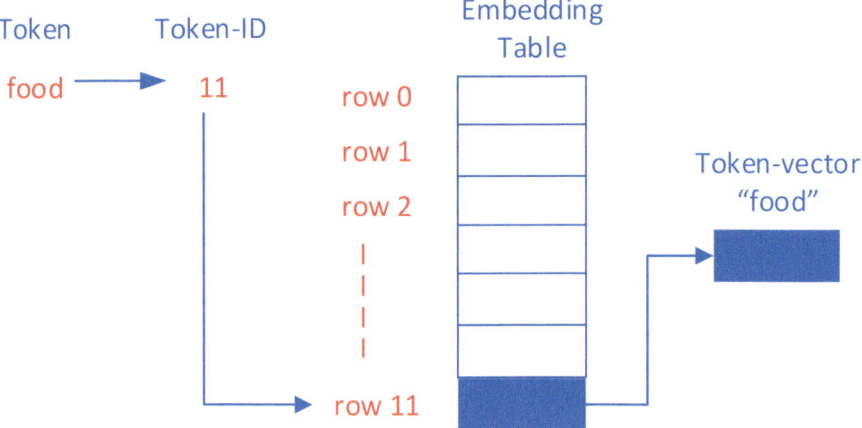

Fig. 15.6 A sample of embedding words into token vectors

4. This token vector is fed to neural network.

There are several steps to transform sentences into vectors as training a neural network is complex.

An LSTM neural network architecture can be designed to perform model training after these preliminary steps.

15.11   Movie Sentiment Analysis with LTSM Using Keras and spaCy

This section will demonstrate the design of LSTM-based RNN text classifier for sentiment analysis with steps below:

1. Data retrieval and preprocessing.
2. Tokenize review utterances with padding.
3. Create utterances pad sequence and put it into input layer.
4. Vectorize each token and verify by token-ID in embedding layer.
5. Input token vectors into LSTM.
6. Train LSTM network.

Let's start by recalling the dataset again.

Step 1: Dataset

IMDB movie reviews identical dataset from sentiment analysis with spaCy section will be used. They had already been processed with *pandas* and condensed into two columns with binary labels.

Reload reviews table and perform data preprocessing as done in previous section to ensure the data is up to date:

```
[61]    import pandas as pd
        import matplotlib.pyplot as plt
        %matplotlib inline
        # Create the movie comment DataFrame and display the
        statistics
        mcommentDF=pd.read_csv('imdb_5000.csv')
        mcommentDF = mcommentDF[['Review','Sentiment']].
        dropna()
        axplot=mcommentDF.Sentiment.value_counts().
        plot(kind='bar', colormap='Paired')
        plt.show()
```

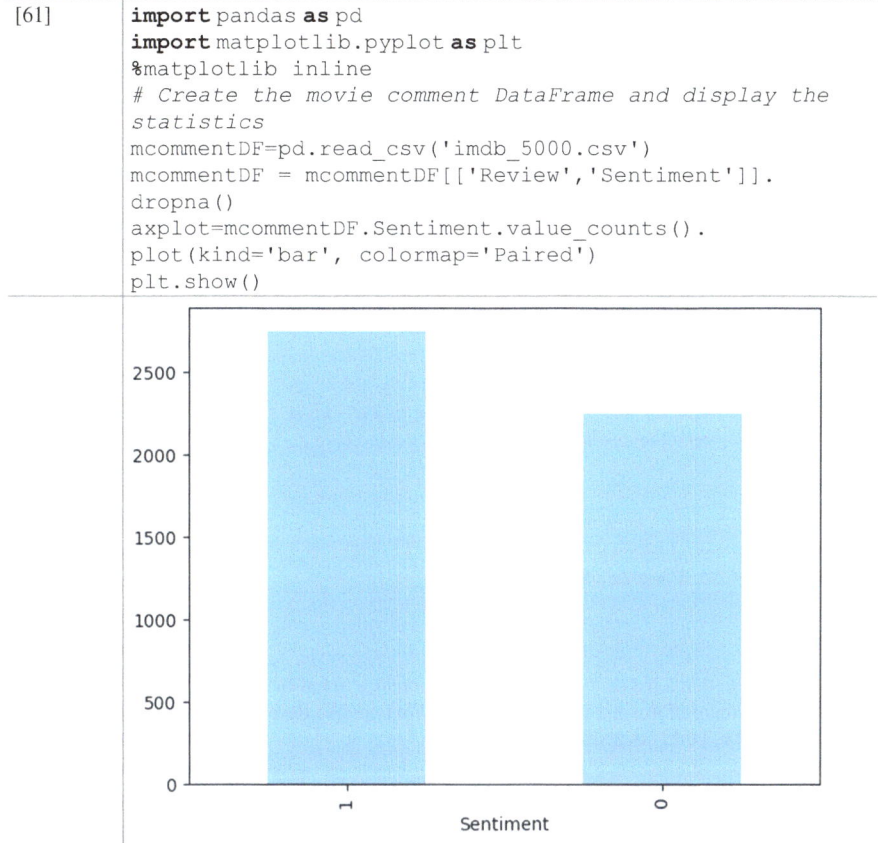

Here is how *mcommentDF* dataset should look:

[62]	mcommentDF.head()	
	Review	**Sentiment**
0	**Possible Spoilers**	0
1	Read the book, forget the movie!	0
2	**Possible Spoilers Ahead**	0
3	What a script, what a story, what a mess!	0
4	I hope this group of film-makers never re-unites.	0

Next, extract review text and review label from each dataset row and add them into Python lists:

[63]
```
# Import spaCy
import spacy
# Load the spaCy NLP model
nlp = spacy.load("en_core_web_md")
```

[64]
```
# Create movie comment sample and categories objects
movie_comment_exp = []
categories = []
# Perform Tokenization
for idx, rw in mcommentDF.iterrows():
    comments = rw["Review"]
    rating = rw["Sentiment"]
    categories.append(rating)
    mtoks = [token.text for token in nlp(comments)]
    movie_comment_exp.append(mtoks)
```

[65]
```
movie_comment_exp[0]
```
['*', '*', 'Possible', 'Spoilers', '*', '*']

Note that a list of words to *movie_comment_exp* has been added, hence each element of this list is a list of tokens. Next, invoke Keras'tokenizer on this tokens list to build vocabulary

Step 2: Data and vocabulary preparation

Since the dataset had already been processed, tokenize dataset sentences and build a vocabulary.

1. Import necessary Python packages.

[66]
```
# Import Tokenizer, pad_sequences
from tensorflow.keras.preprocessing.text import Tokenizer
from tensorflow.keras.preprocessing.sequence import
pad_sequences
import numpy as np
```

2. Feed *ktoken* into token list and convert them into IDs by calling *texts_to_sequences*:

```
[67]   # Create ktoken and perform tokenization
       ktoken = Tokenizer(lower=True)
       ktoken.fit_on_texts(movie_comment_exp)
       # Create utterance sequences object
       seq_utterance = ktoken.texts_to_sequences(movie_
       comment_exp)
```

3. Pad short utterance sequences to a maximum length of 50. This will truncate long reviews to 50 words:

```
[68]   # Set the max length to 50
       MLEN = 50
       # Create pad utterance sequence object
       ps_utterance = pad_sequences(seq_utterance, MLEN,
       padding="post")
```

4. Convert this list of reviews and labels to numpy arrays:

```
[69]   # Convert the ps_utterance into numpy arrays
       ps_utterance = np.array(ps_utterance)
       # Create the category list (catlist)
       catlist = np.array(categories)
```

```
[70]   catlist = catlist.reshape(catlist.shape[0], 1)
```

```
[71]   catlist.shape
       (5000, 1)
```

All basic preparation works are completed at present to create an LSTM network and input data.

Load TensorFlow Keras related modules:

```
[72]   # Import the LSTM model and the optimizers
       from tensorflow.keras.models import Model
       from tensorflow.keras.layers import Input, LSTM, Dense,
       Embedding
       from tensorflow.keras import optimizers
```

Step 3: Implement the Input Layer

```
[73]    utterance_input = Input(shape=(None,))
```

Don't confuse **None** as input shape. Here, *None* means that the dimension can be any scalar number. So, use this expression when Keras infers the input shape

Step 4: Implement the Embedding Layer

Create an Embedding Layer as follows:

```
[74]    # Create the Embedding_Layer
        embedding = Embedding(input_dim = len(ktoken.word_
        index)+1, output_dim = 100)(utterance_input)
```

1. When defining embedding layer, input dimension should always be tokens number in the vocabulary (+1 because the indices start from 1 and not 0. Index 0 is reserved for padding value).
2. Here, 100 is selected as the output shape, hence token vectors for vocabulary tokens will be 100-dimensional. Popular numbers for token vector dimensions are 50, 100, and 200 depending on task complexity.

15.11.1   Step 5: Implement the LSTM Layer

Create LSTM_Layer:

```
[75]    # Create the LSTM_Layer
        LSTM_layer = LSTM(units=256)(embedding)
```

Here, *units* = 256 is the dimension of hidden state. LSTM output shape and hidden state shape are identical due to LSTM architecture.

15.11.2   Step 6: Implement the Output Layer

When a 256-dimensional vector from LSTM layer has obtained, it will be condensed into a 1-dimensional vector (possible values of this vector are 0 and 1, which are class labels):

```
[76]    # Create the Output Layer
        outlayer = Dense(1, activation='sigmoid')(LSTM_layer)
```

A sigmoid function is an S-shaped function used as an activation function to map its input to a [0–1] range in output layer. It is commonly used in many neural networks

15.11.3   Step 7: System Compilation

After the model has defined, it is required to compile with an optimizer, a loss function, and an evaluation metric:

```
[77]    # Create the IMDB User Review LSTM Model (imdb_mdl)
        imdb_mdl = Model(inputs=[utterance_
        input],outputs=[outlayer])
```

Let's look at an *imdb_mdl* model setup:

```
[78]    imdb_mdl.summary()
```

Model: "functional"

Layer (type)	Output Shape	Param #
input_layer (InputLayer)	(None, None)	0
embedding (Embedding)	(None, None, 100)	1,874,400
lstm (LSTM)	(None, 256)	365,568
dense (Dense)	(None, 1)	257

Total params: 2,240,225 (8.55 MB)
Trainable params: 2,240,225 (8.55 MB)
Non-trainable params: 0 (0.00 B)

Next, invoke model compilation:

```
[79]    imdb_mdl.compile(optimizer="adam", loss="binary_
        crossentropy", metrics=["accuracy"])
```

 1. Use adaptive moment estimation (ADAM) as optimizer for LSTM training for imdb_mdl LSTM model.
2. Use binary cross-entropy as loss function.
3. A list of supported performance metrics can be found in Keras official site (Keras 2024).

Step 8: Model Fitting and Experiment Evaluation

Feed *imdb_mdl* model to data with 5 epochs to reduce time:

```
[80]    # Model fitting by using 5 epochs
        imdb_mdl.fit(x=ps_utterance,
             y=catlist,
             batch_size=64,
             epochs=5,
             validation_split=0.3)
```

Epoch 1/5
55/55 _____**5s** 63ms/step - accuracy: 0.5956 - loss: 0.6616 - val_accuracy: 0.7520 - val_loss: 0.5091
Epoch 2/5
55/55 _____ **3s** 60ms/step - accuracy: 0.8932 - loss: 0.2779 - val_accuracy: 0.7907 - val_loss: 0.4508
Epoch 3/5
55/55_____ **3s** 60ms/step - accuracy: 0.9634 - loss: 0.1304 - val_accuracy: 0.7953 - val_loss: 0.4469
Epoch 4/5
55/55 _____ **3s** 61ms/step - accuracy: 0.9791 - loss: 0.0845 - val_accuracy: 0.7587 - val_loss: 0.7780
Epoch 5/5
55/55_____**3s** 61ms/step - accuracy: 0.9955 - loss: 0.0204 - val_accuracy: 0.7507 - val_loss: 0.9288
[51]:
<keras.src.callbacks.history.History at 0x1e7345f4690>

1. x is a list of *ps_utterance* for network training and y is the list of categories (catlist). The epochs parameter is set to 5 to process 5 passes over the data,
2. Data has been processed 5 times in parameter batch_size = 64 means a batch of 64 training utterances are fed into the memory at a time due to memory limitations.
3. The *validation_split* = 0.3 means 70% of the dataset is used for training and 30% is used for system validation.
4. An experiment validation accuracy rate of 0.7793 is acceptable for a basic LSTM network training for 5 epochs only.

WORKSHOP

Workshop 5.2 Further Exploration of LSTM model on Movie Sentiment Analysis
1. Follow Workshop 15.1 logic and use rating (0–10) field of IMDB movie reviews dataset to reconstruct an LSTM for sentiment analysis into three categories: Positive, neutral, and negative.
2. Verify training performance.
3. Experiment with the code further by placing dropout layers at different locations such as after embedding layer or, after LSTM layer.
4. Try different values for embedding dimensions such as 50, 150, and 200 to observe change in accuracy.
5. Experiment with different values instead of 256 at LSTM layer's hidden dimension. Try different parameters for each to perform simulations and see whether the best configuration can be found.

References

Agarwal, B. (2020) Deep Learning-Based Approaches for Sentiment Analysis (Algorithms for Intelligent Systems). Springer.

Albrecht, J., Ramachandran, S. and Winkler, C. (2020) Blueprints for Text Analytics Using Python: Machine Learning-Based Solutions for Common Real World (NLP) Applications. O'Reilly Media.

Altinok, D. (2021) Mastering spaCy: An end-to-end practical guide to implementing NLP applications using the Python ecosystem. Packt Publishing.

Arumugam, R., & Shanmugamani, R. (2018). Hands-on natural language processing with python. Packt Publishing.

Bird, S., Klein, E., and Loper, E. (2009). Natural language processing with python. O'Reilly.

Ekman, M. (2021) Learning Deep Learning: Theory and Practice of Neural Networks, Computer Vision, Natural Language Processing, and Transformers Using TensorFlow. Addison-Wesley Professional.

George, A. (2022) Python Text Mining: Perform Text Processing, Word Embedding, Text Classification and Machine Translation. BPB Publications.

Géron, A. (2019) Hands-On Machine Learning with Scikit-Learn, Keras, and TensorFlow: Concepts, Tools, and Techniques to Build Intelligent Systems. O'Reilly Media.

IMDB (2024) IMDB official site. http://imdb.com. Accessed 17 Dec 2024.

Kaggle (2024) IMDB Large Movie Review Dataset from Kaggle. https://www.kaggle.com/code/nisargchodavadiya/movie-review-analytics-sentiment-ratings/data. Accessed 17 Dec 2024.

Kedia, A. and Rasu, M. (2020) Hands-On Python Natural Language Processing: Explore tools and techniques to analyze and process text with a view to building real-world NLP applications. Packt Publishing.

Keras (2024) Keras official site performance metrics. https://keras.io/api/metrics. Accessed 17 Dec 2024.

Korstanje, J. (2021) Advanced Forecasting with Python: With State-of-the-Art-Models Including LSTMs, Facebook's Prophet, and Amazon's DeepAR. Apress.

NLPGitHub (2024) URL: https://github.com/raymondshtlee/NLP/. Accessed 17 Dec 2024.

Perkins, J. (2014). Python 3 text processing with NLTK 3 cookbook. Packt Publishing Ltd.

Pozzi, F., Fersini, E., Messina, E. and Liu, B. (2016) Sentiment Analysis in Social Networks. Morgan Kaufmann.

SpaCy (2024) spaCy official site. https://spacy.io/. Accessed 17 Dec 2024.

Sarkar, D. (2019) Text Analytics with Python: A Practitioner's Guide to Natural Language Processing. Apress.

Siahaan, V. and Sianipar, R. H. (2022) Text Processing and Sentiment Analysis using Machine Learning and Deep Learning with Python GUI. Balige Publishing.

Srinivasa-Desikan, B. (2018). Natural language processing and computational linguistics: A practical guide to text analysis with python, gensim, SpaCy, and keras. Packt Publishing, Limited.

TensorFlow (2024) TensorFlow official site. https://tensorflow.org/. Accessed 17 Dec 2024.

Vasiliev, Y. (2020) Natural Language Processing with Python and spaCy: A Practical Introduction. No Starch Press.

Chapter 16
Workshop#6 Transformers with spaCy and TensorFlow (Hour 11–12)

16.1 Introduction

In Chap. 8, the basic concept about transfer learning, its motivation, and related background knowledge such as *RNN* with *Transformer Technology* and *BERT model* are introduced.

This workshop will learn about the latest topic *Transformers* in NLP, and how to use them with TensorFlow and spaCy. First, will learn about Transformers and Transfer learning. Second, will learn about a commonly used Transformer architecture—Bidirectional *Encoder Representations from Transformers (BERT)* as well as how BERT Tokenizer and WordPiece algorithms work.

Further, will learn how to start with pre-trained transformer models of HuggingFace library (HuggingFace 2024) and practice to fine-tune HuggingFace Transformers with TensorFlow and Keras (TensorFlow 2024; Keras 2024) followed by how spaCy v3.0 (spaCy 2024) integrates transformer models as pre-trained pipelines. These techniques and tools will be used in the last workshop for building a Q&A chatbot.

Hence, this workshop will cover the following topics:

- Transformers and Transfer Learning
- Understanding BERT
- Transformers and TensorFlow
- Transformers and spaCy

© The Author(s), under exclusive license to Springer Nature Singapore Pte Ltd. 2025
R. Lee, *Natural Language Processing*,
https://doi.org/10.1007/978-981-96-3208-4_16

16.2   Technical Requirements

In this workshop, Transformers, TensorFlow, and spaCy (TensorFlow 2024; spaCy 2024) are to be installed in own PC/notebook computer. Please ensure that the following Python packages are installed before starting the workshop:

- Python (demo version 3.11.9)
- spacy (demo version 3.4.4)
- keras (demo version 3.3.3)
- transformers (demo version 4.44.2)
- tensorflow (demo version 2.17.0)
- tf-keras (demo version 2.17.0)
- torch (demo version 2.4.1)
- torchvision (demo version 0.19.1)
- spacy-transformers (demo version 1.3.5)
- numPy (demo version 1.26.4)
- pandas (demo version 2.2.2)
- matplotlib (demo version 3.9.2)

If these packages are not installed on PC/laptop, use pip install xxx command. The detailed requirements list and Python package version used in this workshop can be found in the requirements.txt file stored in the NLP GitHub repository (NLPGitHub 2024).

16.3   Transformers and Transfer Learning in a Nutshell

Transformer in NLP is an innovative idea which aims to solve sequential modeling tasks and target problems introduced by Long-Short-Term-Memory (LSTM) architecture (Ekman 2021; Korstanje 2021).

It is a contemporary machine learning concept and architecture introduced by Vaswani et al. (2017) in a research paper *Attention Is All You Need*. It explained that "The Transformer is the first transduction model relying entirely on self-attention to compute representations of its input and output without using sequence-aligned RNNs or convolution."

Transduction in this context means transforming input words to output words by transforming input words and sentences into vectors. A transformer is trained on a large corpus such as Wiki or news. These vectors will be used to convey information regarding word semantics, sentence structures, and sentence semantics for downstream tasks.

Word vectors like Glove and FastText are already trained on Wikipedia corpus that can be used in semantic similarity calculations, hence, Transfer Learning means to import knowledge from pre-trained word vectors or pre-trained statistical models.

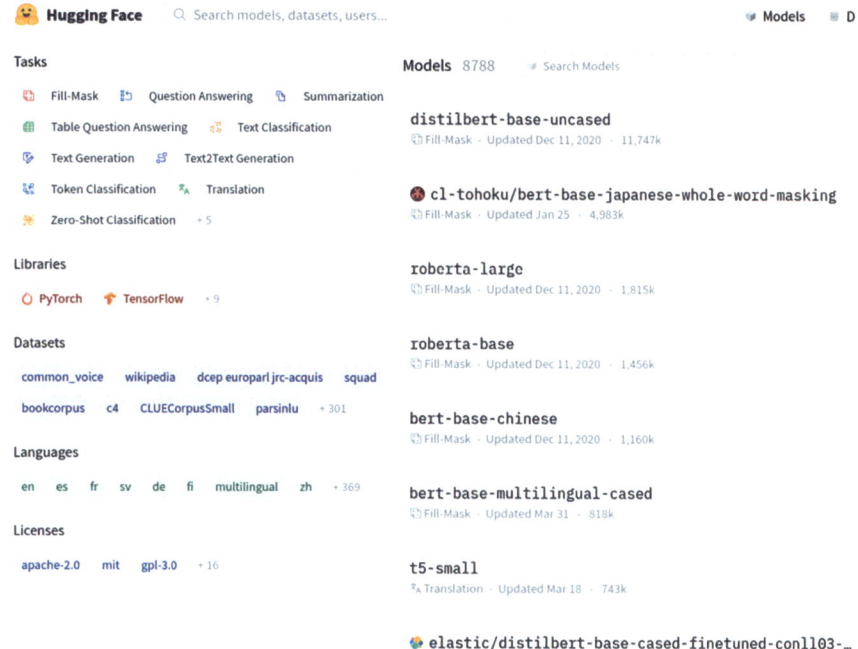

Fig. 16.1 Sample Input Texts and their corresponding Output Class Labels

Transformers offer many pre-trained models to perform NLP tasks such as text classification, text summarization, question answering, machine translation, and natural language generation in over 100 languages. It aims to make state-of-the-art NLP accessible to everyone (Bansal 2021; Rothman 2022; Tunstall et al. 2022; Yıldırım and Asgari-Chenaghlu 2021).

A list of Transformer models provided by HuggingFace (2024) is shown in Fig. 16.1. Each model is named with a combination of architecture names such as BERT or DistilBert, possibly a language code, i.e., en, de, multilingual, which is located at the left side of the figure, and information regarding whether the model is cased or uncased, i.e., distinguish between uppercase and lowercase characters.

Task names are also listed on the left-hand side. Each model is labeled with a task name such as text classification or machine translation for the Q&A chatbot.

16.4   Why Transformers?

Let's review text classification with spaCy in LSTM architecture.

LSTMs work for modeling text effectively, but there are shortcomings:

- LSTM architecture has difficulties in learning long texts sometimes. Statistical dependencies in a long text have problems represented by LSTM because it can fail to recall words processed earlier as time steps progress.
- LSTMs are sequential which means that a single word can process at each time step but is impossible to parallelize learning process causing bottleneck.

Transformers address these problems by not using recurrent layers at all; their architecture is different from LSTM architecture (Bansal 2021; Rothman 2022; Tunstall et al. 2022; Yıldırım and Asgari-Chenaghlu 2021). A Transformer architecture has an input encoder block at the left, called encoder, and an output decoder at the right, called decoder as shown in Fig. 16.2.

The architecture is catered for a machine translation task, input is a sequence of words from source language, and output is a sequence of words in the target language. Encoder generates a vector representation of input words and passes them to decoder where the word vector transfer is represented by an arrow from encoder

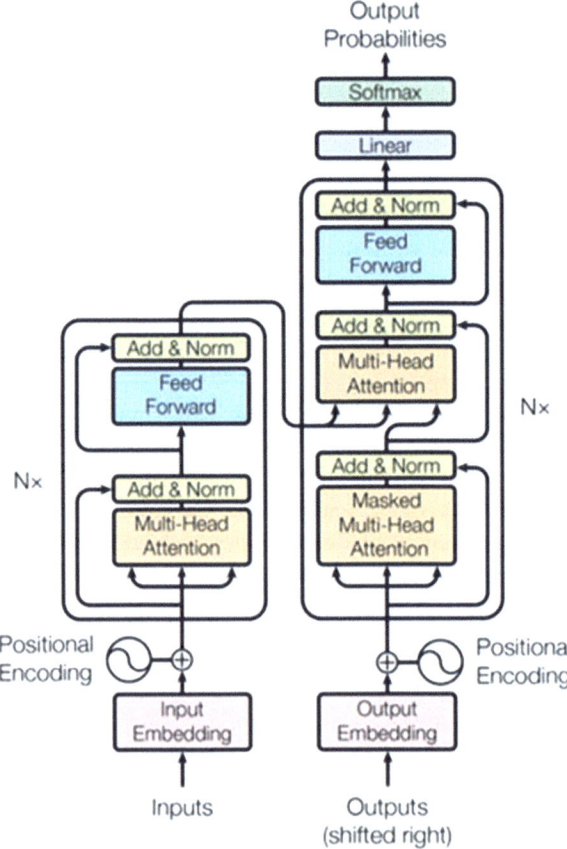

Fig. 16.2 Transformer architecture

block to decoder block direction. The decoder extracts input word vectors, transforms output words into word vectors, and generates the probability of each output word.

There are feedforward layers, which are dense layers in encoder and decoder blocks used for text classification with spaCy. The innovative transformers can place in a Multi-Head Attention block to create a dense representation for each word with self-attention mechanism. This mechanism relates each word in input sentence to other words in the input sentence. Word embedding is calculated by taking a weighted average of other words' embeddings, and each word significance can be calculated in input sentence to enable the architecture focus on each input word sequentially.

A self-attention mechanism of how input words at the left-hand side attend input word *it* at the right-hand side is shown in Fig. 16.3. Dark colors represented

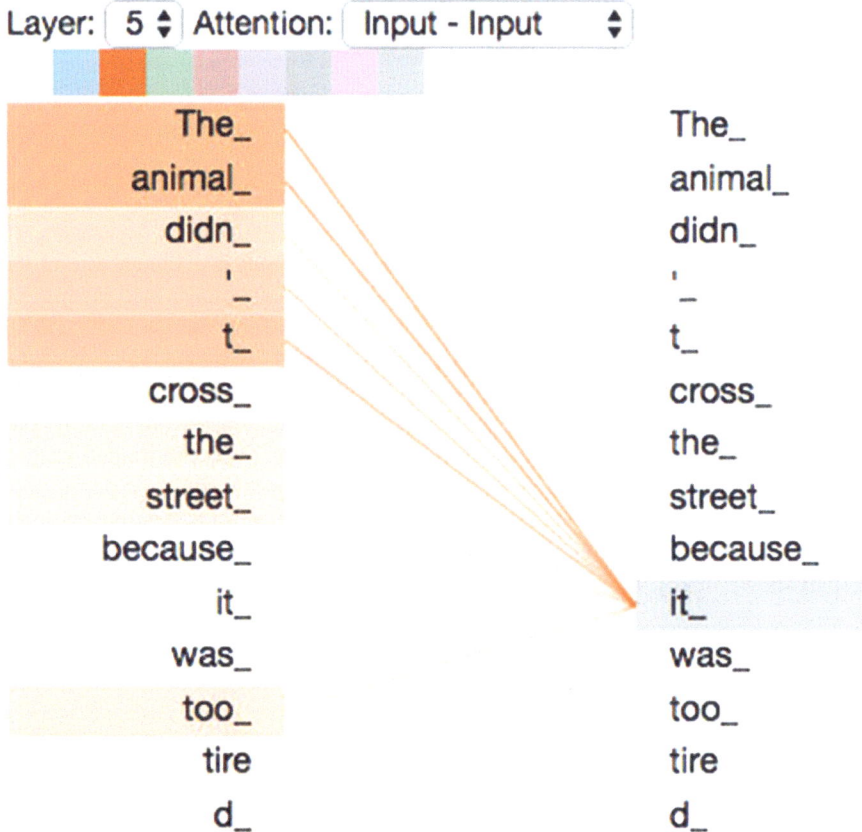

Fig. 16.3 Illustration of the self-attention mechanism

relevance, phrase *the animal* are related to *it* than other words in the sentence. This signified transformer can resolve many semantic dependencies in a sentence and is used in different tasks such as text classification and machine translation since they have several architectures depending on tasks. BERT is a popular architecture to be used.

16.5   An Overview of BERT Technology

16.5.1   What Is BERT?

BERT is introduced in a Google's original research paper published by Devlin et al. (2019), the complete Google BERT model can be downloaded from Google's GitHub archive (GoogleBert 2024).

It has the following output features (Bansal 2021; Rothman 2022; Tunstall et al. 2022; Yıldırım and Asgari-Chenaghlu 2021):

- Bidirectional: Each input sentence text data training is processed from left to right and from right to left.
- Encoder: An encoder encodes input sentence.
- Representations: A representation is a word vector.
- Transformers: A transformer-based architecture.

BERT is a trained transformer encoder stack. The input is a sentence, and the output is a sequence of word vectors. Word vectors are contextual which means that a word vector is assigned to a word based on an input sentence. In short, BERT outputs contextual word representations as shown in Fig. 16.4.

It is noted that word *bank* has different meanings in these two sentences, word vectors are the same because Glove and FastText are static. Each word has only one vector and vectors are saved to a file after training. Then, these pre-trained vectors can be downloaded to our application. BERT word vectors are dynamic on the contrary. It can generate different word vectors for the same word depending on input sentence. Word vectors generated by BERT are shown in Fig. 16.5 against the counterpart shown in Fig. 16.4.

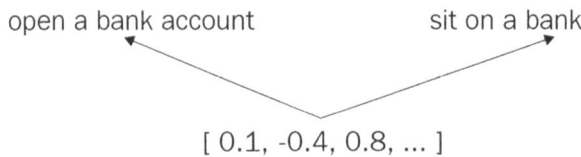

Fig. 16.4 Word vector for the word "bank"

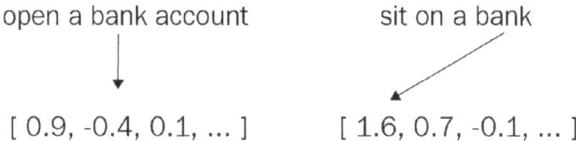

Fig. 16.5 Two distinct word vectors generated by BERT for the same word *bank* in two different contexts

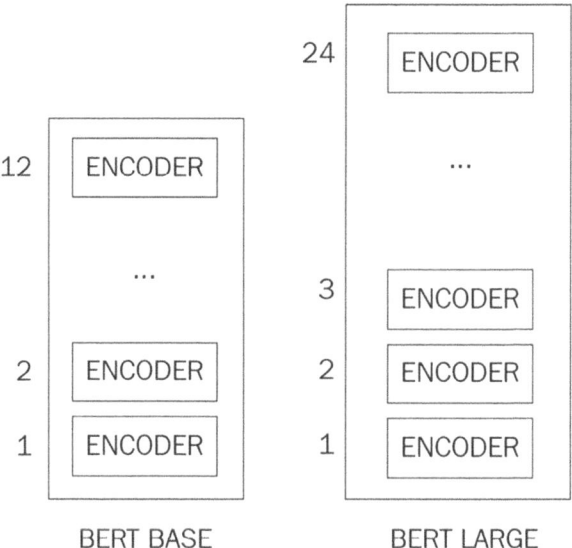

Fig. 16.6   BERT Base and Large architectures (having 12 and 24 encoder layers respectively)

16.5.2   BERT Architecture

BERT is a transformer encoder stack, which means several encoder layers are stacked on top of each other. The first layer initializes word vectors randomly, and then each encoder layer transforms output of the previous encoder layer. Figure 16.6 illustrates two BERT model sizes: BERT Base and BERT Large.

BERT Base and BERT Large have 12 and 24 encoder layers to generate word vectors sizes of 768 and 1024 comparatively.

BERT outputs word vectors for each input word. A high-level overview of BERT inputs and outputs is illustrated in Fig. 16.7. It showed that BERT input should be in a special format to include special tokens such as CLS.

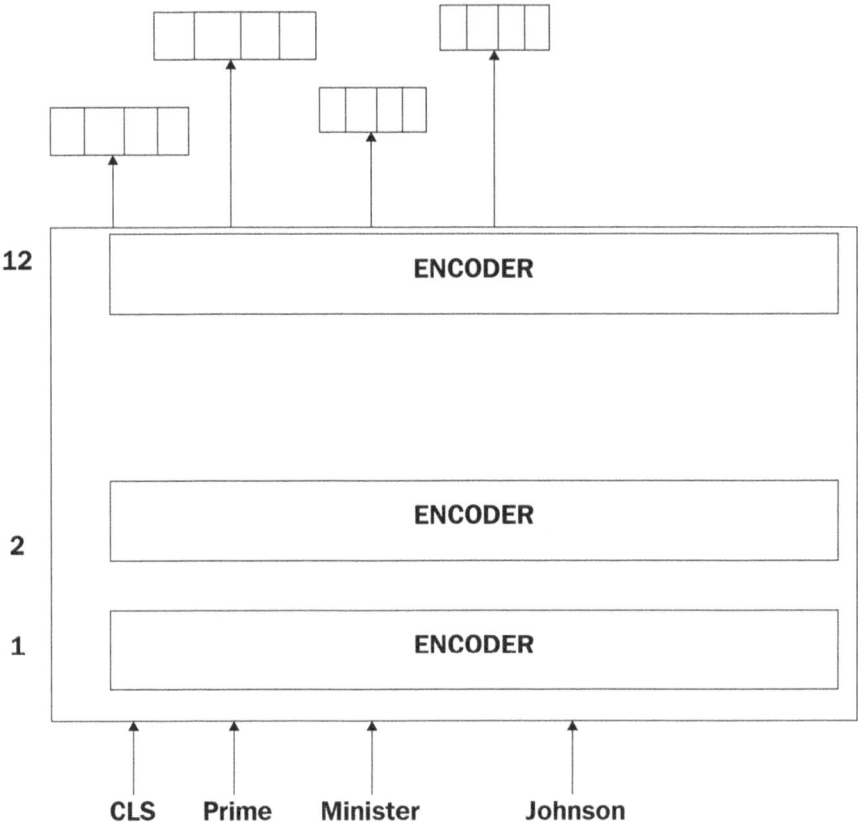

Fig. 16.7 BERT model input word and output word vectors

16.5.3 BERT Input Format

After learning BERT basic architecture, let's look at how to generate output vectors using BERT.

BERT input format can represent a single sentence and a pair of sentences in a single sequence of tokens (for tasks such as question answering and semantic similarity, we input two sentences to the model).

BERT works with a special tokens class and a special tokenization algorithm called WordPiece.

There are several types of special tokens [CLS], [SEP], and [PAD]:

- [CLS] is the first special token type for every input sequence. This token is a quantity of input sentences for classification tasks but disregard non-classification tasks.

- [SEP] is a sentence separator. If the input is a single sentence, this token will be placed at the end of sentence, i.e., [CLS] sentence [SEP], or to separate two sentences, i.e., [CLS] sentence1 [SEP] sentence2 [SEP].
- [PAD] is a special token for padding. The padding values can generate sentences from dataset with equal length. BERT receives sentences with fixed length only, hence, padding short sentences is required prior feeding to BERT. The tokens maximum length can feed to BERT is 512.

It was learnt that a sentence can feed to Keras model one word at a time, input sentences can be tokenized into words using spaCy tokenizer, but BERT works differently as it uses WordPiece tokenization. A *word piece* is literally a piece of a word.

WordPiece algorithm breaks down words into several subwords, its logic behind is to break down complex/long tokens into tokens, e.g., the word *playing* is tokenized as play + ##ing. A ## character is placed before every word piece to indicate that this token is not a word from language's vocabulary but is a word piece.

Let's look at some examples:

playing play, ##ing

played play, ##ed

going go, ##ing

vocabulary = [play,go, ##ing, ##ed]

It can concise language vocabulary as WordPiece groups common subwords.

WordPiece tokenization can divide rare/unseen words into their subwords.

After input sentence is tokenized and special tokens are added, each token is converted to its ID and feed token ID sequences to BERT.

An input sentence transformed into BERT input format is illustrated in Fig. 16.8.

BERT Tokenizer has several methods to perform above tasks, but it has an encoding method that combines these steps into a single step.

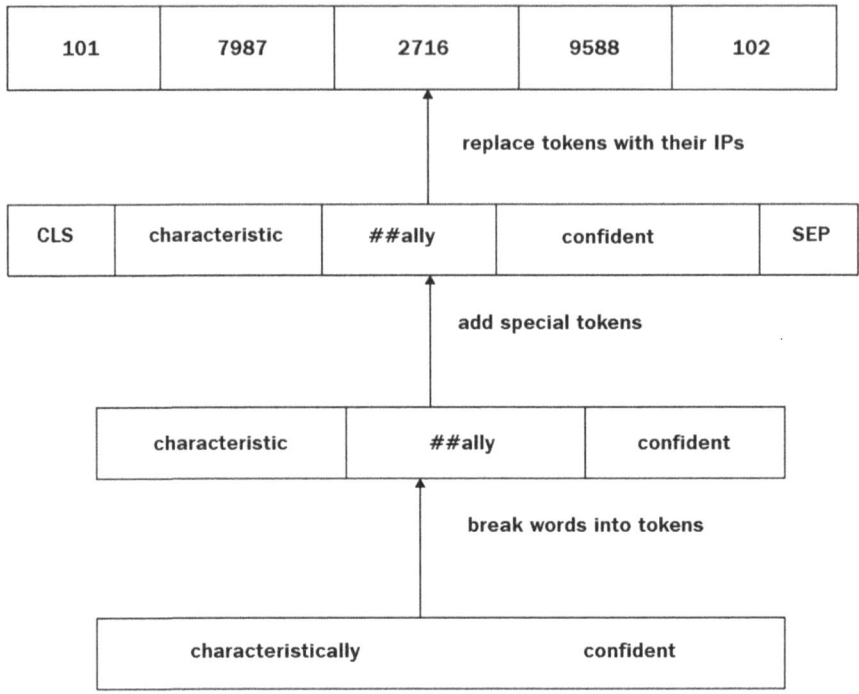

Fig. 16.8 Transforming an input sentence into BERT input format

16.5.4 How to Train BERT?

BERT originators stated that "We then train a large model (12-layers to 24-layers Transformer) on a large corpus (Wikipedia + BookCorpus) for a long time (1 M update steps), and that's BERT." in Google Research's BERT GitHub repository (GoogleBert 2024).

BERT is trained by masked language model (MLM) and NSP.

Language modeling is the task of predicting the next token given the sequence of previous tokens. For example, given the sequence of words *Yesterday I visited*, a language model can predict the next token as one of the tokens *church*, *hospital*, *school*, and so on.

MLM is different. A percentage of tokens are masked randomly to replace a [MASK] token and presume MLM predicts the masked words.

BERT's MLM is implemented as follows:

1. Select 15 input tokens randomly.
2. About 80% of selected tokens are replaced by [MASK].
3. About 10% of selected tokens are replaced by another token from vocabulary.
4. About 10% remain unchanged.

A training sentence to LMM example is as follows:

[CLS] Yesterday I [MASK] my friend at [MASK] house [SEP]

NSP is the task of predicting the next sentence given by an input sentence. There are two sentences fed to BERT and presume BERT predicts sentences order if second sentence is followed by first sentence.

An input of two sentences separated by a [SEP] token to NSP example is as follows:

[CLS] A man robbed a [MASK] yesterday [MASK] 8 o'clock [SEP] He [MASK] the bank with 6 million dollars [SEP]

Label = IsNext

It showed that the second sentence can follow the first sentence; hence, the predicted label is IsNext.

Here is another example:

[CLS] Rabbits like to [MASK] carrots and [MASK] leaves [SEP] [MASK] Schwarzenegger is elected as the governor of [MASK] [SEP]

Label= NotNext

This example showed that the pair of sentences generate a *NotNext* label without contextual or semantical relevance.

16.6 Transformers with TensorFlow

Pre-trained transformer models are provided to program developers in open sources by many organizations including Google (GoogleBert 2024), Facebook (Facebook-transformer 2024), and HuggingFace (HuggingFace-transformer 2024).

HuggingFace is an AI company that focuses on NLP apportioned to open source.

These pre-trained models and agreeable interfaces can integrate transformers into Python code, as interfaces are compatible with either PyTorch or TensorFlow or both. HuggingFace's pre-trained transformers and their TensorFlow interface to transformer models will be used in this workshop.

16.6.1   HuggingFace Transformers

This section will explore HuggingFace's pre-trained models, TensorFlow interface, and its conventions. HuggingFace offers several models as in Fig. 16.1. Each model is dedicated to tasks such as text classification, question answering, and sequence-to-sequence modeling.

A HuggingFace documentation of a distilbert-base-uncased-distilled-squad model is shown in Fig. 16.9. A Question Answering task tag is assigned to the upper left corner in the documentation followed by supporting deep learning libraries PyTorch, TensorFlow, TFLite, TFSavedModel, training dataset, e.g., squad, model language, e.g., en for English; the license and base model's name, e.g., DistilBERT.

Some models are trained with similar algorithms that belong to an identical model family. For example, the DistilBERT family has many models such as distilbert-base-uncased and distilbert-multilingual-cased. Each model name includes information such as casing to distinguish uppercase/lowercase or model language such as en, de, or multilingual.

HuggingFace documentation provides information about each model family with individual model's API in detail. Lists of available models and BERT model architecture variations are shown in Fig. 16.10.

BERT model has many task variations such as text classification, question answering, and NSP.

Each of these models is obtained by placing extra layers atop of BERT output as these outputs are a sequence of word vectors for each word of input sentences.

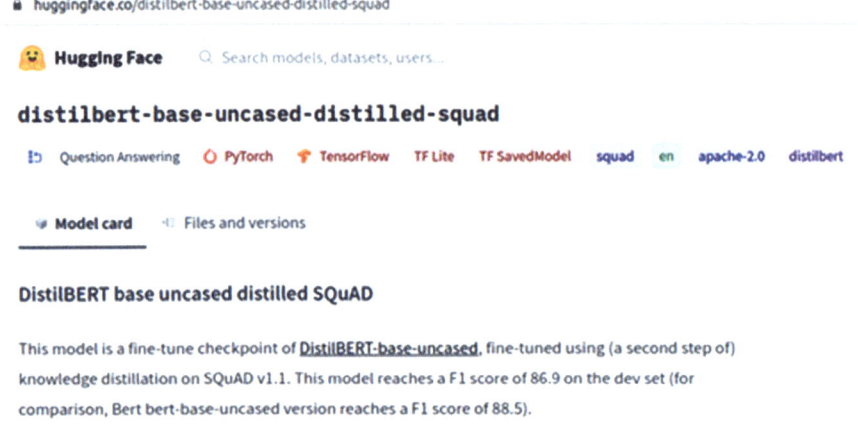

Fig. 16.9 Documentation of the distilbert-base-uncased-distilled-squad model

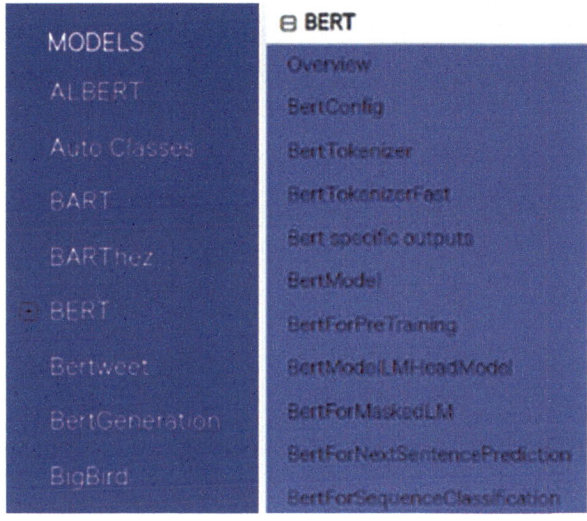

Fig. 16.10 Lists of the available models (left-hand side) and BERT model variations (right-hand side)

For example, a BERTForSequenceClassification model is obtained by placing a dense layer atop of BERT word vectors.

16.6.2   Using the BERT Tokenizer

BERT uses the WordPiece algorithm for tokenization to ensure that each input word is divided into subwords.

Let's look at how to prepare input data with HuggingFace library.

```
[1]    # Import transformer package
       from transformers import BertTokenizer
       # Create bert_tokenizer and sample utterance (utt1) and
       tokens (tok1)
       btokenizer
       = BertTokenizer.from_pretrained('bert-base-uncased')
       utt1 = "He lived characteristically idle and romantic."
       utt1 = "[CLS] " + utt1 + " [SEP]"
       tok1 = btokenizer.tokenize(utt1)
```

```
[2]    # Display bert tokens
       tok1
```

```
['[CLS]',
'he',
'lived',
'characteristic',
'##ally',
'idle',
'and',
'romantic',
'.',
'[SEP]']
```

[3]
```
# Convert bert tokens to ids (id1)
id1 = btokenizer.convert_tokens_to_ids(tok1)
id1
```

[101, 2002, 2973, 8281, 3973, 18,373, 1998, 6298, 1012, 102]

1. Import BertTokenizer. Note that different models have different tokenizers, e.g., XLNet model's tokenizer is called XLNetTokenizer.
2. Call from_pretrained method on tokenizer object and provide model's name. Needless to download pre-trained bert-base-uncased (or model) as this method downloads model by itself.
3. Call tokenize method. It tokenizes sentences by dividing all words into subwords.
4. Print tokens to examine subwords. The words *he*, *lived*, *idle*, exist in Tokenizer's vocabulary are to be remained. *Characteristically* is a rare word that does not exist in Tokenizer's vocabulary. Thus, tokenizer splits this word into subwords *characteristic* and *##ally*. Notice that *##ally* starts with characters *##* to emphasize that this is a piece of word.
5. Call convert_tokens_to_ids.

Since [CLS] and [SEP] tokens must add to the beginning and end of input sentence, it is required to add them manually for the preceding code, but these preprocessing steps can perform in a single step.

BERT provides a method called encode that can:

- add CLS and SEP tokens to input sentence
- tokenize sentence by dividing tokens into subwords
- converts tokens to their token IDs

Call encode method on input sentence directly as follows:

| [4] | ```
from transformers import BertTokenizer
btokenizer
= BertTokenizer.from_pretrained('bert-base-uncased')
utt2 = "He lived characteristically idle and romantic."
id2 = btokenizer.encode(utt2)
print(id2)
``` |
|---|---|
| | [101, 2002, 2973, 8281, 3973, 18,373, 1998, 6298, 1012, 102] |

This code segment outputs token IDs in a single step instead of step-by-step. The result is a python list

Since all input sentences in a dataset must have equal length because BERT cannot process variable-length sentences, padding the longest sentence from dataset into short sentences is required using the parameter "padding='longest'".

Writeup conversion codes are also required if a TensorFlow tensor is used instead of a plain list. HuggingFace library provides *encode_plus* to combine all these steps into the single method as follows:

| [5] | ```
from transformers import BertTokenizer
                     btokenizer
= BertTokenizer.from_pretrained('bert-base-uncased')
utt3 = "He lived characteristically idle and romantic."
encoded = btokenizer.encode_plus(
                    text=utt3,
        add_special_tokens=True,
            padding='longest',
            return_tensors="tf"
                           )
    id3 = encoded["input_ids"]
                print(id3)
``` |
|---|---|
| | tf.Tensor([[101 2002 2973 8281 3973 18373 1998 6298 1012 102]], shape=(1, 10), dtype=int32) |

Call *encode_plus* to input sentence directly. It is padded to a length of 10 including special tokens [CLS] and [SEP]. The output is a direct TensorFlow tensor with token IDs.

Verify parameter list of *encode_plus( )* by:

[6]	`btokenizer.encode_plus?`

	Signature:
	btokenizer.encode_plus(
	    text: Union[str, List[str], List[int]],
	    text_pair: Union[str, List[str], List[int], NoneType] = **None,**
	    add_special_tokens: bool = **True,**
	    padding: Union[bool, str, transformers.utils.generic.PaddingStrategy] = **False,**
	    truncation: Union[bool, str, transformers.tokenization_utils_
	  base.TruncationStrategy] = **None,**
	    max_length: Optional[int] = **None,**
	    stride: int = **0,**
	    is_split_into_words: bool = **False,**
	    pad_to_multiple_of: Optional[int] = **None,**
	    return_tensors: Union[str, transformers.utils.generic.TensorType, NoneType]
	  = **None,**
	    return_token_type_ids: Optional[bool] = **None,**
	    return_attention_mask: Optional[bool] = **None,**
	    return_overflowing_tokens: bool = **False,**
	    return_special_tokens_mask: bool = **False,**
	    return_offsets_mapping: bool = **False,**
	    return_length: bool = **False,**
	    verbose: bool = **True,**
	    **kwargs,
	) -> transformers.tokenization_utils_base.BatchEncoding
	…

BERT tokenizer provides several methods on input sentences. Data preparation is not straightforward, but practice makes perfect. Try out code examples with own text

It is ready to process transformed input sentences to BERT model and obtain BERT word vectors.

16.6.3 Word Vectors in BERT

This section will examine BERT model output as they are a sequence of word vectors assigned by one vector per input word. BERT has a special output format. Let's look at the code first.

```
[7]    from transformers import BertTokenizer, TFBertModel
       btokenizer
       = BertTokenizer.from_pretrained('bert-base-uncased')
       bmodel
       = TFBertModel.from_pretrained("bert-base-uncased")
       utt4 = "He was idle."
       encoded = btokenizer.encode_plus(
           text=utt4,
           add_special_tokens=True,
           padding='longest',
           max_length=10,
           return_attention_mask=True,
           return_tensors="tf"
       )
       id4 = encoded["input_ids"]
       outputs = bmodel(id4)
```

- Import TFBertModel.
- Initialize BERT model with a BERT-base-uncased pre-trained model.
- Transform input sentence to BERT input format with encode_plus, and capture result tf.Tensor in the input variable.
- Feed sentence to BERT model and capture output with the output's variables.

BERT model output is a tuple of two elements. Let's print the shapes of output pair:

```
[8]    print(outputs[0].shape)
       (1, 6, 768)
```

```
[9]    print(outputs[1].shape)
       (1, 768)
```

1. Shape, i.e., batch size, sequence length, and hidden size is the first element of output. A batch size is the number of sentences that can feed to model instantly. When one sentence is fed, the batch size is 1. The sequence length is 10 because sentence is fed max_length=10 to the tokenizer and padded to length of 10. hidden_size is a BERT parameter. BERT architecture has 768 hidden layers size to produce word vectors with 768 dimensions. Hence, the first output element contains 768-dimensional vectors per word means it contains 10 words x 768-dimensional vectors.

2. The second output is only one vector of 768-dimension. This vector is the word embedding of [CLS] token. Since [CLS] token is an aggregate of the whole sentence, this token embedding is regarded as embeddings pooled version of all words in the sentence. The shape of output tuple is always the batch size, hidden_size. It is to collect [CLS] token's embedding per input sentence basically.

When BERT embeddings are extracted, they can be used to train text classification model with TensorFlow and tf.keras.

16.7   Revisit Text Classification Using BERT

Some of the codes will be used from the previous workshop, but this time the code is shorter because the embedding and LSTM layers will be replaced by BERT to train a binary text classifier and tf.keras.

This section will use an email log dataset *emails.csv* for spam mail classification found in the NLP Workshop6 GitHub repository (NLPGitHub 2024).

16.7.1   Data Preparation

Before Text Classification model using BERT is created, let's prepare the data first just like being learnt in the previous workshop:

1. Import related modules:

```
[10]    import pandas as pd
        import numpy as np
        import tensorflow
        from tensorflow.keras.layers import Dense, Input
        from tensorflow.keras.models import Model
```

2. Read eamils.csv datafile.

```
[11]    emails=pd.read_csv("emails.csv",encoding='ISO-8859-1')
        emails.head()
```

		text	spam
0	Subject: naturally irresistible your corporate...		1
1	Subject: the stock trading gunslinger fanny i...		1
2	Subject: unbelievable new homes made easy im ...		1
3	Subject: 4 color printing special request add...		1
4	Subject: do not have money , get software cds ...		1

3. Use dropna() to remove records with missing contents.

[12]
```
emails=emails.dropna()
emails=emails.reset_index(drop=True)
emails.columns = ['text','label']
emails.head()
```

		text	label
0	Subject: naturally irresistible your corporate...		1
1	Subject: the stock trading gunslinger fanny i...		1
2	Subject: unbelievable new homes made easy im ...		1
3	Subject: 4 color printing special request add...		1
4	Subject: do not have money , get software cds ...		1

16.7.2   Start the BERT Model Construction

4. Import BERT models and tokenizer:

[13]
```
from transformers import BertTokenizer, TFBertModel,
BertConfig, TFBertForSequenceClassification
bert_tokenizer = BertTokenizer.from_pretrained
("bert-base-uncased")
bmodel
= TFBertModel.from_pretrained("bert-base-uncased")
```

1. Import BertTokenizer and BERT model, TFBertModel.
2. Initialize both tokenizer and BERT model with a pre-trained bert-base-uncased model. Note that model's name starts with TF as names of all HuggingFace pre-trained models for TensorFlow start with TF. Please pay attention to this when using other transformer models.

5. Process input data with BertTokenizer:

[14]	emails.head()

	text	label
0	Subject: naturally irresistible your corporate...	1
1	Subject: the stock trading gunslinger fanny i...	1
2	Subject: unbelievable new homes made easy im ...	1
3	Subject: 4 color printing special request add...	1
4	Subject: do not have money , get software cds ...	1

6. Double check databank to see whether data has:

[15]	messages=emails['text'] labels=emails['label'] len(messages),len(labels)
	(5728, 5728)

7. Use BERT Tokenizer:

[16]	input_ids = []
	attention_masks = []
	for msg in messages:
	    bert_inp = bert_tokenizer.encode_plus(
	        msg,
	        add_special_tokens=True,
	        max_length=64,
	        padding='max_length',
	        truncation=True,
	max_length
	        return_attention_mask=True
	    )
	    input_ids.append(bert_inp['input_ids'])
	    attention_masks.append(bert_inp['attention_mask'])
	input_ids = np.asarray(input_ids)
	attention_masks = np.array(attention_masks)
	labels = np.array(labels)

Note: This snippet will generate token IDs for each input sentence of the dataset and append them to a list. Tags are a list of category labels, consisting of 0 and 1. We then convert the python lists, input_ids, and labels into numpy arrays to feed them to the Keras model

8. Define Keras model using the following lines:

```
[17]    # Custom BERT Layer to handle input/output properly
        class BertLayer(Layer):
            def __init__(self, bert_model):
                super(BertLayer, self).__init__()
                self.bert_model = bert_model
            def call(self, inputs):
                input_ids, attention_mask = inputs
                return self.bert_model(input_ids=input_ids,
          attention_mask=attention_mask)[1] # Pooled output
          (CLS token)
        # Define the model architecture
        input_ids_layer = Input(shape=(64,), dtype=tf.int32,
        name="input_ids")
        attention_mask_layer = Input(shape=(64,), dtype=tf.
        int32, name="attention_mask")
        # Pass inputs to the custom BERT layer
        bert_outputs = BertLayer(bmodel)([input_ids_layer,
        attention_mask_layer])
        # Add a classification layer
        outputs
        = Dense(units=1, activation="sigmoid")(bert_outputs)
        # Create the model
        model = Model(inputs=[input_ids_layer, attention_mask_
        layer], outputs=outputs)
        # Compile the model
        adam = tf.keras.optimizers.Adam(learning_rate=2e-5,
        epsilon=1e-08)
        model.compile(loss="binary_crossentropy",
        metrics=["accuracy"], optimizer=adam)
```

9. Perform model fitting and use 1 epoch to save time:

```
[18]    # Summary of the model
        model.summary()
        # Train the model
        model.fit([input_ids, attention_masks], labels,
        epochs=1, batch_size=1)
```

```
Model: "functional"
```

Layer (type)	Output Shape	Param #	Connected to
input_ids (InputLayer)	(None, 64)	0	-
attention_mask (InputLayer)	(None, 64)	0	-
bert_layer (BertLayer)	(None, 768)	0	input_ids[0][0], attention_mask[0][0]
dense (Dense)	(None, 1)	769	bert_layer[0][0]

```
Total params: 769 (3.00 KB)
Trainable params: 769 (3.00 KB)
Non-trainable params: 0 (0.00 B)
5728/5728 ──────────── 223s 38ms/step - accuracy: 0.7612 - loss: 0.5341
<keras.src.callbacks.history.History at 0x1edeeb594d0>
```

A BERT-based text classifier using less than 10 lines of code is to:
1. Define input layer to input sentences to model. The shape is 64 because each input sentence has 64 tokens in length. Pad each sentence to 64 tokens when encode_plus method is called.
2. Feed input sentences to BERT model.
3. Extract second output of BERT output at the third line. Since BERT model's output is a tuple, the first element of output tuple is a sequence of word vectors, and the second element is a single vector that represents the whole sentence called pooled output vector. Bert[1] extracts pooled output vector which is a vector of shape (1, 768).
4. Squash pooled output vector to a vector of shape 1 by a sigmoid function which is the class label.
5. Define Keras model with inputs and outputs.
6. Compile model.
7. Fit Keras model.

BERT model accepts one line only but can transfer enormous knowledge of Wiki corpus to model. This model obtains an accuracy of 0.96 at the end of the training. A single epoch is usually fitted to the model due to BERT overfits a moderate size corpus.

The rest of the code handles compiling and fitting Keras model as BERT has a huge memory requirement as can be seen by RAM requirements of Google Research's GitHub archive (GoogleBert-Memory, 2024).

The training code operates for about an hour on a local machine, where bigger datasets require more time even for one epoch.

This section will learn how to train a Keras model with BERT from scratch.

16.8   Transformer Pipeline Technology

HuggingFace Transformers library provides pipelines to assist program developers and benefit from transformer code immediately without custom training. A pipeline is a combination of a tokenizer and a pre-trained model.

HuggingFace provides models for various NLP tasks, its HuggingFace pipelines offer:

- Sentiment analysis (Agarwal 2020; Siahaan and Sianipar 2022).
- Question answering (Rothman 2022; Tunstall et al. 2022).
- Text summarization (Albrecht et al. 2020; Kedia and Rasu 2020).
- Translation (Arumugam and Shanmugamani 2018; Géron 2019).

This section will explore pipelines for sentiment analysis and question answering.

16.8.1 Transformer Pipeline for Sentiment Analysis

Let's start examples on sentiment analysis:

```
[19]    from transformers import pipeline
        nlp = pipeline("sentiment-analysis")
        utt5 = "I hate I am being a worker in the desert."
        utt6 = "I like you who are beautiful and kind."
                    result1 = nlp(utt5)
                    result2 = nlp(utt6)
```

The following steps are taken in the preceding code snippet:
1. Import pipeline function from transformers' library. This function creates pipeline objects with task name given as a parameter. Hence, a sentiment analysis pipeline object nlp is created by calling this function on the second line.
2. Define two example sentences with negative and positive sentiments. Then feed these sentences to the pipeline object nlp.

Check outputs:

```
[20]    result1
        [{'label': 'NEGATIVE', 'score': 0.9276903867721558}]
```

```
[21]    result2
        [{'label': 'POSITIVE', 'score': 0.9998767375946045}]
```

16.8.2 Transformer Pipeline for QA System

Next, will perform a question answering. Let's see the code:

| [22] | ```
from transformers import pipeline
nlp = pipeline("question-answering")
res = nlp({
    'question': 'What is the name of this book ?',
    'context': "I'll publish my new book Natural
 Language Processing soon."
})
print(res)
``` |
|---|---|
| | {'score': 0.9857430458068848, 'start': 25, 'end': 52, 'answer': 'Natural Language Processing'} |

Again, import pipeline function to create a pipeline object nlp. A context which has identical background information for the model is required for question-answering tasks to the model
• Request the model about this book's name after giving information of *this new publication will be available soon*.
• The answer is *natural language processing*, as expected.
• Try your own examples as simple exercise.

HuggingFace transformers studies are completed. Let's move on to the final section to see what spaCy offers on transformers.

Workshop 6.1 Revisit Sentiment Analysis using Transformer Technology
1. Use either previous workshop databank or another to import databank for sentiment analysis.
2. Try to implement sentiment analysis using previous and Transformer technology learnt in this workshop.
3. Compare performances and analysis (bonus).

16.9   Transformer and spaCy

SpaCy v3.0 had released new features and components. It has integrated transformers into spaCy NLP pipeline to introduce one more pipeline component called Transformer. This component allows users to use all HuggingFace models with spaCy pipelines. A spaCy NLP pipeline without transformers is illustrated in Fig. 16.11.

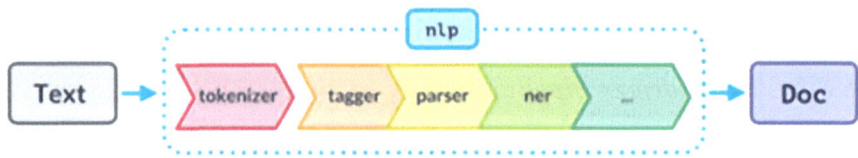

Fig. 16.11 Vector-based spaCy pipeline components
```

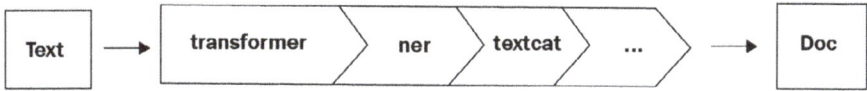

Fig. 16.12 Transformed-based spaCy pipeline components

en_core_web_trf

English transformer pipeline (roberta-base). Components: transformer, tagger, parser, ner, attribute_ruler, lemmatizer.

LANGUAGE	**EN** English
TYPE	**CORE** Vocabulary, syntax, entities, vectors
GENRE	**WEB** written text (blogs, news, comments)
SIZE	**TRF** 438 MB
COMPONENTS ⑦	transformer, tagger, parser, ner, attribute_ruler, lemmatizer
PIPELINE ⑦	transformer, tagger, parser, ner, attribute_ruler, lemmatizer
VECTORS ⑦	0 keys, 0 unique vectors (0 dimensions)
SOURCES ⑦	OntoNotes 5
AUTHOR	Explosion
LICENSE	MIT

Fig. 16.13 spaCy English transformer-based language models

A transformer-based pipeline component is illustrated in Fig. 16.12.

Transformer-based models and v2 style models are listed under Models page of the documentation (spaCy-model 2024) in English model for each supported language. Transformer-based models have various sizes and pipeline components like v2 style models. Also, each model has corpus and genre information like v2 style models. An example of an English transformer-based language model from Models page is shown in Fig. 16.13.

It showed that the first pipeline component is a transformer that generates word representations and deals with WordPiece algorithm to tokenize words into subwords. Word vectors are fed to the rest of the pipeline.

Downloading, loading, and using transformer-based models are identical to v2 style models.

English has two pre-trained transformer-based models, en_core_web_trf and en_core_web_lg currently. Let's start by downloading the en_core_web_trf model:

```
python3 -m spacy download en_core_web_trf
```

Import spaCy module and transformer-based model:

| [23] | ```
import spacy
import torch
import spacy_transformers
nlp = spacy.load("en_core_web_trf")
``` |
|---|---|

After loading model and initializing pipeline, use this model the same way as in v2 style models:

| [24] | ```
utt7 = nlp("I visited my friend Betty at her house.")
utt7.ents
``` |
|---|---|
| | (Betty,) |

| [25] | ```
for word in utt7:
    print(word.pos_, word.lemma_)
``` |
|---|---|
	PRON I
	VERB visit
	PRON my
	NOUN friend
	PROPN Betty
	ADP at
	PRON her
	NOUN house
	PUNCT .

These features related to the transformer component can be accessed by ._trf_data.trf_data which contain word pieces, input ids, and vectors generated by the transformer.

Let's examine the features one by one:

| [26] | ```
utt8 = nlp("It went there unwillingly.")
``` |
|---|---|

[27]	`utt8._.trf_data.wordpieces`
	WordpieceBatch(strings=[['<s>', 'It', 'Ġwent', 'Ġthere', 'Ġunw', 'ill', 'ingly', '.', '</s>']], input_ids=array([[0, 243, 439, 89, 10963, 1873, 7790, 4, 2]]), attention_mask=array([[1., 1., 1., 1., 1., 1., 1., 1., 1.]], dtype=float32), lengths=[9], token_type_ids=None)

There are five elements: word pieces, input IDs, attention masks, lengths, and token type IDs in the preceding output.

Word pieces are subwords generated by WordPiece algorithm. The word pieces of this sentence are as follows:

```
<s>
It
Gwent
Gthere
Gunw
Ill
ingly
.
</s>
```

The first and last tokens are special tokens used at the beginning and end of the sentence. The word *unwillingly* is divided into three subwords—*unw*, *ill*, and *ingly*. A G character is used to mark word boundaries. Tokens without G are subwords, such as *ill* and *ingly* in the preceding word piece list, except first word in the sentence marked by <'s' > .

Input IDs have identical meanings which are subword IDs assigned by the transformer's tokenizer.

The attention mask is a list of 0 s and 1 s for pointing the transformer to tokens it should notice. 0 corresponds to PAD tokens, while all other tokens should have a corresponding 1.

Lengths refer to the length of sentence after dividing it into subwords. Here is 9 but notice that len(doc) outputs is 5, while spaCy always operates on linguistic words.

token_type_ids are used by transformer tokenizers to mark sentence boundaries of two sentences input tasks such as question and answering. Since there is only one text provided, this feature is inapplicable.

Token vectors are generated by transformer, doc._.trf_data.tensors which contain transformer output, a sequence of word vectors per word, and the pooled output vector. Please refer to Obtaining BERT word vectors section if necessary.

[28]	`utt8._.trf_data.tensors[0].shape`
	(1, 9, 768)

[29]	`utt8._.trf_data.tensors[1].shape`
	(1, 768)

The first element of tuple is the vectors for tokens. Each vector is 768-dimensional; hence 9 words produce 9 × 768-dimensional vectors. The second element of tuple is the pooled output vector which is an aggregate representation for input sentence, and the shape is 1 × 768

spaCy provides user-friendly API and packaging for complicated models such as transformers. Transformer integration is a validation of using spaCy for NLP.

References

Agarwal, B. (2020) Deep Learning-Based Approaches for Sentiment Analysis (Algorithms for Intelligent Systems). Springer.

Albrecht, J., Ramachandran, S. and Winkler, C. (2020) Blueprints for Text Analytics Using Python: Machine Learning-Based Solutions for Common Real World (NLP) Applications. O'Reilly Media.

Arumugam, R., & Shanmugamani, R. (2018). Hands-on natural language processing with python. Packt Publishing.

Bansal, A. (2021) Advanced Natural Language Processing with TensorFlow 2: Build effective real-world NLP applications using NER, RNNs, seq2seq models, Transformers, and more. Packt Publishing.

Devlin, J., Chang, M. W., Lee, K. and Toutanova, K. (2019). BERT: Pre-training of Deep Bidirectional Transformers for Language Understanding. Archive: https://arxiv.org/pdf/1810.04805.pdf.

Ekman, M. (2021) Learning Deep Learning: Theory and Practice of Neural Networks, Computer Vision, Natural Language Processing, and Transformers Using TensorFlow. Addison-Wesley Professional.

Facebook-transformer (2024) Facebook Transformer Model archive. https://github.com/pytorch/fairseq/blob/master/examples/language_model/README.md. Accessed 17 Dec 2024.

Géron, A. (2019) Hands-On Machine Learning with Scikit-Learn, Keras, and TensorFlow: Concepts, Tools, and Techniques to Build Intelligent Systems. O'Reilly Media.

GoogleBert (2024) Google Bert Model Github archive. https://github.com/google-research/bert. Accessed 17 Dec 2024.

GoogleBert-Memory (2024) GoogleBert Memory Requirement. https://github.com/google-research/bert#out-of-memory-issues. Accessed 17 Dec 2024.

HuggingFace (2024) Hugging Face official site. https://huggingface.co/. Accessed 17 Dec 2024.

HuggingFace_transformer (2024) HuggingFace Transformer Model archive. https://github.com/huggingface/transformers. Accessed 17 Dec 2024.

Kedia, A. and Rasu, M. (2020) Hands-On Python Natural Language Processing: Explore tools and techniques to analyze and process text with a view to building real-world NLP applications. Packt Publishing.

Keras (2024) Keras official site. https://keras.io/. Accessed 17 Dec 2024.

Korstanje, J. (2021) Advanced Forecasting with Python: With State-of-the-Art-Models Including LSTMs, Facebook's Prophet, and Amazon's DeepAR. Apress.

NLPGitHub (2024) URL: https://github.com/raymondshtlee/NLP/. Accessed 17 Dec 2024.

Rothman, D. (2022) Transformers for Natural Language Processing: Build, train, and fine-tune deep neural network architectures for NLP with Python, PyTorch, TensorFlow, BERT, and GPT-3. Packt Publishing.

SpaCy (2024) spaCy official site. https://spacy.io/. Accessed 17 Dec 2024.

SpaCy-model (2024) spaCy English Pipeline Model. https://spacy.io/models/en. Accessed 17 Dec 2024.

Siahaan, V. and Sianipar, R. H. (2022) Text Processing and Sentiment Analysis using Machine Learning and Deep Learning with Python GUI. Balige Publishing.

TensorFlow (2024) TensorFlow official site. https://tensorflow.org /. Accessed 17 Dec 2024.

Tunstall, L, Werra, L. and Wolf, T. (2022) Natural Language Processing with Transformers: Building Language Applications with Hugging Face. O'Reilly Media.

Vaswani, A., Shazeer, N., Parmar, N., Uszkoreit, J., Jones, L., Gomez, A. N., … & Polosukhin, I. (2017). Attention is all you need. Advances in neural information processing systems, 30. https://arxiv.org/abs/1706.03762.

Yıldırım, S and Asgari-Chenaghlu, M. (2021) Mastering Transformers: Build state-of-the-art models from scratch with advanced natural language processing techniques. Packt Publishing.

Chapter 17
Workshop#7 Building Chatbot with TensorFlow and Transformer Technology (Hour 13–14)

17.1 Introduction

In the previous 6 NLP workshops, we studied NLP implementation tools and techniques ranging from tokenization, N-gram generation to semantic and sentiment analysis with various key NLP Python enabling technologies: NLTK, spaCy, TensorFlow and contemporary Transformer Technology. This final workshop will explore how to integrate them for the design and implementation of a domain-based chatbot system on a movie domain.

This workshop will explore:

1. Technical requirements for chatbot system.
2. Knowledge domain—the Cornell Large Movie Conversation Dataset is a well-known conversation dataset with over 200,000 movie dialogues of 10,000+ movie characters (Cornell 2024; Cornell_Movie_Corpus 2024).
3. A step-by-step Movie Chatbot system implementation which involve movie dialogue preprocessing, model construction, attention learning, system integration with spaCy, TensorFlow, Keras and Transformer Technology, an important tool in NLP system implementation (Bansal 2021; Devlin et al. 2019; Géron 2019; Rothman 2022; Tunstall et al. 2022; Yıldırım and Asgari-Chenaghlu 2021).
4. Evaluation metrics with real-time chat examples.

17.2 Technical Requirements

In this workshop, transformers, TensorFlow, and spaCy (TensorFlow 2024; spaCy 2024) are to be installed in PC/notebook computer. Please ensure that the following Python packages are installed before starting the workshop:

© The Author(s), under exclusive license to Springer Nature Singapore Pte Ltd. 2025
R. Lee, *Natural Language Processing*, https://doi.org/10.1007/978-981-96-3208-4_17

- Python (demo version 3.11.9)
- spacy (demo version 3.4.4)
- keras (demo version 3.3.3)
- transformers (demo version 4.44.2)
- tensorflow (demo version 2.17.0)
- tensorflow_datasets (demo version 4.9.6)
- tf-keras (demo version 2.17.0)
- numPy (demo version 1.26.4)
- pandas (demo version 2.2.2)
- matplotlib (demo version 3.9.2)
- pydot (demo version 3.0.1)
- graphviz (demo version 0.20.3)
- pydot-ng (demo version 2.0.0)

If these packages are not installed on PC/laptop, use pip install xxx command. The detailed requirements list and Python package version used in this workshop can be found in the requirements.txt file stored in the NLP GitHub repository (NLPGitHub 2024).

17.3   AI Chatbot in a Nutshell

17.3.1   What Is a Chatbot?

Conversational artificial intelligence (conversational AI) is a field of machine learning that aims to create technology and enables users to have text or speech-based interactions with machines. Chatbots, virtual assistants, and voice assistants are typical conversational AI products (Batish 2018; Freed 2021; Janarthanam 2017; Raj 2018).

A chatbot is a software application designed to make conversations with humans.

Chatbots are widely used in human resources, marketing and sales, banking, healthcare, and non-commercial areas such as personal conversations. They include:

- Amazon Alexa is a voice-based virtual assistant to perform tasks per user requests or inquiries, i.e., play music, podcasts, set alarms, read audiobooks, provide real-time weather, traffic, other information, etc. Alexa Home can connect smart home devices to oversee premises, electrical appliances, etc.
- Facebook Messenger and Telegram instant messaging services provide interfaces and API documentations (Facebook 2024, Telegram 2024) for developers to connect bots.

- Google Assistant provides real-time weather, flight, traffic information, send and receive text messages, email services, device information, set alarms, and integrate with smart home devices, etc. available on Google Maps, Google Search, and standalone Android and iOS applications.
- IKEA provides customer service chatbot called Anna, AccuWeather, and FAQ chatbots.
- Sephora has virtual make-up artist and customer service chatbots at Facebook messenger.
- Siri integrates with iPhone, iPad, iPod, and macOS to initiate, answer calls, send, receive text messages and WhatsApp messages on iPhone.

Other virtual assistants include AllGenie, Bixby, Celia, Cortana, Duer, and Xiaowei.

17.3.2 What Is a Wake Word in Chatbot?

A wake word is the gateway between user and user's digital assistant/Chatbot. Voice assistants such as Alexa and Siri are powered by AI with word detection abilities to queries response and commands, as illustrated in Fig. 17.1.

Common wake words include Hey, Google, Alexa, and Hey Siri.

Today's wake word performance and speech recognition are operated by machine learning or AI with cloud processing.

Sensory's wake word and phrase recognition engines use deep neural networks to provide an embedded or on-device wake word and phrase recognition engine.

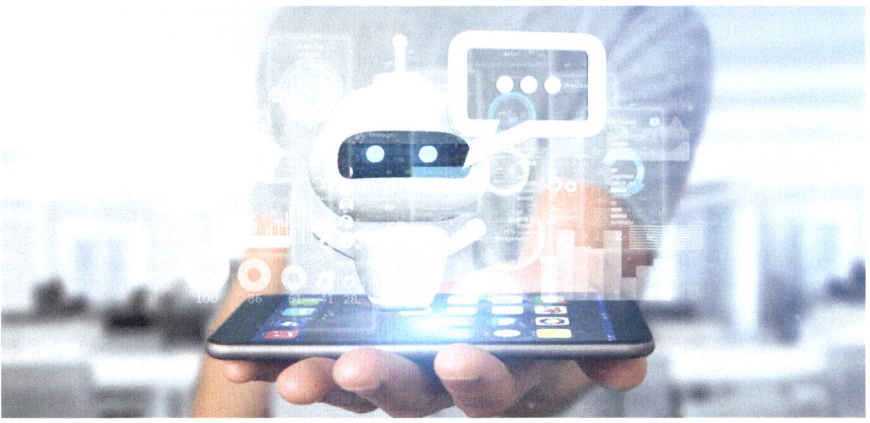

Fig. 17.1 Wake word to invoke Chatbot (Tuchong 2024)

17.3.2.1   Tailor-Made Wake Word

Wake words like Alexa, Siri, and Google are associated with highly valued and technical products experiences, other companies had created tailor-made wake word and uniqueness to their products, i.e., Hi Toyota had opened a doorway to voice user interface to strengthen the relationship between customers and the brand.

17.3.2.2   Why Embedded Word Detection?

Wake word technology has been used in cases beyond mobile applications. Some battery powered devices like Bluetooth headphones, smart watches, cameras, and emergency alert devices.

Chatbot allow users to utter commands naturally. Queries like *what time is it?* or *how many steps have I taken?* are phrases examples that a chatbot can process zero latency with high accuracy.

Wake word technology can integrate with voice recognition applications like touch screen food ordering, voice-control microwaves, or user identification settings at televisions or vehicles.

17.3.3   NLP Components in a Chatbot

A typical chatbot consists of major components:

1. Speech-to-text converts user speech into text. The input is a wav/mp3 file, and the output is a text file containing user's utterance.
2. Conversational NLU performs intent recognition and entity extraction on user's utterance text. The output is the user's intent with a list of entities. Resolving references in the current to previous utterances is processed by this component.
3. Dialogue manager retains conversation memory to generate a meaningful and coherent chat. This component is regarded as dialogue memory in conversational state hitherto entities and intents appeared. Hence, the input is the previous dialogue state for the current user to parse intent and entities to a new dialogue state output.
4. Answer generator gives all inputs from previous stages to generate answers to user's utterance.
5. Text-to-speech generates a speech file (WAV or mp3) from system's answers.

Each of these components is trained and evaluated separately, e.g., speech-to-text training is performed by speech files and corresponding transcriptions on an annotated speech corpus.

17.4 Building Movie Chatbot by Using TensorFlow and Transformer Technology

This workshop will integrate the learnt technologies including: TensorFlow (Bansal 2021; Ekman 2021; TensorFlow 2024), Keras (Géron 2019; Keras 2024a), Transformer technology with Attention Learning Scheme (Ekman 2021; Kedia and Rasu 2020; Rothman 2022; Tunstall et al. 2022; Vaswani et al. 2017; Yıldırım and Asgari-Chenaghlu 2021) to build a domain-based chatbot system. The Cornell Large Movie Dialog Corpus (Cornell 2024) will be used as conversation dataset for system training. The movie dataset can be downloaded from either Cornell databank (2024) or Kaggle's Cornell Movie Corpus archive (2024).

Use pip install command to invoke TensorFlow package and install its dataset:

```
[1]     import tensorflow as tflow
        tflow.random.set_seed(1234)
        # !pip install tensorflow-datasets==1.2.0
        import tensorflow_datasets as tflowDS
        import re
        import matplotlib.pyplot as pyplt
```

1. Install and import tensorflow-datasets in addition to TensorFlow package. Please use pip install command as script if not installed already.
2. Use *random.set_seed( )* method to set all random seeds required to replicate TensorFlow codes.

17.4.1 The Chatbot Dataset

The Cornell Movie Dialogues corpus is used in this project. This dataset, movie_conversations.txt contains lists of conversation IDs and movie_lines.txt associative conversation ID. It has generated 220,579 conversations and 10,292 movie characters among movies.

17.4.2 Movie Dialogue Preprocessing

The maximum numbers of conversations (MAX_CONV) and the maximum length of utterance (MLEN) are set for 50,000 and 40 for system training respectively.

Preprocessing data procedure (PP) involves the following steps:

1. Obtain 50,000 movie dialogue pairs from dataset.
2. PP each utterance by special and control characters removal.
3. Construct tokenizer.
4. Tokenize each utterance.
5. Cap the max utterance length to MLEN.
6. Filter and pad utterances.

```
[2]    # Set the maximum number of training conversation
       MAX_CONV = 50000
       # Preprocess all utterances
       def pp_utterance(utterance):
           utterance = utterance.lower().strip()
           # Add a space to the following special characters
           utterance = re.sub(r"([?.!,])", r" \1 ", utterance)
           # Delete extrac spaces
           utterance = re.sub(r'[" "]+', " ", utterance)
           # Other than below characters, the other character
       replace by spaces
           utterance = re.sub(r"[^a-zA-Z?.,!]+", " ",
       utterance)
           utterance = utterance.strip()
           return utterance
       def get_dialogs():
           # Create the dialog object (dlogs)
           id2dlogs = {}
           # Open the movie_lines text file
           with open('data/movie_lines.txt', encoding = 'utf-8',
       errors = 'ignore') as f_dlogs:
               dlogs = f_dlogs.readlines()
           for dlog in dlogs:
               sections = dlog.replace('\n', '').split(' +++$+++
       ')
               id2dlogs[sections[0]] = sections[4]
           query, ans = [], []
           with open('data/movie_conversations.txt',
                   encoding = 'utf-8', errors = 'ignore') as
       f_conv:
               convs = f_conv.readlines()
           for conv in convs:
               sections = conv.replace('\n', '').split(' +++$+++
       ')
           # Create movie conservation object m_conv as a list
           m_conv = [conv[1:-1] for conv in sections[3][1:-1].
       split(', ')]
             for i in range(len(m_conv) - 1):
               query.append(pp_utterance(id2dlogs[m_conv[i]]))
               ans.append(pp_utterance(id2dlogs[m_conv[i + 1]]))
               if len(query) >= MAX_CONV:
               return query, ans
           return query, ans
       queries, responses = get_dialogs()
```

Select query 13 and verify response:

| [3] | ```python
print('Query 13: {}'.format(queries[13]))
print('Response 13: {}'.format(responses[13]))
``` |
|---|---|
| | Query 13: that s because it s such a nice one.
Response 13: forget french. |

Select query 100 and verify response:

| [4] | ```python
print('Query 100: {}'.format(queries[100]))
print('Response 100: {}'.format(responses[100]))
``` |
|---|---|
| | Query 100: you set me up.
Response 100: i just wanted |

Verify queries (responses) size to see whether it situates within MAX_CONV:

| [5] | ```python
len(queries)
``` |
|---|---|
| | 50000 |

| [6] | ```python
Len(responses)
``` |
|---|---|
| | 50000 |

1. After max 50,000 movie conversations had obtained to perform basic preprocessing, it is sufficient for model training.
2. Perform tokenization procedure to add START and END tokens using commands below.

17.4.3   Tokenization of Movie Conversation

| [7] | ```python
# Define the Movie Token object
m_token =
tflowDS.
deprecated.text.SubwordTextEncoder.build_from_corpus
(queries + responses, target_vocab_size = 2**13)
# Define the Start and End tokens
START_TOKEN, END_TOKEN =
[m_token.vocab_size], [m_token.vocab_size + 1]
# Define the size of Vocab (SVCAB)
SVCAB = m_token.vocab_size + 2
``` |
|---|---|
```

Verify movie token lists for conv 13 and 100:

[8]	`print('The movie token of conv 13: {}'.format(m_token.`
	`encode`
	`(queries[13])))`

	The movie token of conv 13: [15, 8, 151, 12, 8, 354, 10, 347, 188, 1]

[9]	`print('The movie token of conv 100: {}'.format(m_token.`
	`encode`
	`(queries[100])))`

	The movie token of conv 100: [5, 539, 36, 119, 1]

17.4.4   Filtering and Padding Process

Cap utterance max length (MLEN) to 40, perform filtering and padding:

| [10] | ```
# Set the maximum length of each utterance MLEN to 40
MLEN = 40
# Performs the filtering and padding of each utterance
def filter_pad (qq, aa):
    m_token_qq, m_token_aa = [], []
    for (utterance1, utterance2) in zip(qq, aa):
        utterance1 = START_TOKEN + m_token.
encode(utterance1) + END_TOKEN
        utterance2 = START_TOKEN + m_token.
encode(utterance2) + END_TOKEN
        if len(utterance1) <= MLEN and len(utterance2) <=
MLEN:
        m_token_qq.append(utterance1)
        m_token_aa.append(utterance2)
    # pad tokenized sentences
    m_token_qq = tflow.keras.preprocessing.sequence.
pad_sequences(m_token_qq, maxlen=MLEN, padding =
'post')
    m_token_aa = tflow.keras.preprocessing.sequence.
pad_sequences(m_token_aa, maxlen=MLEN, padding =
'post')
    return m_token_qq, m_token_aa
queries, responses = filter_pad (queries, responses)
``` |
| --- | --- |

Review the size of movie vocab (SVCAB) and total number of conversation (conv):

[11]	`print('Size of vocab: {}'.format(SVCAB))` `print('Total number of conv: {}'.format(len(queries)))`
```

| Size of vocab: 8333 |
| Total number of conv: 44095 |

1. Note that the total number of conversations after the filtering and padding process is 44,095 which is less than the previous max conv size of 50,000 as some conversations are filtered out.
2. SVCAB size is around 8000 which makes sense as the total numbers of conversation is around 44,000 lines, and the number of vocabulary used is between 5000 and 10,000.

17.4.5   Creation of TensorFlow Movie Dataset Object (mDS)

TensorFlow dataset object is created by using Dataset.from_tensor_slices() method of TensorFlow Data class as below:

```
[12]    tflow.data.Dataset.from_tensor_slices?
```

```
[13]    # Define the Batch and Buffer size
        sBatch = 64
        sBuffer = 20000
        # Create mDS object from TensorFlow class
        mDS
        =
        tflow.
        data.Dataset.from_tensor_slices(({'inNodes':queries,
        'decNodes':responses[:, :-1]},{'outNodes':responses[:,
        1:]}))
        mDS = mDS.cache()
        mDS = mDS.shuffle(sBuffer)
        mDS = mDS.batch(sBatch)
        mDS = mDS.prefetch(tflow.data.experimental.AUTOTUNE)
```

1. Create a TensorFlow dataset object first to define batch and buffer size
2. Define three layers of Transformer model:
a. input node layer (inNodes)—Queries
b. decoder input node layer (decNodes)—Responses
c. output node layer (outNodes)—Responses
3. Define prefetch scheme—AUTOTUNE in our project.

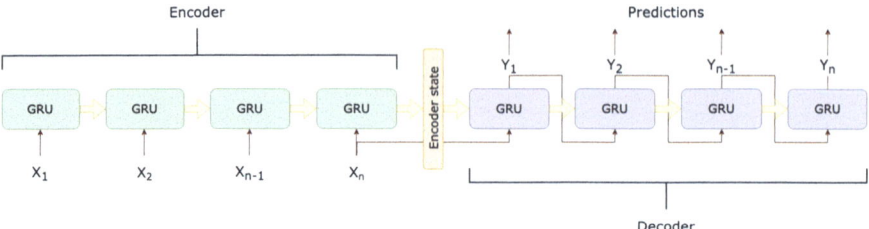

Fig. 17.2 Attention learning with transformer technology

17.4.6   Calculate Attention Learning Weights

The main concept of transformer technology is the Attention Learning technique, which aimed at network capability to focus *attention* to various parts of the training sequence during recurrent network learning. AI chatbot corresponds to *self-attention* learning on movie dialogs, in which the network has attention ability to different positions of dialogue token sequences to compute utterances representation. A system architecture of the Attention Learning model with Transformer technology is illustrated in Fig. 17.2. Implement Attention Equation to calculate the attention weight is given by:

$$Attention(Q,K,V) = softmax\left(\frac{QK^T}{\sqrt{d_k}}\right) \qquad (17.1)$$

Attention Equation is a typical scaled-dot-product attention function in transformer object Query (Q), K (Key), and V (Value) Value and Python implementation is given below:

```
[14]    # Calculate the Attention Weight, Query (q), Key(k),
        Value(v), Mask(m)
        def calc_attention(q, k, v, m):
            qk = tflow.matmul(q, k, transpose_b = True)
            dep = tflow.cast(tflow.shape(k)[-1], tflow.float32)
            mlogs = qk / tflow.math.sqrt(dep)
            # Use the masking for padding
            if m is not None:
                mlogs += (m * -1e9)
            # Apply softmax on the final axis of the utterance
        sequence
            att_wts = tflow.nn.softmax(mlogs, axis = -1)
            # Apply matmul() operation
            out_wts = tflow.matmul(att_wts, v)
            return out_wts
```

17.4.7   Multi-Head-Attention (MHAttention)

Multi-Head-Attention (MHAttention) consists of the following steps:

1. Construct linear layers
2. Perform head-splitting
3. Calculate attention weights
4. Combine heads
5. Condense layers

MHAttention is implemented as follows:

```
[15]   class MHAttention(tflow.keras.layers.Layer):
       def __init__(self, dm, nhd, name="MHAttention"):
         super(MHAttention, self).__init__(name=name)
         self.nhd = nhd
         self.dm = dm
         assert dm % self.nhd == 0
         self.dep = dm // self.nhd
         self.qdes = tflow.keras.layers.Dense(units=dm)
         self.kdes = tflow.keras.layers.Dense(units=dm)
         self.vdes = tflow.keras.layers.Dense(units=dm)
         self.des = tflow.keras.layers.Dense(units=dm)
       def sheads(self, inNodes, bsize):
         inNodes = tflow.reshape(
             inNodes, shape=(bsize, -1, self.nhd, self.dep))
         return tflow.transpose(inNodes, perm=[0, 2, 1, 3])
       def call(self, inNodes):
         q, k, v, m = inNodes['q'], inNodes['k'],
       inNodes['v'], inNodes['m']
         bsize = tflow.shape(q)[0]
         # 1. Construct Linear-layers
         q = self.qdes(q)
         k = self.kdes(k)
         v = self.vdes(v)
         # 2. Perform Head-splitting
         q = self.sheads(q, bsize)
         k = self.sheads(k, bsize)
         v = self.sheads(v, bsize)
         # 3. Calculate Attention Weights
         sattention = calc_attention(q, k, v, m)
         sattention = tflow.transpose(sattention, perm=[0, 2,
       1, 3])
         # 4. Head Combining
         cattention = tflow.reshape(sattention,
             (bsize, -1, self.dm))
         # 5. Layer Condensation
         outNodes = self.des(cattention)
         return outNodes
```

17.4.8   System Implementation

Step 1. Implement Masking

Implement (1) Padding Mask and (2) Look_ahead Mask to mask token sequences.

| [16] | ```
# Generate Padding Mask (gen_pmask)
def gen_pmask(p):
    pmask = tflow.cast(tflow.math.equal(p, 0), tflow.
float32)
    return pmask[:, tflow.newaxis, tflow.newaxis, :]
``` |
| --- | --- |

| [17] | ```
# Generate Look_Ahead Mask (gen_lamask)
def gen_lamask(x):
    slen = tflow.shape(x)[1]
    lamask = 1- tflow.linalg.band_part(tflow.ones((slen,
slen)), -1, 0)
    pmask = gen_pmask(x)
    return tflow.maximum(lamask, pmask)
``` |
| --- | --- |

Review *lamask* with a sample matrix:

| [18] | ```
print(gen_lamask(tflow.constant([[1, 2, 0, 4, 5]])))
``` |
| --- | --- |
| | tf.Tensor(
 [[[[0. 1. 1. 1. 1.]
    [0. 0. 1. 1. 1.]
    [0. 0. 1. 1. 1.]
    [0. 0. 1. 0. 1.]
    [0. 0. 1. 0. 0.]]]], shape=(1, 1, 5, 5), dtype=float32) |

Step 2. Implement Positional Encoding

The main function of positional encoding is to provide model with information about the relative position of word tokens within utterance for attention learning given by the following formula:

$$PE_{(pos,2i)} = \sin\left(pos / 10000^{2i/d_{model}} \right)$$
$$PE_{(pos,2i+1)} = \cos\left(pos / 10000^{2i/d_{model}} \right)$$

(17.2)
```

```
[19]   # Implementation of Positional Encoding Class
       (PEncoding)
       class PEncoding(tflow.keras.layers.Layer):
          def __init__(self, pos, dm):
             super(PEncoding, self).__init__()
             self.pencode = self.pencods(pos, dm)
          def gdeg(self, pos, i, dm):
             deg = 1 / tflow.pow(10000,(2 * (i // 2)) /
       tflow.cast(dm, tflow.float32))
             return pos * deg
          def pencods(self, pos, dm):
             deg_rads = self.gdeg(pos = tflow.range(pos,
          dtype=tflow.float32)[:, tflow.newaxis], i=tflow.range(dm,
          dtype=tflow.float32)[tflow.newaxis, :], dm = dm)
             m_sin = tflow.math.sin(deg_rads[:, 1::2])
             m_cos = tflow.math.cos(deg_rads[:, 1::2])
             pencode = tflow.concat([m_sin, m_cos], axis = -1)
             pencode = pencode[tflow.newaxis,…]
             return tflow.cast(pencode, tflow.float32)
          def call(self, inNodes):
             # Convert SparseTensor to DenseTensor if
       necessary
             if isinstance(inNodes, tflow.sparse.SparseTensor):
             inNodes = tflow.sparse.to_dense(inNodes)
             # Add positional encoding to input nodes
             return inNodes + self.pencode[:, :tflow.
       shape(inNodes)[1], :]
```

Try to plot *PositionalEncoding* diagram:

```
[20]   # Create PositionalEncoding Sample
       pencoding_sample = PEncoding(50, 512)
       pyplt.pcolormesh(pencoding_sample.pencode.numpy()[0],
       cmap = 'RdBu')
       pyplt.xlabel('Depth')
       pyplt.xlim((0, 512))
       pyplt.ylabel('Position')
       pyplt.colorbar()
       pyplt.show()
```

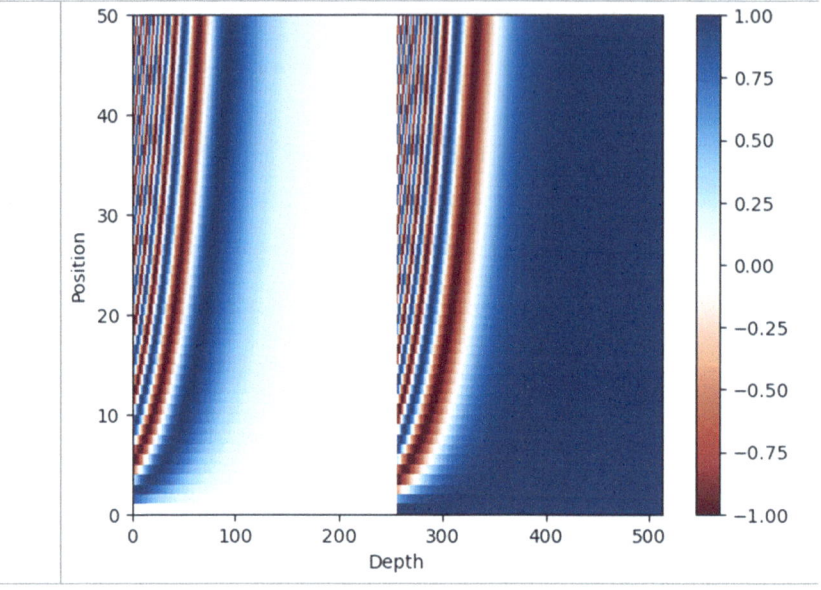

Step 3. Implement Encoder Layer

Encoder Layer (enclayer) implementation involves:

1. Create MHAttention object
2. Two dense layers

Details as shown below:

```
[21]    # Implementation of Encoder Layer (enclayer)
        def enclayer(i, dm, nhd, drop, name="enclayer"):
        inNodes = tflow.keras.Input(shape=(None, dm),
        name="inNodes")
        pmask = tflow.keras.Input(shape=(1, 1, None),
        name="pmask")
        att = MHAttention(
            dm, nhd, name="att")({
                'q': inNodes,
                'k': inNodes,
                'v': inNodes,
                'm': pmask
            })
        att = tflow.keras.layers.Dropout(rate=drop)(att)
        att = tflow.keras.layers.LayerNormalization(
            epsilon=1e-6)(inNodes + att)
        outNodes = tflow.keras.layers.Dense(units=i,
        activation='relu')(att)
        outNodes = tflow.keras.layers.Dense(units=dm)(outNodes)
        outNodes = tflow.keras.layers.Dropout(rate=drop)
        (outNodes)
        outNodes = tflow.keras.layers.LayerNormalization(
            epsilon=1e-6)(att + outNodes)
        return tflow.keras.Model(
            inputs=[inNodes, pmask], outputs=outNodes,
          name=name)
```

1. An attention learning object is defined and used at encoder layer implementation class.
2. *relu* function is used as default setting for encoder layer activation function. Current research includes the modification (or change) of activation function for system enhancement.

Try to display a sample encoder layer using Keras plot model():

```
[22]    # Create a sample Encoder Layer and display object
        diagram
        enclayer_sample = enclayer(i = 512, dm = 128, nhd = 4,
        drop = 0.3, name = "enclayer_sample")
        tflow.keras.utils.plot_model(enclayer_sample, to_file =
        'enclayer.png', show_shapes = True)
```

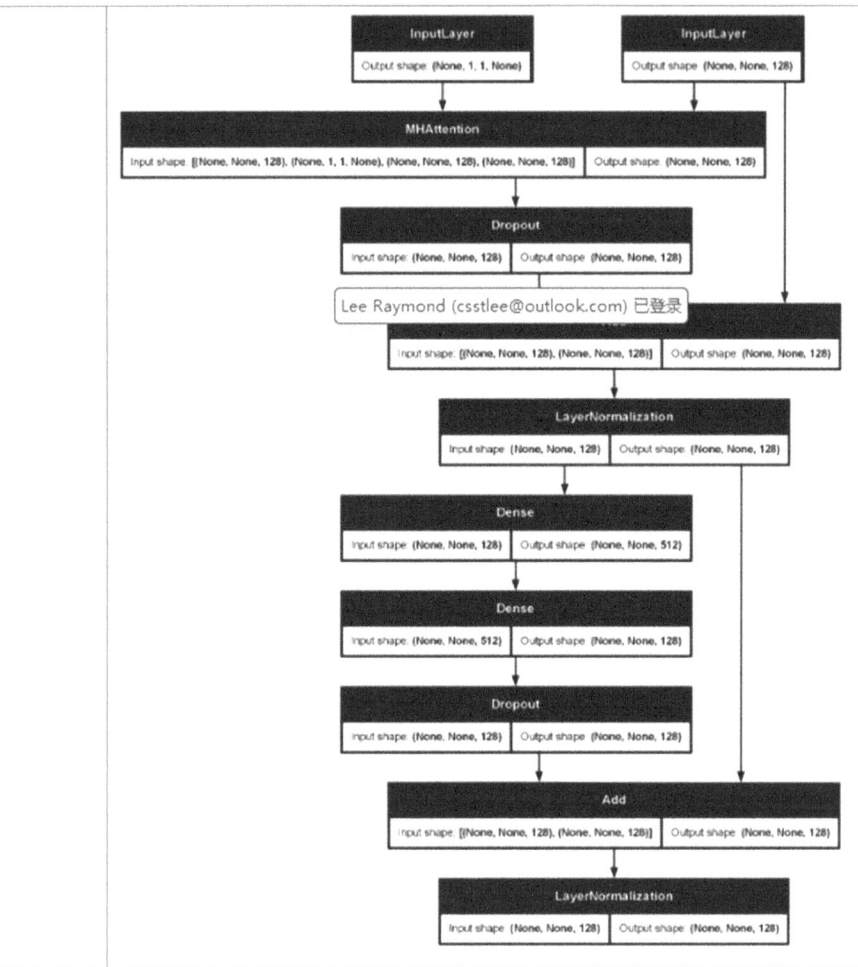

Step 4. Implement Encoder

Encoder implementation involves the following processes:

1. Embed inputs
2. Perform positional encoding scheme
3. Encode Num Layers

```
[23]    # Implementation of Encoder Class (encoder)
        def encoder(svcab,
              nlayers,
              x,
              dm,
              nhd,
              drop,
              name="encoder"):
        inNodes = tflow.keras.Input(shape=(None,),
        name="inNodes")
        pmask = tflow.keras.Input(shape=(1, 1, None),
        name="pmask")
        embeddings = tflow.keras.layers.Embedding(svcab, dm)
        (inNodes)
        embeddings *= tflow.math.sqrt(tflow.cast(dm, tflow.float32))
        embeddings = PEncoding(svcab, dm)(embeddings)
        outNodes = tflow.keras.layers.Dropout(rate=drop)
        (embeddings)
        for i in range(nlayers):
            outNodes = enclayer(
                i=x,
                dm=dm,
                nhd=nhd,
                drop=drop,
                name="enclayer_{}".format(i),
            )([outNodes, pmask])
        return tflow.keras.Model(
            inputs=[inNodes, pmask], outputs=outNodes,
          name=name)
```

Display a sample encoder using Keras plot model:

```
[24]    # Create a sample Encoder Sample and display object
        diagram
        encoder_sample = encoder(svcab = 8192,
              nlayers = 2,
              x = 512,
              dm = 128,
              nhd = 4,
              drop = 0.3,
              name = "encoder_sample")
        tflow.keras.utils.plot_model
        (encoder_sample, to_file='encoder_sample.png', show_
        shapes = True)
```

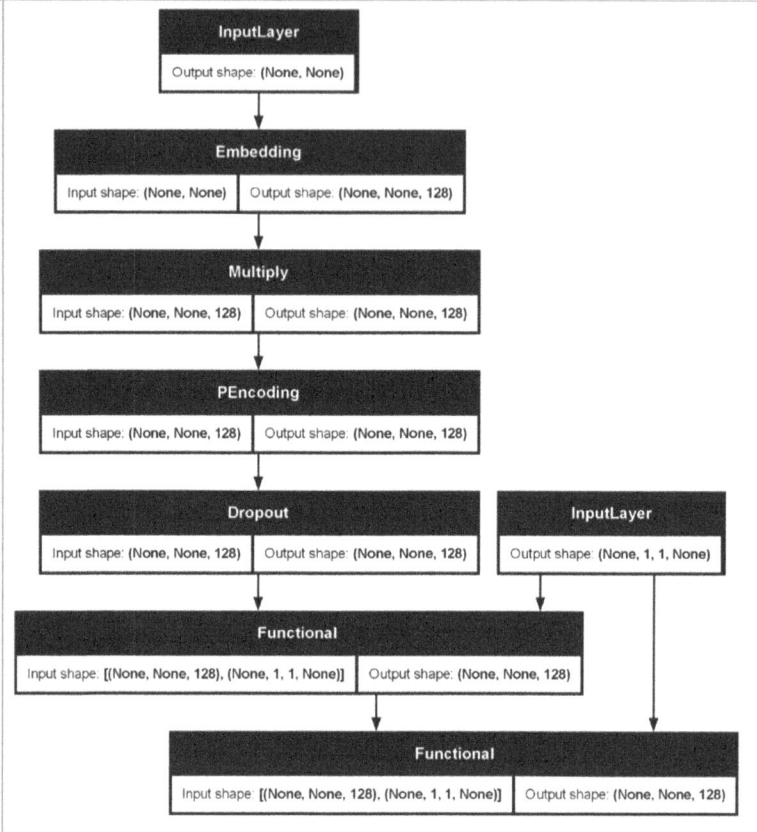

Step 5. Implement Decoder Layer

Decoder Layer implementation involves the following steps:

1. MHAttention
2. 2 Dense Decoder Layers with dropout

```
[25]   # Implementation of Decoder Layer (declayer)
       def declayer(i, dm, nhd, drop, name = "declayer"):
         inNodes = tflow.keras.Input(shape=(None, dm), name
       ="inNodes")
         encouts = tflow.keras.Input(shape=(None, dm),
       name="encouts")
         lamask = tflow.keras.Input(shape=(1, None, None),
       name = "lamask")
         pmask = tflow.keras.Input(shape=(1, 1, None), name =
       "pmask")
         att1 = MHAttention(dm, nhd, name="att1")
       (inNodes={'q':inNodes,
                 'k':inNodes,
                 'v':inNodes,
                 'm':lamask})
         att1 = tflow.keras.layers.LayerNormalization(epsilon
       =1e-6)
       (att1 + inNodes)
         att2 = MHAttention(dm,nhd, name = "att2")
       (inNodes={'q':att1,
                 'k':encouts,
                 'v':encouts,
                 'm':pmask})
         att2 = tflow.keras.layers.Dropout(rate=drop)(att2)
         att2 = tflow.keras.layers.LayerNormalization(epsilon
       = 1e-6)(att2 + att1)
         outNodes = tflow.keras.layers.Dense(units=i,
       activation='relu')(att2)
         outNodes = tflow.keras.layers.Dense(units=dm)
       (outNodes)
         outNodes = tflow.keras.layers.Dropout(rate=drop)
       (outNodes)
         outNodes = tflow.keras.layers.
       LayerNormalization(epsilon=1e-6)(outNodes + att2)
         return tflow.keras.Model(inputs=[inNodes, encouts,
       lamask, pmask],
             outputs = outNodes,
             name = name)
```

1. Encoder layer implements single attention learning object, and decoder layer implements two attention learning objects att1 and att2 according to transformer learning model.
2. Again, *relu* function is used as activation function. It can modify or adopt different activation function to improve network performance as studied in Sect. 17.1.

Display sample decoder layer using Keras *plot_model()*:

[26]

```
# Create a decoder layer sample and show object
association diagram
declayer_sample = declayer(i = 512, dm = 128, nhd = 4,
drop = 0.3,
name = "declayer_sample")
tflow.keras.utils.plot_model
(declayer_sample, to_file='declayer_sample.png', show_
shapes=True)
```

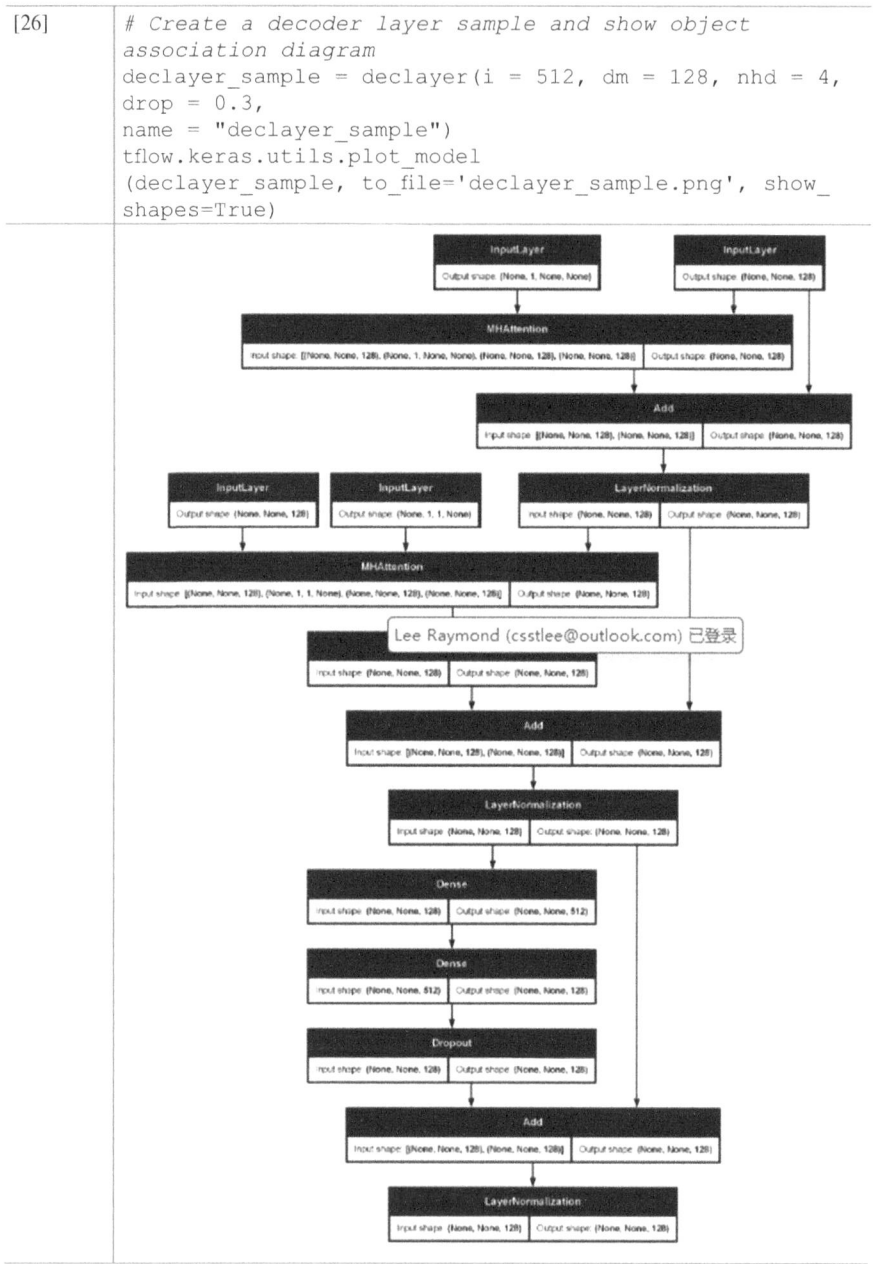

Step 6. Implement Decoder

Decoder implementation involves the following processes:

1. Embed network outputs

2. Look ahead and pad masking
3. Positional encoding scheme
4. Perform N-decoder layers

```
[27]  # Implementation of Decoder class (decoder)
      def decoder(svcab,
            nlayers,
            x,
            dm,
            nhd,
            drop,
            name='decoder'):
        inNodes = tflow.keras.Input(shape=(None,), name="inNodes")
        encouts = tflow.keras.Input(shape=(None, dm),
      name="encouts")
        lamask = tflow.keras.Input(shape=(1, None, None),
      name="lamask")
        pmask = tflow.keras.Input(shape=(1, 1, None),
      name="pmask")
        embeddings = tflow.keras.layers.Embedding(svcab, dm)
      (inNodes)
        embeddings *= tflow.math.sqrt(tflow.cast(dm, tflow.float32))
        embeddings = PEncoding(svcab, dm)(embeddings)
        outNodes = tflow.keras.layers.Dropout(rate=drop)
      (embeddings)
        for i in range(nlayers):
            outNodes = declayer(i = x,
                dm=dm,
                nhd=nhd,
                drop=drop,
                name = 'declayer_{}'.format(i),)(inputs=[outNodes,
      encouts, lamask, pmask])
            return tflow.keras.Model(inputs=[inNodes, encouts,
      lamask, pmask],
                outputs = outNodes,
                name = name)
```

Display sample decoder using Keras plot_model:

```
[28]  # Create a decoder sample and show object association
      diagram
      decoder_sample = decoder(svcab=8192,
            nlayers=2,
            x = 512,
            dm = 128,
            nhd = 4,
            drop = 0.3,
            name = "decoder_sample")
      tflow.keras.utils.plot_model(decoder_sample,
      to_file='decoder_sample.png', show_shapes =True)
```

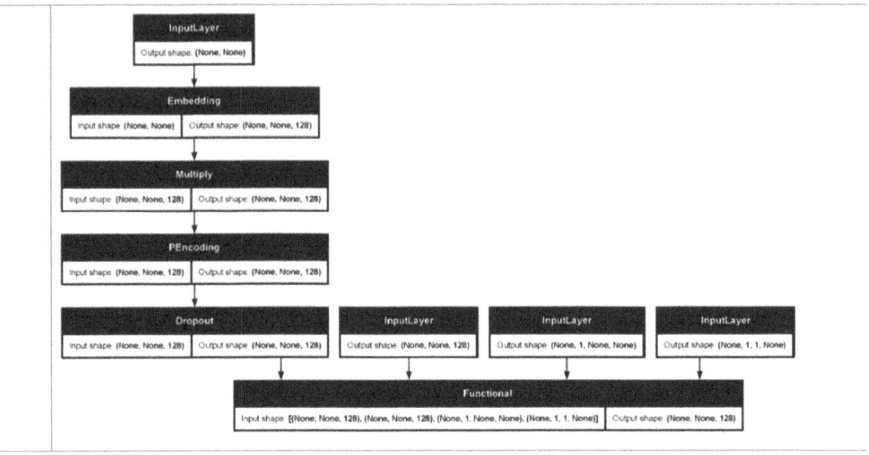

Step 7. Implement Transformer

Transformer involves implementing encoder, decoder, and the final linear layer.

Transformer decoder output is input to linear layer as a recurrent neural network (RNN) and output model is returned.

```
[29]   # Implementation of Transformer Class
       def transformer(svcab, nlayers, x, dm, nhd, drop,
       name="transformer"):
           queries = tflow.keras.Input(shape=(None,), name="inNodes")
           dec_queries = tflow.keras.Input(shape=(None,),
         name="decNodes")
           enc_pmask = tflow.keras.layers.Lambda(
           gen_pmask, output_shape=(1, 1, None),
           name="enc_pmask")(queries)
           # Perform Look Ahead Masking for Decoder Input for the
         Att1
           lamask = tflow.keras.layers.Lambda(gen_lamask,
                   output_shape=(1, None, None),
                   name = "lamask")(dec_queries)
           # Perform Padding Masking for Encoder Output for the Att2
           dec_pmask = tflow.keras.layers.Lambda(gen_pmask,
                   output_shape=(1, 1, None),
                   name="dec_pmask")(queries)
           encouts = encoder(svcab=svcab,
               nlayers = nlayers,
               x = x,
               dm = dm,
               nhd = nhd,
               drop = drop,)(inputs = [queries, enc_pmask])
           decouts = decoder(svcab=svcab,
               nlayers = nlayers,
               x = x,
               dm = dm,
               nhd = nhd,
               drop=drop,)(inputs=[dec_queries, encouts, lamask,
         dec_pmask])
           responses =
       tflow.keras.layers.Dense(units=svcab, name="outNodes")
       (decouts)
           return tflow.keras.Model(inputs=[queries, dec_queries],
       outputs=responses, name=name)
```

Display sample transformer object using Keras plot_model:

```
[30]   # Create a transformer sample and display object diagram
       transformer_sample = transformer(svcab=8192, nlayers=4,
       x=512,
           dm=128, nhd = 4, drop=0.3, name="transformer_sample")
       tflow.keras.utils.plot_model(transformer_sample,
       to_file="transformer_sample.png", show_shapes=True)
```

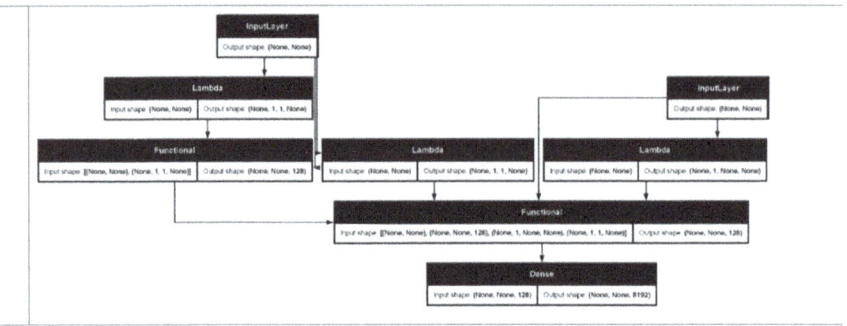

Step 8. Model Training

Parameters for nLayers, dm and units (x) had reduced to speed up training process.

[31]	# Create Transformer Model
	tflow.keras.backend.clear_session()
	model = transformer(svcab = SVCAB,
	                 nlayers=2,
	                 x=512,
	                 dm=256,
	                 nhd=8,
	                 drop=0.1)

1. A movie Chatbot Transformer model consists of two layers with 512 units, data-model size 256, head number 8, and dropout rate 0.1 according to transformer model as in Fig. 17.2.
2. It is recommended to modify these parameter settings to improve network performance as discussed in Sect. 17.1.

Step 9. Implement Model Evaluation Function

A loss function is implemented for system evaluation. It is important to use a padding mask when calculating the loss since target sequences are padded.

| [32] | # *Implementation of Evaluation Function (Loss Function)*
def Eval_function(xtrue, xpred):
        # Reshape xtrue
        xtrue = tflow.reshape(xtrue, shape=(-1, MLEN - 1))
        # Compute sparse categorical crossentropy loss
        loss_val = tflow.keras.losses.SparseCategoricalCrossentropy(
                from_logits=True, reduction='none')(xtrue, xpred)
        # Mask padding values (assuming 0 is the padding value)
        mask_val = tflow.cast(tflow.not_equal(xtrue, 0), tflow.float32)
        loss_val = tflow.multiply(loss_val, mask_val)
        return tflow.reduce_mean(loss_val) |

Step 10. Implement Customized Learning Rate

Adam_Optimizer with a customized learning rate is used with the formula below:

$$l_{\text{rate}} = d_{\text{model}}^{-5} * \min\left(\text{step\_num}^{-0.5}, \text{step\_num} * \text{warmup\_steps}^{-1.5}\right) \quad (17.3)$$

| [33] | # *Implementation of Customized Learning Rate*
class CLearning(tflow.keras.optimizers.schedules.LearningRateSchedule):
        def __init__(self, dm, warmup_steps=4000):
                super(CLearning, self).__init__()
                self.dm = dm
                self.dm = tflow.cast(self.dm, tflow.float32)
                self.warmup_steps = warmup_steps
        def __call__(self, step):
        # *arg1 = tflow.math.rsqrt(step)*
                arg1 = tflow.math.rsqrt(tflow.cast(step, tflow.float32))
                arg2 = tflow.cast(step, tflow.float32) * (tflow.cast(self.warmup_steps,
tflow.float32)**-1.5)
                return tflow.math.rsqrt(self.dm) * tflow.math.minimum(arg1, arg2) |

Plot customized learning rate:

| [34] | ```
# Create customized learning rate object and display
performance
CLearning_sample = CLearning(dm=128)
pyplt.plot(CLearning_sample(tflow.range(200000,
dtype=tflow.float32)))
pyplt.ylabel("Learning Rate")
pyplt.xlabel("Train Step")
``` |
```

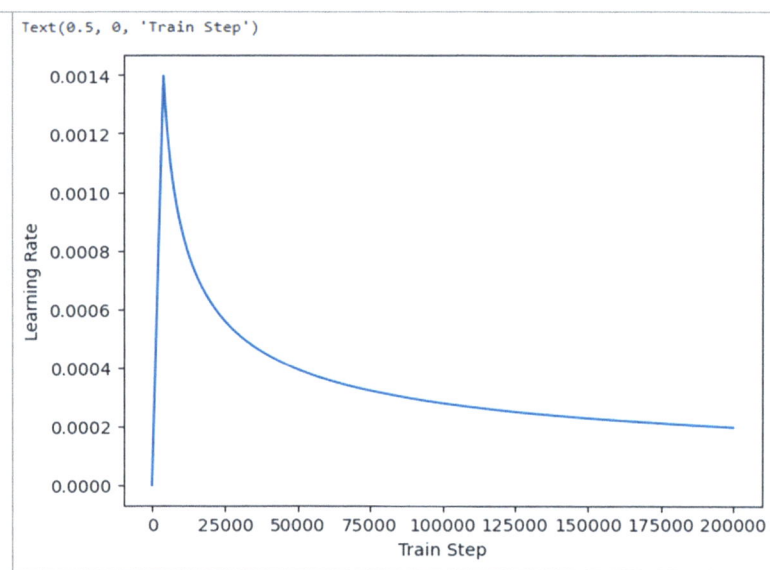

Step 11. Compile Chatbot Model

Step 12. System Training (Model Fitting)

```
[35]    # Compile Movie Chatbot Model
        # Set the Customized Learning Rate
        cLRate = CLearning(256)
        # Set Adam Optimizers
        optimizer = tflow.keras.optimizers.Adam(learning_
        rate=cLRate, beta_1=0.9, beta_2=0.98, epsilon=1e-9)
        # Implement Accuracy Evaluation Scheme
        def accuracy(xtrue, xpred):
            xtrue = tflow.reshape(xtrue, shape=(-1, MLEN - 1))
            xpred = xpred[:, :tflow.shape(xtrue)[1], :] # Slice
          to match time steps
            return tflow.keras.metrics.sparse_categorical_
          accuracy(xtrue, xpred)
        # Compile Chatbot Model
        model.compile(optimizer=optimizer, loss=Eval_function,
        metrics=[accuracy])
```

Train Chatbot transformer model by calling *model.fit( )* for 20 epochs to save time.

```
[36]    EPOCHS = 20
        model.fit(mDS, epochs = EPOCHS)
```

```
Epoch 1/20
689/689 ___ 211s 299ms/step - accuracy: 0.0240 - loss: 2.5178
Epoch 2/20
689/689 ___ 249s 361ms/step - accuracy: 0.0765 - loss: 1.5446
Epoch 3/20
689/689 ___ 257s 373ms/step - accuracy: 0.0857 - loss: 1.4137
Epoch 4/20
689/689 ___ 255s 371ms/step - accuracy: 0.0903 - loss: 1.3390
Epoch 5/20
689/689 ___ 161s 233ms/step - accuracy: 0.0951 - loss: 1.2668
Epoch 6/20
689/689 ___ 162s 235ms/step - accuracy: 0.0986 - loss: 1.2131
Epoch 7/20
689/689 ___ 161s 234ms/step - accuracy: 0.1031 - loss: 1.1636
Epoch 8/20
689/689 ___ 162s 235ms/step - accuracy: 0.1088 - loss: 1.0969
Epoch 9/20
689/689 ___ 179s 260ms/step - accuracy: 0.1148 - loss: 1.0374
Epoch 10/20
689/689 ___ 180s 261ms/step - accuracy: 0.1202 - loss: 0.9825
Epoch 11/20
689/689 ___ 178s 258ms/step - accuracy: 0.1268 - loss: 0.9476
Epoch 12/20
689/689 ___ 176s 256ms/step - accuracy: 0.1326 - loss: 0.9058
Epoch 13/20
689/689 ___ 177s 257ms/step - accuracy: 0.1381 - loss: 0.8731
Epoch 14/20
689/689 ___ 177s 256ms/step - accuracy: 0.1428 - loss: 0.8378
Epoch 15/20
689/689 ___ 176s 256ms/step - accuracy: 0.1476 - loss: 0.8064
Epoch 16/20
689/689 ___ 182s 264ms/step - accuracy: 0.1522 - loss: 0.7809
Epoch 17/20
689/689 ___ 179s 260ms/step - accuracy: 0.1556 - loss: 0.7551
Epoch 18/20
689/689 ___ 175s 254ms/step - accuracy: 0.1609 - loss: 0.7331
Epoch 19/20
689/689 ___ 174s 253ms/step - accuracy: 0.1644 - loss: 0.7117
Epoch 20/20
689/689 ___ 175s 254ms/step - accuracy: 0.1663 - loss: 0.6878
[36]: <keras.src.callbacks.history.History at 0x22eeb101b50>
```

Step 13. System Evaluation and Live Chatting

System evaluation and live chatting implementation involve the following steps:

1. Create Mining() method by performing data preprocessing of all utterances.
2. Perform tokenization of utterances and padded with START and END tokens.
3. Perform LookAhead and Padding Masks.
4. Construct Transformer model with attention learning.
5. Implement chatting() method by decoder scheme.
6. Combine chatted word sequences to decoder input.

7. Use transformer model for system to predict responses based on previous training epochs.

```
[37]    # Implementation of Movie Chatting class - mchat
        def mchat(utterance):
            # Utterance Preprocessing and add START AND END
            TOKENS
            utterance = pp_utterance(utterance)
            utterance = tflow.expand_dims(START_TOKEN +
        m_token.encode(utterance) + END_TOKEN, axis = 0)
            # Create response object
            response = tflow.expand_dims(START_TOKEN, 0)
            for i in range(MLEN):
                chatting = model(inputs = [utterance, response],
            training = False)
            # Choose last_word from token sequence
            chatting = chatting[:, -1:, :]
            chatted_id = tflow.cast(tflow.argmax(chatting, axis=-
        1), tflow.int32)
            # Return with chattedID with ENDTOKEN
            if tflow.equal(chatted_id, END_TOKEN[0]):
                break
            # Combine CHATTEDID with utterance response
            response = tflow.concat([response, chatted_id],
        axis=-1)
            return tflow.squeeze(response, axis = 0)
        #
        Implementation
        of main class for Movie Chatting - mchatting
        def mchatting(utterance):
            mchatting = mchat(utterance)
            chatted_utterance =
        m_token.decode([i for i in mchatting if i < m_token.
        vocab_size])
            print('Query: {}'.format(utterance))
            print('Response: {}'.format(chatted_utterance))
            return chatted_utterance
```

Try some movie conversations to see whether it works:

```
[38]    output = mchatting("Where have you been?")
        Query: Where have you been?
        Response: i m going to see you.
```

```
[39]    output = mchatting("It's a trap")
        Query: It's a trap
        Response: you re not going anywhere !
```

[40]	`output = mchatting("Do you need help?")`
	Query: Do you need help? Response: i don t know.

[41]	`output = mchatting("What do you think?")`
	Query: What do you think? Response: i don t know. i don t know. i m not sure. i just had to see what i m saying.

[42]	`output = mchatting("Are you happy?")`
	Query: Are you happy? Response: yes. i m very happy.

1. Training showed that epochs 1–20 are rather slow but increased in accuracy and decreased in loss rate.

2. Two chatbots experiments with one used 2 epochs and the other used 20 epochs. Results showed that performance on 20 epochs has satisfactory performance than the one with 2 epochs.

3. Increase epochs, say, up to 50 epochs to review whether accuracy has continuous improvement. It is natural to require more time unless there are sufficient GPUs.

Workshop 7.1 Fine-tune Chatbot Model

TensorFlow and Transformer technology are used to develop a domain-based Chatbot system

There are rooms to fine-tune model performance like any AI model. It can be conducted by:

1. Dataset Level

- Enhance preprocessing process.

- Improve data record selection scheme, e.g., sample size, utterance MLEN, etc.

2. Network Model Level

- Fine-tune system parameters, e.g., learning rate and method, etc.

- Fine-tune Transformer Model by modifying Attention Function, etc. Compare performances (MUST) and analysis (bonus).

Fine-tune Movie Chatbot model and compare with the original version

Workshop 7.2 Mini Project - Build a Semantic-Level AI Chatbot System

Extend character-level and word-level NLU to a semantic-level NLU

1. Modify codes of AI Chatbot learnt in this section to implement a semantic-level AI Chatbot system.

2. Compare system performance of this revised system with previous character-level and word-level AI Chatbot system.

17.5   Related Works

This workshop had integrated all NLP related implementation techniques including TensorFlow and Keras with Transformer Technology to design an AI-based NLP application chatbot system. It is a step-by-step implementation consisting of data preprocessing, model construction, system training, testing evaluation process; and Attention Learning and Transformer Technology with TensorFlow and Keras implementation platform easily applied to other chatbot domain and interactive QA systems using Cornell Large Movie dataset with over 200,000 movie conversations with 10,000+ movie characters.

Nevertheless, it is only the dawn of the journey. There are regular new R&D prevalence and usage in NLP applications. Below are lists of renowned domains and resources related to chatbot systems for reference.

Datasets for Chatbot Systems
• Taskmaster from Google Research (Google Research 2024a).
• Simulated Dialogue dataset from Google Research (GoogleResearch 2024b).
• Dialogue Challenge dataset from Microsoft (MicrosoftDialog 2024).
• Dialogue State Tracking Challenge dataset (DSTC 2024).

Keras Modules and Optimizer
• Keras layers (Keras 2024a).
• Keras optimizers (Keras 2024b).
• An overview of optimizers (Ruder 2024).
• Adam optimizer (Adam 2024).

Famous Chatbot System
• Amazon Alexa developer blog (Alexa 2024).
• Apple Siri Developer (AppleSiri 2024).
• Duer from Baidu (Duer 2024).
• Google Assistant (GoogleAssistant 2024).
• Microsoft Cortana Developer (MicrosoftCortana 2024).
• Samsung Bixby Developer (SamsungBixby 2024).
• Xiaowei from Tencent (Xiaowei 2024).

References

Adam (2024) Adam optimizer: https://arxiv.org/abs/1412.6980. Accessed 17 Dec 2024.
Alexa (2024) Amazon Alexa developer blog: https://developer.amazon.com/blogs/home/tag/Alexa. Accessed 17 Dec 2024.
AppleSiri (2024) Apple Siri Developer: https://developer.apple.com/siri/. Accessed 17 Dec 2024.
Bansal, A. (2021) Advanced Natural Language Processing with TensorFlow 2: Build effective real-world NLP applications using NER, RNNs, seq2seq models, Transformers, and more. Packt Publishing.

Batish, R. (2018) Voicebot and Chatbot Design: Flexible conversational interfaces with Amazon Alexa, Google Home, and Facebook Messenger. Packt Publishing.

Cornell (2024) https://www.cs.cornell.edu/~cristian/Chameleons_in_imagined_conversations. html. Accessed 17 Dec 2024.

Cornell_Movie_Corpus (2024) Cornell Movie Corpus archive. https://www.kaggle.com/datasets/ Cornell-University/movie-dialog-corpus. Accessed 17 Dec 2024.

Devlin, J., Chang, M. W., Lee, K. and Toutanova, K. (2019). BERT: Pre-training of Deep Bidirectional Transformers for Language Understanding. Archive: https://arxiv.org/ pdf/1810.04805.pdf.

Duer (2024) Duer Baidu AI Chatbot. http://duer.baidu.com/en/index.html. Accessed 17 Dec 2024.

DSTC (2024) Dialog State Tracking Challenge dataset: https://github.com/matthen/dstc. Accessed 17 Dec 2024.

Ekman, M. (2021) Learning Deep Learning: Theory and Practice of Neural Networks, Computer Vision, Natural Language Processing, and Transformers Using TensorFlow. Addison-Wesley Professional.

Facebook (2024) Facebook Messenger API documentation. https://developers.facebook.com/ docs/messenger-platform/getting-started/quick-start/. Accessed 17 Dec 2024.

Freed, A. (2021) Conversational AI: Chatbots that work. Manning.

Géron, A. (2019) Hands-On Machine Learning with Scikit-Learn, Keras, and TensorFlow: Concepts, Tools, and Techniques to Build Intelligent Systems. O'Reilly Media.

GoogleAssistant (2024) Google Assistant: https://assistant.google.com/. Accessed 17 Dec 2024.

GoogleResearch (2024a) Taskmaster from Google Research. https://github.com/google-research-datasets/Taskmaster/tree/master/TM-1-2019. Accessed 17 Dec 2024.

GoogleResearch (2024b) Simulated Dialogue dataset from Google Research. https://github.com/ google-research-datasets/simulated-dialogue. Accessed 17 Dec 2024.

Janarthanam, S. (2017) Hands-On Chatbots and Conversational UI Development: Build chatbots and voice user interfaces with Chatfuel, Dialogflow, Microsoft Bot Framework, Twilio, and Alexa Skills. Packt Publishing.

Kedia, A. and Rasu, M. (2020) Hands-On Python Natural Language Processing: Explore tools and techniques to analyze and process text with a view to building real-world NLP applications. Packt Publishing.

Keras (2024a) Keras official sites: https://keras.io. Accessed 17 Dec 2024.

Keras (2024b) Keras optimizers: https://keras.io/api/optimizers/. Accessed 17 Dec 2024.

MicrosoftCortana (2024) Microsoft Cortana Developer: https://www.microsoft.com/en-us/cortana/. Accessed 17 Dec 2024.

MicrosoftDialog (2024) https://github.com/xiul-msr/e2e_dialog_challenge. Accessed 17 Dec 2024.

NLPGitHub (2024) URL: https://github.com/raymondshtlee/NLP/. Accessed 17 Dec 2024.

Raj, S. (2018) Building Chatbots with Python: Using Natural Language Processing and Machine Learning. Apress.

Rothman, D. (2022) Transformers for Natural Language Processing: Build, train, and fine-tune deep neural network architectures for NLP with Python, PyTorch, TensorFlow, BERT, and GPT-3. Packt Publishing.

Ruder (2024) An overview of optimizers: https://ruder.io/optimizing-gradient-descent/. Accessed 17 Dec 2024.

SamsungBixby (2024) Samsung Bixby Developer: https://developer.samsung.com/bixby. Accessed 17 Dec 2024.

SpaCy (2024) spaCy official site. https://spacy.io/. Accessed 17 Dec 2024.

Telegram (2024) Telegram bot API documentation: (https://core.telegram.org/bots. Accessed 17 Dec 2024.

TensorFlow (2024) TensorFlow official site> https://tensorflow.org /. Accessed 17 Dec 2024.

Tuchong (2024) Wake word to invoke your Chatbot. https://stock.tuchong.com/image/detail?imag eId=918495180260638796. Accessed 17 Dec 2024.

Tunstall, L, Werra, L. and Wolf, T. (2022) Natural Language Processing with Transformers: Building Language Applications with Hugging Face. O'Reilly Media.

Vaswani, A., Shazeer, N., Parmar, N., Uszkoreit, J., Jones, L., Gomez, A. N., … & Polosukhin, I. (2017). Attention is all you need. Advances in neural information processing systems, 30. https://arxiv.org/abs/1706.03762.

Xiaowei (2024) Xiaowei chatbot system from Tencent. https://xiaowei.tencent.com/. Accessed 17 Dec 2024.

Yıldırım, S and Asgari-Chenaghlu, M. (2021) Mastering Transformers: Build state-of-the-art models from scratch with advanced natural language processing techniques. Packt Publishing.

Index

The manufacturer's authorised representative in the EU is Springer
Nature Customer Service Centre GmbH, Europaplatz 3, 69115 Heidelberg,
Germany. If you have any concerns regarding our products, please
contact ProductSafety@springernature.com

Printed and bound by CPI Group (UK) Ltd, Croydon, CR0 4YY

29/04/2026

02099543-0004